Effect of Biotic and Abiotic Factors on the Physiology of Horticultural Plants

Effect of Biotic and Abiotic Factors on the Physiology of Horticultural Plants

Guest Editors

George Zervoudakis
Filippos Bantis

MDPI

Basel • Beijing • Wuhan • Barcelona • Belgrade • Novi Sad • Cluj • Manchester

Guest Editors
George Zervoudakis
Department of Agriculture
University of Patras
Messolonghi
Greece

Filippos Bantis
Department of Agriculture
University of Western
Macedonia
Florina
Greece

Editorial Office
MDPI AG
Grosspeteranlage 5
4052 Basel, Switzerland

This is a reprint of the Special Issue, published open access by the journal *Plants* (ISSN 2223-7747), freely accessible at: https://www.mdpi.com/journal/plants/special_issues/1175W7ZF9K.

For citation purposes, cite each article independently as indicated on the article page online and as indicated below:

Lastname, A.A.; Lastname, B.B. Article Title. *Journal Name* **Year**, *Volume Number*, Page Range.

ISBN 978-3-7258-7460-6 (Hbk)
ISBN 978-3-7258-7461-3 (PDF)
https://doi.org/10.3390/books978-3-7258-7461-3

Contents

About the Editors

George Zervoudakis

George Zervoudakis is an Associate Professor of Plant Physiology and Biochemistry at the Department of Agriculture, University of Patras, Greece. His research interests focus on plant physiology and oxidative stress, particularly in medicinal and aromatic plants and vegetables. He has contributed to several scientific publications.

Filippos Bantis

Filippos Bantis is an Assistant Professor of Vegetable Science and Greenhouse Production at the Department of Agriculture, University of Western Macedonia, Greece. His research interests focus on plant physiology, plant photomorphogenesis, production, and nutritional quality, particularly in vegetable crops. He has contributed to several scientific publications and participated in a plethora of national and EU research projects.

Preface

This Special Issue focuses on the physiological responses of horticultural plants to biotic and abiotic stress factors, aiming to advance knowledge of plant adaptation and resilience under changing environmental conditions. It brings together original research and review articles addressing key mechanisms of stress tolerance, including antioxidant defense, photosynthetic regulation, and signaling pathways. This collection is intended for researchers, agronomists, and professionals seeking to better understand and manage stress factors in high-value cropping systems.

George Zervoudakis and Filippos Bantis
Guest Editors

plants

MDPI

Editorial

Effect of Biotic and Abiotic Factors on the Physiology of Horticultural Plants

Filippos Bantis [1,*] and George Zervoudakis [2,*]

1 Department of Agriculture, University of Western Macedonia, 53100 Florina, Greece
2 Department of Agriculture, University of Patras, 30200 Messolonghi, Greece
* Correspondence: fbantis@uowm.gr (F.B.); gzerv@upatras.gr (G.Z.)

Horticultural crops comprise a diverse group of intensively cultivated plant species including vegetables, fruits, ornamental plants, and medicinal or aromatic crops, which are typically characterized by high economic value, intensive management, and strong dependence on environmental conditions. These crops play a crucial role in global food production and economic sustainability. However, their productivity and quality are increasingly challenged by environmental pressures, including climate crisis, abiotic constraints such as salinity, extreme temperatures, and water deficit, as well as biotic stresses. Understanding how plants respond to these factors is essential for developing resilient production systems. The Special Issue "Effect of Biotic and Abiotic Factors on the Physiology of Horticultural Plants" was organized to advance knowledge on both plant physiological responses and stress tolerance mechanisms and innovative management strategies.

Another important challenge facing modern horticulture is the genetic narrowing that has resulted from intensive breeding programs primarily targeting high yield, uniformity, and market-related traits. While these efforts have significantly improved productivity and product quality, they have often been accompanied by the loss or underutilization of genetic traits associated with tolerance to environmental stresses. As climate variability intensifies, the reintegration of such traits has become increasingly important for restoring resilience in elite cultivars. These traits are frequently present in landraces, wild relatives, and less intensively bred genotypes. Understanding the physiological basis of plant responses to biotic and abiotic stresses is therefore essential for guiding breeding strategies aimed at improving crop adaptability under changing environmental conditions.

The contributions gathered in this Special Issue explore the diverse interactions between horticultural plants and their environments. Several studies focus on responses to abiotic parameters, highlighting how factors such as salinity [1–4], drought [5], and root-zone temperature [6] induce oxidative damage and metabolic alterations, while also demonstrating the potential of biostimulants or adaptive mechanisms to mitigate stress effects and improve plant performance. Other contributions [7,8] investigate plant–pathogen interactions and defense mechanisms, revealing the complex roles of antioxidant systems, phytohormone signaling, and physiological adjustments in plant resistance against diseases. Another contribution [9] deals with the phenotypic plasticity and adaptive strategies of riparian plants in heterogeneous light habitats.

Soil salinity is a major limitation for strawberry cultivation worldwide, compromising vegetative growth, fruit yield, and marketable quality. Developing integrated strategies that combine physiological understanding with biological interventions and breeding approaches is therefore essential for sustaining production in salt-affected regions. In the review of Passa et al. [1], the authors synthesize current knowledge on salinity-induced physiological, biochemical, and molecular responses in strawberry, while critically evaluating the role of biostimulants, signaling molecules, beneficial microorganisms, and emerging genetic tools in enhancing salt tolerance. They highlight the complexity of osmotic adjustment, antioxidant defense, ion homeostasis, and stress-responsive gene networks underlying plant adaptation. Importantly, the review emphasizes that many identified tolerance mechanisms are characterized at early developmental stages and do not always translate into stable improvements in fruit yield and quality. Significant genotypic variation among cultivars and crop wild relatives is discussed as a valuable resource for breeding, although stress-response traits must be carefully validated for agronomic relevance. Overall, the work underscores the need to bridge mechanistic knowledge with long-term field performance to develop truly salt-resilient strawberry cultivars.

Salinity and alkaline stress also severely constrain maize productivity worldwide, primarily through oxidative damage and impairment of photosynthesis. Identifying biostimulant-based strategies to reinforce physiological resilience under such conditions is therefore of major agronomic importance. In the study of Qi et al. [2], the authors investigated the role of exogenous melatonin (150 μM) in mitigating 100 mM Na_2CO_3 stress in a salt-susceptible maize line (P138) and its salt-resistant EMS mutant, combining physiological measurements with qRT-PCR analysis of antioxidant- and pigment-related genes. They demonstrated that melatonin application improved relative water content, chlorophyll concentration, and gas exchange parameters (Pn, Gs, Tr), while reducing membrane injury and H_2O_2 accumulation. Moreover, key genes associated with antioxidant defense and photosynthetic pigment biosynthesis were upregulated, indicating coordinated transcriptional regulation. Overall, their findings show that melatonin preserves photosynthetic apparatus integrity, enhances ROS scavenging, and strengthens maize seedling tolerance to alkaline salt stress.

Mild salinity can act not only as a stress factor but also as a regulatory signal influencing early plant development, making its precise management highly relevant for seedling production systems. Understanding how low salt concentrations modulate water transport and growth processes is particularly important for controlled environments such as hydroponics and greenhouses. Masepan et al. [3] investigated the effects of low NaCl concentrations (10 and 20 mM) on Chinese white radish seedlings grown in sand culture, evaluating growth traits, ion accumulation, water status, aquaporin gene expression, and physiological parameters four days after sowing. They demonstrated that 10 mM NaCl markedly stimulated cotyledon growth, increasing biomass, area, thickness, and mesophyll cell size. This enhancement was associated with elevated Na^+, Cl^-, K^+, nitrogen, carbon, and proline contents, as well as upregulation of key aquaporin genes (PIP and TIP isoforms), indicating improved water transport capacity. Additionally, maintenance of Hill reaction activity and enhanced antioxidant capacity supported greater cell expansion and succulence. Collectively, the findings suggest that low NaCl application can function as a simple agronomic strategy to promote early seedling vigor through aquaporin-mediated water regulation.

Salinity is also a major constraint in coastal and marginal agricultural systems, affecting both plant productivity and the biosynthesis of valuable secondary metabolites, even in halophytic species. Enhancing salt tolerance while preserving phytochemical quality is therefore essential for the sustainable cultivation of medicinal and aromatic crops such as *Crithmum maritimum*. In the study of Giannakoula et al. [4], the authors evaluated the physiological, antioxidant, and essential oil responses of sea fennel under increasing NaCl salinity (10 and 20 dS m^{-1}) combined with foliar application of two biostimulants, Aquamin and Cultisano, assessing growth traits, oxidative stress markers, photosynthetic performance, and volatile compound profiles (GC–MS). They found that moderate salinity stimulated growth and photosynthesis, whereas high salinity reduced biomass, photosynthetic efficiency, and water use efficiency, accompanied by elevated MDA and H_2O_2 levels. Biostimulant treatments mitigated oxidative damage, improved pigment retention and antioxidant capacity, and partially restored photosynthetic activity under severe salinity. Moreover, they modulated essential oil composition, with Aquamin particularly enhancing p-cymene and stabilizing monoterpenes, while Cultisano promoted specific compounds such as thymol-methyl-ether. Overall, the study demonstrates that targeted biostimulant applications can simultaneously enhance salt resilience and phytochemical value in halophytic crops, supporting their sustainable exploitation in saline environments.

Drought stress is among the most critical constraints to vegetable production, particularly for pepper, a crop highly sensitive to water deficits due to its physiological characteristics. Identifying genotypes that combine yield stability with fruit quality under limited irrigation is therefore essential for sustainable production and breeding programs. Ntanasi et al. [5] evaluated the effects of a 40% reduction in irrigation on yield, macronutrient composition, and fruit quality traits in two landraces ('JO109', 'JO204'), the cultivar 'Yolo Wonder', and the commercial hybrid 'Sammy RZ', under greenhouse conditions. They found that most cultivars suffered significant yield reductions due to decreased fruit number or weight; however, the landrace 'JO109' maintained stable yields under drought stress. This resilience was associated with higher leaf K concentrations and lower Na accumulation in fruits, underlining the protective role of potassium in stress mitigation. Moreover, deficit irrigation enhanced fruit firmness and total soluble solids, particularly in 'JO109', without negatively affecting titratable acidity, indicating that certain genotypes can preserve both productivity and quality under water-limited conditions.

Root-zone temperature is a decisive factor in soilless cultivation systems, directly influencing root activity, nutrient uptake, and ultimately crop productivity and quality. Optimizing this parameter is particularly important for leafy vegetables grown in NFT and aeroponic systems during cool periods. Bantis et al. [6] evaluated mini Romaine lettuce cultivated in nutrient film technique and aeroponics under heated nutrient solutions (14, 18, and 22 °C) compared with ambient conditions (11–12 °C), assessing yield, root growth, photosynthetic performance, pigment content, and nutritional traits. They demonstrated that increasing root-zone temperature, particularly to 22 °C, substantially enhanced leaf biomass and leaf length in both systems, with more pronounced effects in NFT. Suboptimal temperatures limited root development and biomass accumulation despite relatively maintained photosynthetic activity, indicating sink-related growth restrictions. In NFT, moderate heating also reduced nitrate accumulation and increased total soluble solids, while pigment responses varied between systems. Overall, the findings highlight that targeted root-zone heating can significantly improve production efficiency in soilless lettuce cultivation, especially in low- and mid-tech greenhouses during cooler months.

Gray mold, caused by *Botrytis cinerea*, is one of the most destructive diseases in grape production worldwide, posing major challenges for yield and fruit quality. Understanding the molecular interplay between secondary metabolism and plant defense responses is therefore crucial for developing more resistant cultivars. In the study of Hao et al. [7], the authors investigated the role of VvMYBPA1, a key transcription factor regulating proanthocyanidin biosynthesis, using agroinfiltration-mediated transient overexpression in grape berries and subsequent analysis of antioxidant enzymes and ROS-related genes. They found that overexpression of VvMYBPA1 increased susceptibility to *B. cinerea*, despite enhancing proanthocyanidin accumulation. This heightened susceptibility was associated with reduced β-1,3-glucanase and polyphenol oxidase activities, increased expression of respiratory burst oxidase homologs (*VvRBOHs*), decreased peroxidase activity, and excessive H_2O_2 accumulation leading to enhanced cell death. Overall, the study demonstrates

 https://doi.org/10.3390/plants15060961

that VvMYBPA1 negatively regulates grape resistance to gray mold by disturbing ROS homeostasis, providing valuable knowledge for future genome-editing strategies targeting MYB transcription factors to improve disease resistance.

Leaf blight represents a serious threat to ornamental crops, compromising both aesthetic quality and commercial value, particularly in high-value shrubs such as *Ilex verticillata*. A deeper understanding of pathogen identity and host defense mechanisms is essential for developing effective disease management strategies. In this study, Lu et al. [8] identified *Alternaria alternata* as the causal agent of leaf blight through morphological characterization, multi-locus molecular analyses (ITS, TEF1-α, G3PDH, RPB2), and pathogenicity assays fulfilling Koch's postulates, and further explored host responses using transcriptomic, physiological, and phytohormone analyses. They revealed a dynamic, multi-layered defense response characterized by early activation of pattern-recognition receptors and antioxidant enzymes, followed by extensive metabolic reprogramming. An early oxidative burst, indicated by increased malondialdehyde levels, was accompanied by sustained antioxidant activity (SOD, CAT), reflecting adaptive stress mitigation. Moreover, distinct salicylic acid and jasmonic acid crosstalk, with a mid-infection SA peak and a later JA rebound, highlighted coordinated hormonal regulation of defense. Collectively, these findings elucidate the complex antioxidant and phytohormone-mediated responses of *I. verticillata* to *A. alternata*, offering valuable targets for breeding and hormone-based disease management strategies.

Understanding how riparian plants adjust their structural and physiological traits to heterogeneous light environments is essential for interpreting ecological adaptation and energy allocation strategies. In dynamic riverine systems, coordinated variation between twig and leaf traits, together with thermal dissipation capacity, plays a key role in maintaining photosynthetic performance. Li et al. [9] examined populations of *Hippophae rhamnoides* along three canopy light gradients (full light, moderate shade, and canopy cover) in the Taohe River riparian zone, assessing twig–leaf morphological traits and chlorophyll fluorescence parameters to explore their relationship with leaf thermal dissipation. They found significant shifts in twig architecture, specific leaf area (SLA), leaf thickness, and energy partitioning across habitats. In high-light environments, plants developed thicker leaves with lower SLA and allocated a greater proportion of absorbed light energy to heat dissipation, enhancing photoprotection. Under shaded and canopy-covered conditions, morphological adjustments and reduced thermal dissipation reflected strategies to optimize light capture, balance competition, and maintain photosynthetic stability. Overall, the coordinated variation in twig–leaf traits and fluorescence characteristics highlights the phenotypic plasticity and adaptive strategies of riparian plants in heterogeneous light habitats.

This Special Issue underscores the relevance of plant physiology research for sustainable horticulture. Advances in stress physiology can guide the development of cultivation practices, breeding strategies, and innovative inputs aimed at improving crop resilience to produce more with fewer resources. In conclusion, the studies included in this Special Issue contribute to a deeper understanding of how biotic and abiotic factors affect horticultural plant physiology. By bridging fundamental research with practical applications, this collection highlights emerging directions in stress biology and sustainable crop production, offering valuable perspectives for researchers, agronomists, and practitioners working toward resilient horticultural systems.

Funding: This research received no external funding.

Acknowledgments: We gratefully acknowledge all the authors that participated in this Special Issues.

https://doi.org/10.3390/plants15060961

Conflicts of Interest: The authors declare no conflicts of interest.

References

1. Passa, K.; Gerakari, M.; Goufa, M.; Tani, E.; Papasotiropoulos, V. Salinity Stress in Strawberry (*Fragaria* × *ananassa* Duch.): Biological Intervention Strategies and Breeding Approaches for Salt-Tolerant Cultivars. *Plants* **2026**, *15*, 432. [CrossRef] [PubMed]
2. Qi, G.; Zhao, X.; He, F.; Sun, S.; Shi, Z.; Niu, Y. Exogenous Melatonin Reinforces Photosynthesis, Antioxidant Defense and Gene Expression to Ameliorate Na_2CO_3 Stress in Maize. *Plants* **2024**, *13*, 2844. [CrossRef] [PubMed]
3. Masepan, N.; Intarasit, S.; Panya, A.; Jungklang, J. Low NaCl Concentrations Increase Cotyledon Growth in Chinese White Radish (*Raphanus sativus* L. var. *longipinnatus* Bailey) Seedlings via Aquaporin-Mediated Water Transport. *Plants* **2025**, *14*, 1616. [CrossRef] [PubMed]
4. Giannakoula, A.E.; Stefanakis, M.K.; Papaioannou, C.; Tsimpliaraki, S.; Kaftantzi, S.; Niniraki, A.; Gertsis, A. Mitigation of Oxidative Damage Caused by Salinity in the Halophyte *Crithmum maritimum* L. via Biostimulant-Enhanced Antioxidant Activity. *Plants* **2025**, *14*, 3788. [CrossRef] [PubMed]
5. Ntanasi, T.; Karavidas, I.; Savvas, D.; Spyrou, G.P.; Giannothanasis, E.; Consentino, B.B.; Papasotiropoulos, V.; Sabatino, L.; Ntatsi, G. Physiological and Yield Responses of Pepper (*Capsicum annuum* L.) Genotypes to Drought Stress. *Plants* **2025**, *14*, 1934. [CrossRef] [PubMed]
6. Bantis, F.; Tostsidis, N.; Zervoudakis, Z.; Koukounaras, A.; Koulopoulos, A. Root-Zone Heating Boosts the Production of Mini Romaine Lettuce Grown in Nutrient Film Technique and Aeroponics Systems. *Plants* **2026**, *15*, 422. [CrossRef] [PubMed]
7. Hao, L.; Zhang, Y.; Ge, Z.; Meng, X.; Sun, Y.; Yi, H. Transient Overexpression of *VvMYBPA1* in Grape Berries Enhances Susceptibility to *Botrytis cinerea* Through ROS Homeostasis Modulation. *Plants* **2025**, *14*, 2469. [CrossRef] [PubMed]
8. Lu, H.; Zhou, C.; Cheng, P.; Huang, L.; Shen, Q.; Zheng, Y.; Li, Y.; Dai, W.; Zhang, J.; Shen, D.; et al. Leaf Blight in *Ilex verticillata* Caused by *Alternaria alternata*: Mechanisms of Antioxidant Defense, Phytohormone Crosstalk, and Oxidative Stress Responses. *Plants* **2025**, *14*, 3057. [CrossRef] [PubMed]
9. Li, Q.; Ma, M.; Tang, Y.; Zhao, T.; Zhao, C.; Li, B. Correlation Analysis of Twig and Leaf Characteristics and Leaf Thermal Dissipation of *Hippophae rhamnoides* in the Riparian Zone of the Taohe River in Gansu Province, China. *Plants* **2025**, *14*, 282. [CrossRef] [PubMed]

https://doi.org/10.3390/plants15060961

plants

MDPI

Review

Salinity Stress in Strawberry (*Fragaria* × *ananassa* Duch.): Biological Intervention Strategies and Breeding Approaches for Salt-Tolerant Cultivars

Kondylia Passa [1], Maria Gerakari [2], Maria Goufa [2], Eleni Tani [2] and Vasileios Papasotiropoulos [2,*]

[1] Department of Agriculture, University of Patras, Nea Ktiria, 30200 Messolonghi, Greece; k.passa@upatras.gr
[2] Laboratory of Plant Breeding & Biometry, Department of Crop Science, Agricultural University of Athens, Iera Odos 75, 11855 Athens, Greece; mgerakari@aua.gr (M.G.); marog@aua.gr (M.G.); etani@aua.gr (E.T.)
* Correspondence: vpapasot@aua.gr; Tel.: +30-210-5294622

Abstract

Soil salinity is a major constraint to strawberry cultivation, adversely affecting plant growth, yield, and fruit quality. Salinity stress triggers complex physiological and biochemical responses, including osmotic adjustment, antioxidant defense, ion homeostasis, and shifts in metabolite accumulation, ultimately reducing crop productivity and fruit quality. This review synthesizes current knowledge on strawberry responses to salinity, with emphasis on physiological mechanisms, genotypic variation in tolerance, and emerging biologically based approaches, such as biostimulants, small signaling molecules, and beneficial microorganisms, that alleviate salinity stress and enhance plant resilience. In parallel, recent advances in the identification of stress-responsive genes and regulatory pathways are discussed in the context of their relevance for breeding salt-resilient cultivars. This review also identifies critical gaps in current knowledge that, despite significant progress, hinder the translation of mechanistic insights into stable yield and fruit quality under field conditions. By integrating physiological knowledge with advances in biological and breeding-based approaches, together with long-term field validation, this review provides a comprehensive framework for improving strawberry performance under saline conditions and guiding future cultivation and genetic improvement strategies.

Keywords: strawberry; soil salinity; breeding; stress; tolerance; *Fragaria*; biostimulants

1. Introduction

Agriculture, as the backbone of global food production, faces numerous challenges that threaten its sustainability and capacity to meet the nutritional demands of a growing population. Among these challenges, soil salinity has emerged as one of the most pressing abiotic stresses hindering crop productivity worldwide. Soil salinization, defined as the accumulation of soluble salts in the soil profile, represents a major form of land degradation with profound consequences for agriculture, particularly in terms of plant growth, performance, yield, soil health, and long-term sustainability [1,2].

Soil salinity encompasses saline, sodic, and alkaline soil types, each characterized by variable concentrations of salts and exchangeable cations. Saline soils contain high concentrations of soluble salts that reduce soil osmotic potential and impair plant water uptake, whereas sodic soils are dominated by sodium ions (Na^+), leading to soil structural degradation and restricted water movement [1]. Alkaline soils, often associated with carbonate accumulation, exhibit high pH values that reduce nutrient availability and hinder root development [3]. Overall, excessive salts in the rhizosphere disrupt plant–soil–water interactions, adversely affecting crop growth, yield, and quality.

The magnitude of soil salinity as a global agricultural challenge is striking. According to recent estimates by the Food and Agriculture Organization (F.A.O.) [4], approximately 424 million hectares of surface soils and 833 million hectares of subsoils are affected by salinity worldwide. Salinity mainly affects riparian, semi-arid, and arid areas [5,6].

Mediterranean countries, including Greece, are particularly vulnerable to salinity due to their semi-arid climate, dependence on irrigation, and overexploitation of groundwater resources [7,8]. Climatic projections for the region indicate rising temperatures, reduced precipitation, and increased evapotranspiration, largely driven by climate change. These trends are expected to accelerate salt accumulation in irrigated soils, ultimately destabilizing yields and compromising crop quality [9–11].

The impact of salinity stress on plants depends on species and variety, growth stage, environmental conditions, and salt composition [12]. A comprehensive understanding of the physiological, biochemical, and molecular mechanisms governing plant stress responses is therefore essential for the development of effective mitigation strategies, including improved management practices and the breeding and cultivation of stress-resilient crop varieties [13,14]. Evidence from developed countries indicates that the adoption of sustainable cultivation practices combined with salinity-tolerant genotypes has enabled the restoration and productive use of salt-affected soils, whereas in many developing regions, reliance on short-term management approaches has failed to provide lasting solutions, thereby exacerbating food insecurity [15].

2. Salinity Stress as a Constraint in Strawberry Cultivation

2.1. *Impacts of Salinity Stress on Strawberry*

Strawberry represents a particularly important case study for salinity tolerance, due to its high economic value, widespread cultivation, and pronounced sensitivity to soil and water salinity [16,17].

Belonging to the Rosaceae family, strawberries are prized for their unique flavor, nutritional value, and bioactive compounds, including anthocyanins, polyphenols, vitamins, and minerals, which contribute to their recognized health-promoting properties [18,19].

In response to strong market demand, global strawberry production has expanded over the past decade into arid and saline regions where irrigation water quality is often poor [20]. Therefore, secondary salinization has become increasingly prevalent, exposing strawberries to elevated salt stress. Strawberry plants are glycophytes and display a low tolerance to salinity, with yield reductions reported at soil electrical conductivity (ECe) values as low as 1.0 dS m^{-1} [17]. For each additional unit increase in ECe, yield declines by approximately 33%, making salinity one of the most severe constraints on strawberry cultivation in irrigated regions [21].

Research on strawberry salinity tolerance has expanded in recent years, addressing genotypic differences as well as the morphological, physiological, biochemical, and molecular responses underlying stress adaptation [22–28]. Focused on the effects of salinity on fruit quality and yield, strawberry research highlights significant physiological and metabolic adjustments that influence consumer-relevant traits [17,29,30].

Based on research findings, secondary salinization induces pronounced physiological stress in strawberry plants [31,32]. Elevated salt concentrations reduce water uptake through osmotic effects and promote the accumulation of sodium and chloride ions in plant tissues, where they compete with essential nutrients such as potassium and calcium [33]. This ionic imbalance disrupts photosynthesis, stomatal regulation, and enzyme activity. Salt stress also enhances the production of reactive oxygen species (ROS), leading to oxidative

https://doi.org/10.3390/plants15030432

damage of cellular membranes and proteins. Although antioxidant defense mechanisms are present, their capacity is often insufficient under prolonged salinity, resulting in leaf chlorosis, growth inhibition, and reduced yield.

Salinity stress affects not only productivity but also fruit quality by altering the balance of sugars and organic acids and influencing the accumulation of secondary metabolites, which are critical determinants of fruit flavor and nutritional value. As a result, key quality traits such as sweetness, acidity, firmness, shelf life, and nutritional attributes are compromised, ultimately reducing market value and consumer acceptability [17,29,34].

An overview of the effects of salt stress on strawberry growth, development, and productivity is provided in Figure 1.

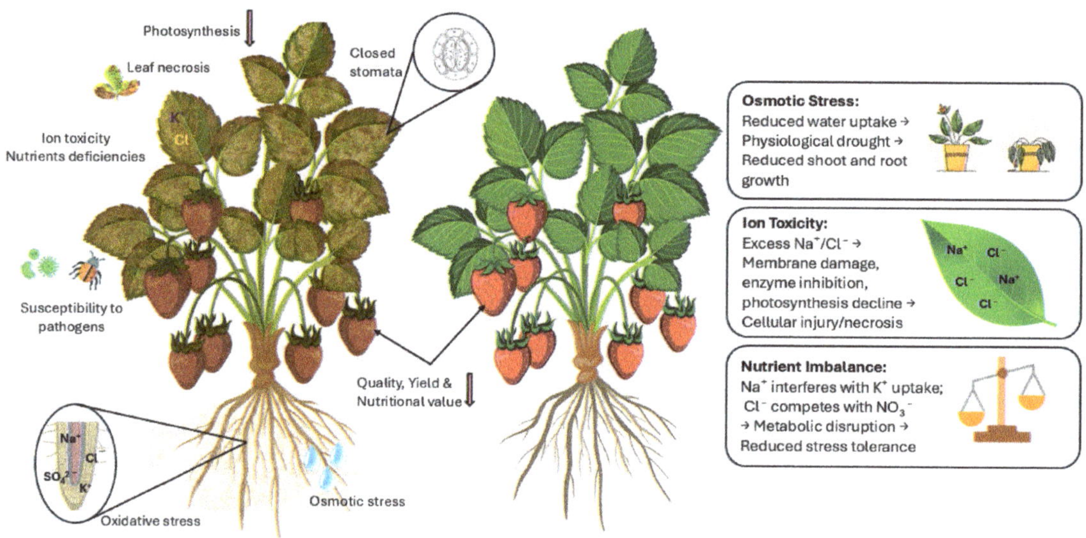

Figure 1. Overview of the physiological and agronomic implications of salt stress in strawberry plants.

The severity of yield and quality losses can escalate when salinity is combined with shallow groundwater and waterlogging, which together intensify stress conditions and limit root aeration [31,32]. For strawberry cultivation, strategies such as soil flushing, adoption of salt-tolerant cultivars, and precision irrigation techniques can help mitigate stress and sustain production.

Salinity stress negatively affects strawberry growth, yield, and fruit quality through a range of physiological and metabolic disturbances. The extent of these effects varies depending on plant material and management conditions. Improving salinity tolerance therefore relies on the identification and development of tolerant cultivars, alongside the adoption of management practices that support soil and water sustainability. A clearer understanding of strawberry responses to salinity can contribute to the development of more resilient production systems under increasingly challenging environmental conditions.

The aim of this review is to provide a critical synthesis of physiological, biochemical, and molecular mechanisms, biological interventions, and breeding-based approaches for improving strawberry tolerance to salinity stress, including biostimulants, stress-related genes, and existing variation among strawberry cultivars and crop wild relatives. This review identifies key knowledge gaps and research priorities to support strawberry cultivation and genetic improvement under saline conditions.

 https://doi.org/10.3390/plants15030432

2.2. Variation in Salinity Responses Across Different Strawberry Species and Cultivars

Strawberry species and their wild ancestors exhibit distinct responses to salinity stress, largely due to differences in genetic background and genome complexity. The genus *Fragaria* comprises 16 priority crop wild relatives (CWRs) that can be exploited in plant breeding to enhance cultivated strawberries, offering opportunities to improve fruit quality, tolerance to abiotic stresses, and resistance to pests and diseases [35]. Comparative analyses of salinity tolerance mechanisms across *Fragaria* species are therefore essential for identifying key regulatory nodes and molecular pathways underpinning adaptive responses to salinity stress.

Among these CWRs, the Chilean strawberry (*F. chiloensis* Duch.), an octoploid wild ancestor of the cultivated strawberry comprises four subspecies: *F. chiloensis* subsp. *chiloensis*, subsp. *lucida*, subsp. *pacifica*, and subsp. *sandwicensis*. This species is widely regarded as tolerant to abiotic stresses, particularly salinity and drought, and thus represents a valuable genetic resource for strawberry breeding [36–38]. Consistent with this view, Nikoloudi [39] demonstrated pronounced differences in salinity stress responses among *Fragaria* CWRs, showing that although *F. virginiana* and *F. chiloensis* are wild progenitors of the cultivated strawberry, *F. chiloensis* exhibits a substantially greater capacity to withstand saline conditions.

Subsequent comparative studies involving cultivated varieties and CWRs further substantiated these findings. Garriga et al. [40] evaluated salinity effects on growth performance, physiological responses, productivity, and fruit quality in three strawberry genotypes: the commercial cultivar 'Camarosa', the cultivated Chilean strawberry 'Bau' (*F. chiloensis* subsp. *chiloensis* f. *chiloensis*), and the wild Chilean strawberry 'Cucao' (*F. chiloensis* subsp. *chiloensis* f. *patagonica*). Notably, *F. chiloensis* f. *patagonica* was largely unaffected by salt stress regarding yield losses, indicating superior salinity tolerance among the genotypes examined.

Towards this end, Kortekamp et al. [41] assessed salinity tolerance in an F_2 population derived from a cross between the salt-sensitive cultivar 'Senga Sengana' and a salt-tolerant *F. chiloensis* ssp. *lucida* accession. Under greenhouse salt stress treatments, F_2 individuals exhibited variable necrotic damage and growth reduction, revealing genetic variation in salt tolerance and identifying genotypes that could serve as a valuable starting material for breeding strawberries with enhanced stress resilience.

These findings underscore the importance of exploiting CWR diversity in strawberry improvement programs, as many researchers have emphasized that the incorporation of resilient wild genotypes into breeding strategies is essential for the development of new varieties capable of minimizing yield losses in saline soils.

A wide range of strawberry commercial cultivars has been evaluated for tolerance to salinity stress [20,22,23,25,29,33,41–54]. Nevertheless, only a limited number of studies have explicitly investigated the relationship between salinity tolerance and yield penalties (Table 1).

Most studies focus on the response of strawberry cultivars to salt stress through the assessment of agronomic, physiological, and biochemical traits, often evaluated before plants reach the fruiting stage. Accordingly, vegetative growth is commonly used as an early indicator of salt sensitivity, as it is generally faster and, in many cases, more reliable than yield-related traits. Supplementary Table S1 summarizes the most representative studies.

https://doi.org/10.3390/plants15030432

Table 1. Genetic material with salinity tolerance in relation to yield reduction.

Genotype	Salinity Tolerance Level	Experimental Setup/Salinity Treatment	Key Physiological & Biochemical Traits	Yield & Survival Performance	References
Chilean strawberry genotypes 'Bau' (*F. chiloensis* ssp. *chiloensis* f. *chiloensis*)-cultivated and the wild 'Cucao' (*F. chiloensis* ssp. *chiloensis* f. *patagonica*)	Tolerant	The experiments were conducted in a polycarbonate greenhouse with natural light and temperature. Stress levels were 0, 30, and 60 mM NaCl solution (EC = 1.6, 3.4, and 5.7 dS m^{-1}, respectively)	Leaf relative water content (LRWC), pigment concentrations, proline, malondialdehyde (MDA) concentration, total soluble sugars (TSS), and titratable acidity (TA)	No significant yield losses compared to "Camarosa" cultivar (25% yield loss at severe stress). Moreover, no differences were detected in fruit diameter (mm) or fruit length (mm)	[40]
F. × *ananassa* 'Korona'	Relatively tolerant	Strawberry plants grown in a sandy medium and watered with solutions of moderate (ECe = 3.9 dS/m), and excessive (ECe = 7.5 dS/m) NaCl concentrations	Significant increase in reduced glutathione. Maintained high fruit quality and better taste under stress	Higher yield stability compared to the more sensitive cultivar 'Elsanta' under equivalent stress (fresh weight reduction of 26% versus 46%)	[33]
F. × *ananassa* 'Elsanta'	Relatively tolerant	Strawberry plants grown in growth chamber in pots watered with 10, 20, or 40 mM NaCl	Low stomatal density/reduced transpiration/delay in the accumulation of Cl^- ions in the shoot	No significant yield reduction under all NaCl treatments	[43]
F. × *ananassa* 'Albion'	Relatively tolerant	USDA-ARS U.S. Salinity Laboratory CA, using 12 raised beds to evaluate five strawberry cultivars under four salinity levels (0.7, 1.0, 1.5, and 2.5 dS m^{-1}	Superior chloride exclusion from roots and petioles. High total soluble sugars (Brix%) regardless of salinity treatment	It exhibited the least mean relative reduction in fruit yield, marketable fruit size survival (94%) while maintaining its fruit taste compared to 'San Andreas', 'Benicia', 'Ventana', and 'Monterey'	[17]
F. × *ananassa* 'Rociera' and 'Camarosa'	'Rociera': relatively higher tolerance 'Camarosa': relatively lower tolerance	Greenhouse, NaCl solution with EC = 2 dS m^{-1}; and (c) NaCl solution with EC = 4 dS m^{-1}	Above-ground plant mass (AGPM), total plant fresh and dry biomass, root fresh and dry weight, length, weight, and diameter of each individual fruit, overall fruit production, and various antioxidants	'Rociera': reduction in the plant growth parameters, the number of leaves and plant water content under severe salinity conditions. 'Camarosa': low salinity tolerance index, low plant water content and growth parameters, but high fruit sucrose, fructose, glucose, and total sugar concentration, as well as sweetness index	[22]
F. × *ananassa* 'Camarosa'	'Camarosa': tolerant	Greenhouse Seedlings grown in 9 L pots containing a mixture of soil and vermiculite Salt solution of 40 mmol/L NaCl in distilled water); L2 (stress level 2—salt solution of 80 mmol/L NaCl in distilled water) for a week	Yield of fruit, fresh biomass, and root biomass CO_2 assimilation Content of sodium (Na) and chloride (Cl) in leaves Phenylalanine ammonia lyase (PAL), peroxidase (POD), and polyphenoloxidase (PPO) activity Total content and the content of reducing sugars Various antioxidants	Mild salt stress did not affect fruit yield The lower level of mild salt stress positively affected photosynthesis and fresh biomass Mild salt stress improved (a) the content of antioxidant compounds and sucrose in the fruit. (b) the content of anthocyanins and sucrose. Higher levels of salt stress induced root growth and the accumulation of phenolic compounds	[20]

Heterogeneity in experimental design, including differences in NaCl concentrations, methods of salinity imposition (acute stress versus gradual increases in NaCl doses), and the duration of stress treatments, has hindered the ability to draw definitive conclusions. For instance, 'Camarosa' is often reported to be more tolerant than other cultivars under mild salinity conditions [47,54]. Nevertheless, substantial yield reductions have been documented in other studies [22,50], indicating that its response is strongly influenced by the cultivation system (hydroponics versus soil) and by the manner in which salt stress is imposed, including practices such as intermediate leaching.

Studies that have explicitly monitored yield responses have consistently reported substantial yield reductions, even in cultivars classified as relatively tolerant. For example, Ferreira et al. [17] reported significant yield reductions in the cultivar 'Albion' under moderate salinity (2.5 dS m^{-1}) in field experiments, although it was classified as the cultivar

 https://doi.org/10.3390/plants15030432

exhibiting the lowest reductions in fruit yield, commercial fruit size, and plant survival. Similarly, under elevated salt stress (80 mmol NaCl L^{-1}), Keutgen and Pawelzik [33] observed fresh weight reductions of 26% in the 'tolerant' cultivar 'Korona' in greenhouse experiments. Denaxa et al. [22] also documented yield losses in the relatively tolerant cv. 'Rociera', with average reductions of 17.3% and 22.7% under moderate and high saline conditions (2 dS m^{-1} or 4 dS m^{-1}, respectively) in greenhouse experiments.

Collectively, these findings highlight considerable variability in the magnitude of yield loss among cultivars under saline conditions; however, they also clearly demonstrate that no cultivar is exempt from yield penalties under salinity stress. The genetic material (both wild and cultivated) exhibiting the lowest yield losses under salinity treatments is summarized in Table 1.

Interestingly, it has been reported by Alnayef et al. [48] that mild salinity (10 mM) can be utilized to improve fruit quality by increasing antioxidant capacity and phenolic accumulation in 'Elsanta' and 'Elsinore', with a relatively low impact on yield (approximately 14–16% loss) when combined with appropriate fertilization. A finding also observed by Galli et al. [20], who reported that mild salt stress did not affect fruit yield in 'Camarosa', while low salt levels enhanced photosynthesis, vegetative growth, and anthocyanin and sucrose contents. Therefore, application of mild salt stress may be an effective biofortification strategy.

Taken together, these studies underscore that improving salt tolerance while preserving yield stability and fruit quality should be a central objective of strawberry breeding programs. However, the availability and identification of genetic material combining these traits remain limited, highlighting a critical gap in the current literature and the availability of breeding resources. Traits such as efficient ionic regulation, enhanced antioxidant capacity, osmotic adjustment, modified stomatal behavior, the ability to balance Na^+ exclusion with Cl^- compartmentalization, and the maintenance of relative water content under moderate salinity are widely considered important determinants of salinity tolerance. However, effective breeding for saline environments must also prioritize yield stability and the preservation of fruit quality traits, as physiological tolerance alone does not necessarily prevent yield or quality penalties under stress conditions.

3. Mitigation Strategies to Enhance Strawberry Performance Under Salinity Stress

Salinity stress represents a significant and growing challenge for strawberry cultivation worldwide. In response, a range of mitigation strategies has been explored, including the use of various biostimulants and other exogenous signaling compounds that modulate plant physiological and molecular stress responses, practices related to supplemental lighting, and the exploitation of beneficial microorganisms and plant–microbe interactions that enhance nutrient acquisition, ion homeostasis, and structural resilience. A deeper understanding of the mechanisms and effects of these complementary approaches is essential for supporting resilient and sustainable strawberry production under increasingly saline and environmentally challenging conditions.

3.1. The Role of Biostimulants and Small Signaling Molecules in Mitigating Salinity Stress

The external application of classical biostimulants and of small signaling molecules has been shown to significantly mitigate the adverse effects of various abiotic stresses, including salinity [55]. Compounds such as 5-aminolevulinic acid (ALA), melatonin (N-acetyl-5-methoxy-tryptamine) (MEL), salicylic acid (SA), humic acid (HA), acetic acid, methyl jasmonate (MeJA), zinc oxide (ZnO) and silicon nanoparticles (NPs), γ-aminobutyric acid

 https://doi.org/10.3390/plants15030432

(GABA), and related substances have been investigated for their ability to alleviate salinity-induced damage and enhance stress tolerance in strawberry [56–61].

ALA, a precursor of tetrapyrroles including chlorophyll, phytochrome, and heme, has been shown to enhance photosynthetic efficiency and antioxidant defenses at *F.* × *ananassa*, cv. Benihonpe plants. Root-applied ALA improves PSII and PSI activity, sustains electron transport, reduces photo-inhibition, and promotes osmotic adjustment under both NaCl-induced salinity (100 mM) and osmotic stress (PEG) in experiments conducted under greenhouse and controlled growth chamber conditions, respectively [62,63]. At the molecular level, ALA modulates stress-responsive genes and promotes H_2O_2-mediated signaling that enhances root Na^+ retention while limiting leaf Na^+ accumulation [63]. Evidence suggests that 10 mg L^{-1} ALA is generally effective, although optimization may depend on developmental stage and environmental conditions [64]. Despite these benefits, direct effects of ALA application on fruit yield remain largely unexplored.

Furthermore, MEL is another important small signaling molecule that plays a vital role as an antioxidant in plant stress responses. Under salt stress, MEL usually accumulates in plant tissues, functioning as a ROS scavenger and enhancing both enzymatic antioxidant capacity and the maintenance of photosynthetic pigments [60,65–67]. The application of MEL in strawberry plants remains poorly documented. In a greenhouse experiment, foliar application of MEL at concentrations of approximately 100–200 µM in 'Camarosa' was associated with improvements in fruit yield and quality under moderate to high NaCl stress (40–80 mM), including higher soluble solids, ascorbic acid, total antioxidant capacity, phenolic content, and sugar levels. MEL has also been suggested to modulate abscisic acid (ABA) signaling, potentially contributing to enhanced stress resilience [68]. However, the underlying metabolic and regulatory responses, as well as the consistency and extent of yield-related effects in strawberries, remain insufficiently characterized, highlighting the need for further investigation.

Moreover, SA, a phenolic signaling molecule, has been found to improve growth, chlorophyll content, water status, nutrient uptake, and antioxidant defenses in strawberry under salinity stress [69]. On the strawberry cultivars 'Camarosa', 'Selva', and 'Gaviota', pot experiments conducted under greenhouse and field conditions have shown that both foliar and pre-treatment applications can mitigate oxidative stress and help maintain physiological performance under salinity. The timing of application appears to be critical, with pre-treatment prior to salt exposure often associated with more pronounced beneficial effects. Based on the existing literature, an optimal combination of SA and salinity stress for enhancing antioxidant capacity, photosynthetic performance, nutrient retention, and, importantly, fruit yield and quality is approximately 0.5–0.9 mM (30–90 ppm) SA, depending on cultivar and growth stage, combined with moderate salinity (~40 mM NaCl). Foliar application one week before stress onset is generally recommended for best growth and physiological outcomes [70–74]. However, while several studies report improvements in photosynthetic performance and stress-related indices, only a limited number have demonstrated tangible increases in fruit yield and quality.

In addition, exogenous GABA application has been shown to enhance strawberry tolerance to salinity by coordinating antioxidant defense, osmotic adjustment, and photosynthetic protection. Transcriptomic analyses showed that foliar GABA (0–20 mM) upregulated chlorophyll metabolism and photosynthesis-related genes, with WRKY, AP2/ERF, and MYB transcription factors contributing to stress resilience [75]. Complementarily, a similar study demonstrated that 25 mM GABA reduced ROS and MDA, improved membrane stability, and increased expression of stress-responsive genes (*DREB*, *cAPX*, *MnSOD*, *GST*), also acting as a priming signal under non-stress conditions [76]. Together, these

https://doi.org/10.3390/plants15030432

findings indicate that GABA functions as both a rapid protective agent and a transcriptional regulator, highlighting its potential for breeding salinity-resilient strawberry varieties.

HA has also been reported to confer beneficial effects under salinity stress in various plant species by improving soil physicochemical properties, as well as by enhancing plant morphological traits, physiological processes, and overall biochemical responses [77–79]. To date, evidence for HA application in strawberry is limited to a single study conducted in Iran on the cultivars 'Kurdistan' and 'Paros', in which root application of HA was associated with reduced Na^+ accumulation, enhanced K^+ uptake, improved photosynthetic performance, and partial recovery of growth and yield. Importantly, the experiments were carried out over multiple seasons, and the authors reported consistent beneficial effects across cultivars [80]. Nevertheless, comprehensive field-scale validation and direct evidence linking HA application to stable commercial yield improvements in strawberry remain limited.

Furthermore, in hydroponic greenhouse experiments, foliar spraying of exogenous acetic acid (1–2 mM) [81] and foliar or nutrient solution applications of MeJA and silicon- NPs [82] have been shown to improve salt tolerance in the strawberry cultivar 'Paros' by enhancing antioxidant defenses, promoting osmolyte accumulation, maintaining photosynthetic performance, and regulating stress-responsive gene expression under moderate salinity (40–50 mM NaCl). These compounds may act synergistically to mitigate oxidative damage and preserve membrane stability, highlighting their potential for combined biostimulant strategies.

Complementary evidence from controlled in vitro experiments indicates that ZnO NPs can also alleviate salinity stress in strawberries. Zeid et al. [83] reported cultivar-dependent responses in *F.* × *ananassa*, with moderate ZnO-NP applications reducing Na^+ accumulation, improving K^+ uptake, and enhancing antioxidant activity under NaCl stress. However, whether these nanoparticle-mediated effects translate into improved whole-plant performance under field conditions remains to be established. Small signaling molecules and biostimulants can serve to enhance strawberry tolerance to salinity by protecting photosynthesis, strengthening antioxidant defenses, maintaining osmotic and ion balance, and regulating stress-responsive genes. Root- or foliar-applied ALA, MEL, SA, HA, GABA, acetic acid, MeJA, and NPs consistently improve these physiological and molecular responses. Despite these beneficial findings, most studies focus on early vegetative growth or physiological and biochemical markers, with limited data on fruit yield, quality, or field performance. Treatment protocols vary widely, and large-scale field validations are scarce. Future research should prioritize standardized, long-term trials linking molecular responses to reproductive performance to fully realize the agronomic potential of these biostimulants.

3.2. *Microorganisms and Plant–Microbe Interactions to Alleviate Salinity Stress*

In recent years, there has been increasing interest in the role of beneficial soil microorganisms, particularly arbuscular mycorrhizal fungi (AMF), in enhancing plant tolerance to salinity. These symbiotic fungi interact with plant roots, facilitating improved water and nutrient acquisition, especially phosphorus and micronutrients, and modulating physiological responses under salt stress [84,85]. In a greenhouse experiment, inoculation of two-month-old strawberry plants (*F.* × *ananassa* cv. Tochiotome) with *Gigaspora margarita* under high salinity (200 mM NaCl) significantly improved shoot and root biomass, maintained chlorophyll content, and reduced leaf browning. AMF symbiosis also promotes ion homeostasis by reducing Na^+ accumulation and decreasing the Na^+/K^+ ratio in both roots and shoots [86]. AMF appears to enhance both physiological performance and structural adaptations, providing a multifaceted protective mechanism for strawberry plants under salinity stress. However, further experimentation is required to confirm the beneficial

 https://doi.org/10.3390/plants15030432

effects of AMF on strawberry performance under salinity conditions, as well as to elucidate their potential impact on yield and yield-related traits.

Field- and greenhouse-based studies provide increasing evidence for the role of plant growth-promoting rhizobacteria (PGPR) in mitigating salinity stress in strawberries. Under naturally saline soil conditions, inoculation with halotolerant PGPR strains significantly improved plant performance in strawberry cultivation, resulting in enhanced vegetative growth, higher chlorophyll content, improved nutrient acquisition (including N, P, K, Ca, Mg, and micronutrients), reduced electrolyte leakage, increased leaf relative water content, and substantial yield gains, with some strains increasing fruit yield by approximately 46–48%; however, the effectiveness of such approaches is likely to depend on strain specificity, environmental conditions, and cultivar background [87]. Similarly, PGPR consortia isolated from saline environments have been shown to enhance biomass accumulation, photosynthetic performance, and overall stress tolerance under moderate salinity (85 mM NaCl), as well as under climate change–simulated conditions combining elevated CO_2 and temperature, indicating potential relevance for future production systems for combined environmental stresses anticipated under climate change scenarios [87–89].

Further evidence suggests that PGPR effects may be amplified when combined with organic soil amendments. Under saline conditions, the joint application of PGPR with humic- and fulvic acid–based fertilizers resulted in marked improvements in strawberry fruit weight and total yield, with combined treatments generally outperforming organic amendments applied alone [88]. It has also been reported that in strawberry, the efficacy of PGPR under salinity depends more on strain–plant–environment compatibility than on microbial diversity per se, supporting the targeted use of native PGPR for improving crop productivity in saline environments [90]. These results indicate synergistic interactions between organic soil promoters and carefully selected PGPR, which not only enhance nutrient availability and root function but also translate into consistent yield gains and improved yield stability under salt stress.

Microbial and microbe-derived interventions enhance strawberry salinity tolerance by maintaining ion homeostasis, improving nutrient uptake, reinforcing root structure, boosting antioxidant defenses and photosynthesis, and activating stress-responsive genes and metabolic pathways. Despite these benefits, most studies focus on vegetative growth under controlled conditions, with limited data on fruit yield, quality, or long-term field performance. Furthermore, the interactions of salinity with other abiotic stresses and the detailed molecular mechanisms underlying microbe-mediated tolerance remain poorly understood. Future research should prioritize standardized, field-based trials that link physiological and molecular responses to reproductive performance and commercial yield to fully realize the agronomic potential of these strategies.

3.3. Supplemental Light Application

Supplemental lighting has been explored to modulate plant physiological responses to salinity stress. Recent studies demonstrate that manipulation of light spectral quality can effectively mitigate the adverse effects of salinity and alkalinity stress in strawberry by targeting photosynthetic and physiological processes [91–93]. Esmaeilizadeh et al. [91] investigated the effects of different spectra of supplemental light on strawberry (cv. Camarosa) under salt and alkalinity stress conditions. The authors demonstrated that red and red/blue light had a positive effect on CO_2 assimilation, while blue/red light increased intrinsic water use efficiency (WUEi) under both stress conditions. Adjustment of light spectra, especially red light, increased photosystem II (PSII) performance indices and quantum yield parameters.

Moreover, Malekzadeh Shamsabad et al. [92] extended the analysis and showed that plants adopt different strategies against abiotic stress depending on light quality. Supplemental blue/red and white/yellow light partially restored vegetative growth, chlorophyll content, relative water content, and mineral nutrient balance, while limiting Na accumulation in plant tissues. Importantly, these light treatments increased the number of inflorescences and fruits and enhanced early fruiting, demonstrating that improvements in the performance of the photosynthetic apparatus can translate into improved reproductive output under stress conditions. More recent work by Malekzadeh Shamsabad et al. [93] further showed that spectral light improves not only early fruit yield but also fruit biochemical quality. Supplemental light spectra enhanced the accumulation of anthocyanins, total phenolic compounds, and antioxidant activity, while modulating titratable acidity in strawberry fruits.

Collectively, these studies provide a mechanistic framework linking light-regulated photosynthetic apparatus performance, ion homeostasis, and metabolic allocation to improved reproductive performance and fruit nutritional quality in strawberries cultivated under saline environments.

Finally, the application of hydrogen-rich water (HRW), a stress modifier, to strawberry seedlings under salinity stress enhances root biomass, promotes K^+ uptake, strengthens antioxidant defenses, and reduces Na^+ accumulation. Transcriptomic and metabolomic analyses indicate activation of genes and pathways involved in ion transport, antioxidant defense, and cell wall biosynthesis, with phenylpropanoid biosynthesis and amino/nucleoside sugar metabolism playing particularly important roles in salt stress mitigation [94].

The roles of biostimulants, signaling molecules, HRW, and plant–microbe interactions in enhancing salinity tolerance are shown in Figure 2.

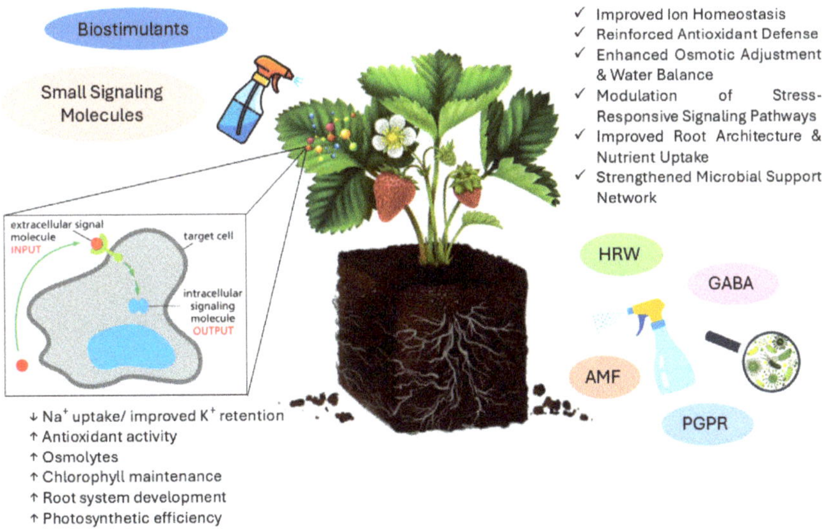

Figure 2. Overview of the beneficial effects through the exogenous application of small signaling molecules, biostimulants, and plant–microbe interactions (AMF: arbuscular mycorrhizal fungi; PGPR: plant growth-promoting rhizobacteria; HRW: hydrogen-rich water; GABA: γ-aminobutyric acid) in enhancing salinity tolerance in strawberry.

The above-discussed research findings of Section 3 are presented in detail in Supplementary Table S2, including information on the experimental setting, plant material and developmental stage, growth conditions, salinity stress level and duration, and parameters analyzed.

https://doi.org/10.3390/plants15030432

4. Genetic and Molecular Approaches for Improving Salinity Tolerance in Strawberry

4.1. Gene Identification and Functional Annotation for Breeding Targets

Salinity tolerance in strawberry involves the coordinated regulation of ion homeostasis, osmotic adjustment, oxidative stress detoxification, and transcriptional control, as evidenced by recent physiological and molecular studies. Gene identification and functional annotation provide a critical framework for understanding the molecular mechanisms underlying complex stress-adaptive traits, enabling the discovery of candidate genes that can be targeted to enhance stress tolerance through breeding [95]. However, consistent with broader trends in crop salinity research, most strawberry studies assess tolerance primarily through physiological, biochemical, and transcriptional responses associated with vegetative growth and survival, whereas direct evidence linking these mechanisms to fruit yield and quality remains limited [96,97]. To enable systematic comparison across studies, stress-responsive genes reported in the literature were organized into major functional categories reflecting their roles in key processes underlying salinity tolerance. The genes were grouped according to their primary molecular functions, while regulatory components such as transcription factors and signaling proteins may contribute to multiple physiological stress-response pathways (Table 2). This framework facilitates evaluation of their contributions to stress resilience and relevance for breeding. In addition, the experimental settings, plant developmental stages, key growth conditions, and salinity stress protocols used across all studies summarized in Table 2 are compiled in Supplementary Table S3.

Table 2. Functional categories of key genes involved in strawberry salinity and their roles in abiotic stress responses.

Functional Category	Genes/Gene Families	Functional Relevance in Abiotic Stress	Reference
Ion homeostasis and transport	*SOS1, SOS2, SOS3, NHX1, NHX2, CLC_C, CLC_G, KUP6, KUP7, SLAH3, ALMT12*, ABCB/C/G transporters	Regulation of Na^+ and Cl^- transport, vacuolar sequestration, and K^+ balance, contributing to salt tolerance and ionic homeostasis	[98–100]
Osmotic adjustment and water balance	Osmotin (*PR-5*), *LEA3, P5CS1, P5CS2, ProDH2, P5CDH, NCED1, NCED3, SnRK2.4, SnRK2.6*, Aquaporins (*PIP, TIP, NIP, SIP*)	Maintenance of cellular turgor and water potential through compatible solute accumulation, ABA-mediated regulation, and water transport	[98,100–105]
Oxidative stress regulation and redox balance	*FaCSDs, FaFSDs, FaMSDs* (*FaMSD5*), *CAT, POD, APX, FaG6PDH1–FaG6PDH19, GSTU5, CHS, CHI, F3H, DFR, ANS*	Detoxification of reactive oxygen species (ROS), maintenance of redox homeostasis, and antioxidant protection under stress	[99,106–109]
Transcriptional control of stress responses	*FvNAC01–37, FaNAC2, FvNAC29, FaTINY2* (AP2/ERF), *FvMYB24, FvMYB82, MYB1, MYB5, MYB30, MYB33, MYB101, MYB108, FaZAT10* (C2H2-ZFP), *FvGRF1–10, FaTEDT1L* (HD-ZIP IV)	Regulation of downstream gene networks controlling ion homeostasis, osmotic adjustment, oxidative stress responses, growth–stress balance, and ABA signaling	[99,100,103–105,108–112]
Signal transduction and post-translational regulation	*FaCDPK1–FaCDPK11*, RBOH (CDPK-linked), FaU-box E3 ligases (*FaU-box98, FaU-box127, FaU-box136*, etc.)	Calcium- and ROS-mediated signaling and ubiquitin-dependent protein turnover modulating stress perception and response intensity	[113,114]
Cell wall remodeling and structural maintenance	*FvLAC1–57* (notably *FvLAC24, FvLAC32, FvLAC51*), *CesA, XHT, EXP, PL*, lignin- and cellulose-associated DEGs	Modification of cell wall composition and mechanical properties contributing to stress endurance and cultivar-dependent tolerance	[115,116]
Epigenetic regulation	Stress-associated DMRs (differentially methylated regions)	DNA methylation changes affect transcriptional plasticity and stress memory	[117]

https://doi.org/10.3390/plants15030432

4.1.1. Ion Homeostasis- and Transport-Related Genes

Among the functional categories, ion homeostasis- and transport-related genes show the clearest association between molecular responses and agronomic performance. In strawberry, key genes in Na^+ (*SOS1*, *SOS2*, *SOS3*, *NHX1/NHX2*) and Cl^- (*CLC*-G, *CLC*-C) transport are consistently associated with reduced ionic toxicity, improved ion partitioning, and superior whole-plant performance under saline irrigation [98,100,112]. Importantly, these associations are supported by experimental studies conducted directly in strawberry, including continuous saline irrigation under greenhouse conditions, where genotypic differences in Cl^- accumulation and partitioning correlate strongly with biomass maintenance and stress tolerance indices [98].

The relevance of these findings is reinforced by evidence from rice, where functional alleles of *OsSOS1* and *OsNHX1* contribute to improved biomass and grain yield under moderate salinity by limiting Na^+ accumulation in photosynthetically active tissues [96,97]. The consistency between strawberry-based physiological evidence and yield-validated rice examples supports the conclusion that strawberry orthologs of *SOS* and *NHX* genes represent realistic breeding targets, particularly when selection focuses on ion exclusion and compartmentation phenotypes.

In addition, several transcriptional regulators are functionally linked to ion homeostasis in strawberry, including FvMYB24 and FaTEDT1L, for which functional assays—often involving heterologous expression of the corresponding genes—indicate activation of conserved ion-regulatory pathways and enhanced salt tolerance [99,100]. FvMYB24, an R2R3-MYB transcription factor from *F. vesca*, conferred enhanced salinity tolerance when overexpressed in *Arabidopsis thaliana*, as evidenced by improved germination, root elongation, fresh biomass, and chlorophyll retention under salt stress [99]. At the molecular level, FvMYB24 is directly bound to the promoter of *SOS1* and positively regulates the expression of key salt-responsive genes, including *SOS1*, *NHX1*, and *LEA3*, thereby linking transcriptional control to Na^+ transport and ion homeostasis. Comparable regulatory behavior was observed for the HD-ZIP IV transcription factor FaTEDT1L, whose overexpression in transgenic Arabidopsis enhanced salt tolerance via increased root growth, reduced water loss, and activation of SOS pathway genes (*SOS1*, *SOS2*, *SOS3*) [100]. Despite this clear mechanistic role, these findings remain largely derived from heterologous systems and early developmental stages, limiting direct inference regarding fruit yield or long-term growth performance in strawberries. Accordingly, their breeding utility is likely to depend on allelic variation or context-specific regulation rather than constitutive expression, given that transcription factors controlling ionic responses may introduce growth penalties when misregulated [118,119].

4.1.2. Osmotic Adjustment and Water Balance

Osmotic adjustment and water balance constitute a central component of strawberry salinity responses (Table 2). The accumulation of osmotin (PR-5) under low water potential represents one of the earliest documented osmoprotective mechanisms and has been demonstrated directly in strawberry tissues [101,102]. More recent studies identify transcription factors such as FaNAC2, FvNAC29, FaTINY2 (AP2/ERF), and FaTEDT1L (HD-ZIP IV) as regulators of proline accumulation, ABA signaling, and transpiration control [100,103–105].

For many of these regulatory genes, functional validation relies heavily on heterologous expression in *Arabidopsis*, where altered survival, root growth, and water-loss phenotypes provide strong mechanistic support. For example, FaNAC2 from 'Benihoppe' strawberry, a member of the ATAF subfamily, was shown to be induced by multiple abi-

 https://doi.org/10.3390/plants15030432

otic stresses and, when overexpressed in *Nicotiana benthamiana*, enhanced salt, cold, and drought tolerance through upregulation of proline and ABA biosynthesis pathways [103].

Additional evidence comes from transgenic approaches in strawberries. Husaini et al. [102] reported that plantlets overexpressing osmotin displayed enhanced salt tolerance in vitro, including higher proline and soluble protein accumulation, improved chlorophyll retention, and survival under NaCl concentrations up to 150 mM. However, these lines also exhibited slightly reduced growth under non-stress conditions, indicating a potential fitness cost associated with constitutive overexpression. Crucially, it remains unknown whether elevated osmotin expression translates into yield or fruit-quality benefits under greenhouse or field conditions.

Similarly, heterologous expression of the woodland strawberry gene *FvNAC29* in Arabidopsis resulted in reduced ROS and malondialdehyde accumulation, increased proline and chlorophyll content, and enhanced activities of antioxidant enzymes (CAT, SOD, POD), indicating coordinated regulation of osmoprotective and oxidative stress responses under salinity [104]. FaTEDT1L has been associated with reduced water loss and enhanced root growth under salinity in transgenic assays, suggesting a contribution to water-balance regulation alongside its effects on ionic pathways [100].

In addition, overexpression of the AP2/ERF transcription factor FaTINY2 from octoploid strawberry was associated with enhanced antioxidant defense, osmotic adjustment, and reduced oxidative damage, as reflected by elevated proline and chlorophyll levels and increased CAT, SOD, and POD activities in transgenic lines [105,120]. Even so, such assays do not capture strawberries' perennial habit or the way resources are partitioned between leaves, roots, and developing fruits. Consistent with observations in rice and wheat—where NAC and AP2/ERF transcription factors enhance drought or salt tolerance but exhibit variable yield effects—osmotic adjustment genes in strawberry should currently be regarded as contributors to stress resilience rather than validated yield-stability determinants [119,120].

4.1.3. Oxidative Stress Regulation and Redox Balance

Salinity-induced oxidative stress is consistently observed in strawberry, with enhanced activity of SODs (FaCSDs, FaMSDs), catalases, peroxidases, and increased NADPH production via FaG6PDHs contributing to reduced ROS accumulation and membrane damage [106,107]. These responses are measured directly in strawberry tissues and correlate with improved physiological performance under salt stress. At the genetic level, genome-wide identification and expression analyses of the G6PDH family suggest that multiple members contribute to redox homeostasis through enhanced NADPH production under salinity [107]. Further support for the role of antioxidant systems in strawberry comes from a targeted analysis of *superoxide dismutase* (*SOD*) genes in octoploid cv. Benihoppe. Zhang et al. [106] identified salinity-induced expression of specific isoforms, including FaMSOD5, highlighting their contribution to ROS detoxification under salt stress. The study also revealed distinct expression patterns of several *FaCSD* genes (e.g., *FaCSD1*, *FaCSD2*, *FaCSD7*, *FaCSD10*) across fruit ripening stages. Because ROS function both as damaging agents under salt stress and as signaling molecules during fruit maturation, coordinated SOD activity may indirectly support yield stability by maintaining vegetative performance and normal fruit development. However, functional validation under greenhouse or field conditions is still necessary to confirm the agronomic relevance of these candidates.

In addition, the MYB transcription factor FvMYB82 (Section 4.1.4) was associated with enhanced antioxidant capacity under salinity, as indicated by increased activities of SOD, POD, and CAT in transgenic lines [112].

https://doi.org/10.3390/plants15030432

Regulatory genes such as *FaZAT10* and *MYB5*, which modulate antioxidant and flavonoid biosynthesis pathways, are largely validated through heterologous expression systems, where improved oxidative stress tolerance is readily observed [108,109]. In cultivar-level comparisons, MYB5-associated anthocyanin accumulation has been shown to track with reduced oxidative damage and improved physiological performance under salinity, reinforcing its role as a modulator of antioxidant capacity rather than a direct yield determinant [109]. Evidence from rice and tomato indicates that antioxidant capacity can support yield maintenance under salinity only when ionic stress is effectively controlled [97]. Accordingly, oxidative stress-related genes in strawberry are best interpreted as supportive components of tolerance rather than primary breeding targets.

4.1.4. Transcriptional Regulators

A substantial proportion of strawberry salinity research focuses on transcription factors, including NAC, MYB, AP2/ERF, GRF, HD-ZIP, and zinc-finger families (Table 2) [99,100,104,105,108–112]. These regulators integrate multiple stress-response pathways and therefore provide strong mechanistic entry points for understanding salinity adaptation. Consistent with this integrative role, several transcription factors discussed here also appear in ion homeostasis and osmotic adjustment sections, where their downstream physiological effects are highlighted.

For example, Li et al. [111] characterized ten *FvGRFs* in *F. vesca*, grouped into five subfamilies based on conserved QLQ and WRC domains linked to chromatin remodeling and transcriptional control. Among them, *FvGRF3*, *FvGRF6*, and *FvGRF8* showed notable salt-induced upregulation, with *FvGRF8* displaying broader tissue expression including fruit, suggesting potential roles in both stress signaling and fruit physiology.

Functional characterization of the MYB transcription factor family has revealed that FvMYB82 from *F. vesca*, through coordinated regulation of osmotic and oxidative stress responses, enables transgenic lines expressing *FvMYB82* to accumulate higher proline and chlorophyll and exhibit elevated activities of key antioxidant enzymes (SOD, POD, CAT), consistent with coupling of osmoprotective mechanisms to improved redox homeostasis under salt stress [112]. Moreover, FvMYB24 has already been analyzed in Section 4.1.1. for its role in ion homeostasis.

The HD-ZIP IV factor FaTEDT1L has been reported to enhance salt tolerance in transgenic assays, with phenotypes including improved root growth, reduced water loss, and induction of *SOS* pathway genes (*SOS1*–*SOS3*). These results suggest that FaTEDT1L can couple water-balance regulation with upstream transcriptional modulation of ionic stress responses, although direct validation of ionic outcomes and yield effects in strawberry remains limited [100].

More broadly, these examples indicate that functional evidence for many strawberry transcriptional regulators is derived largely from heterologous expression systems and assays conducted at early developmental stages. While such approaches provide strong mechanistic support, they cannot reliably predict effects on long-term growth, fruit development, or yield. Accordingly, these genes should be interpreted as indicators of regulatory potential rather than immediate breeding targets, particularly given evidence from rice and wheat that transcription factors often exert pleiotropic effects on growth, flowering, and yield [118,119].

4.1.5. Signal Transduction and Post-Translational Regulators

Salinity responses in strawberry are associated with Ca^{2+} signaling, phosphorylation cascades, and ubiquitin-mediated protein turnover. Crizel et al. [113] characterized the FaCDPK family in octoploid strawberry as Ca^{2+}-responsive kinases linking Ca^{2+} signatures

 https://doi.org/10.3390/plants15030432

to downstream stress responses; among the reported genes, *FaCDPK1* showed the strongest induction under salinity (≈15-fold). Responses to ABA and the ABA-biosynthesis inhibitor NDGA further support isoform-specific hormonal regulation rather than a simple ABA-independent salt model. Jiang et al. [114] provided complementary evidence for protein-level regulation by identifying 155 FaU-box (PUB) E3 ubiquitin ligases and profiling their expression during development and under abiotic stresses, including salt exposure at 200 mM NaCl. The study highlighted candidates including a salt-associated gene (*FaU-box3*) and additional members responsive across stress and hormonal contexts. Importantly, transient overexpression assays indicated that FaU-box127 is associated with changes in fruit composition traits, including soluble sugars and organic acids, emphasizing that post-translational regulators can affect both stress responses and fruit-quality processes [114].

From a breeding perspective, *CDPKs* and *U-box E3* ligases are unlikely to be immediate single-gene targets due to pleiotropic effects; accordingly, stress-responsive members such as *FaCDPK1* and salt-associated candidates such as *FaU-box3* are best prioritized for allele mining and QTL/GWAS colocalization, followed by validation under saline irrigation with yield and quality assessment [113,114].

4.1.6. Structural and Epigenetic Contributors

Structural adaptations mediated by cell wall remodeling genes, including laccases (FvLAC family), contribute to cultivar-dependent stress endurance and are supported by expression analyses conducted directly in strawberry (Table 1) [115,116]. Kong et al. [115] also reported divergent functional roles among FvLAC isoforms, with FvLAC51 associated with enhanced tolerance and FvLAC24 and FvLAC32 linked to reduced performance, indicating specialization within this cell-wall remodeling family. Promoter analyses further suggest possible involvement of some *FvLAC* genes in fruit-related developmental processes, although their agronomic relevance under salinity remains to be established. Nevertheless, as observed in barley and maize, links between cell wall modification and yield under salinity remain largely indirect.

Similarly, epigenetic modifications, such as stress-associated DNA methylation changes, provide insight into transcriptional plasticity and stress memory in strawberry [117]. While these mechanisms are biologically compelling, their relevance to breeding depends on the stability and heritability of epigenetic states under agricultural conditions.

4.2. Breeding Implications

Overall, the reviewed evidence reveals a clear hierarchy in the breeding relevance of molecular mechanisms underlying salinity tolerance in strawberries. Genes directly controlling ion homeostasis—particularly members of the SOS and NHX families—emerge as the most credible near-term breeding targets, supported by sustained salinity experiments conducted in strawberry and by yield-validated examples from rice [96–98]. In contrast, genes involved in osmotic adjustment and oxidative stress regulation provide essential physiological support but lack consistent validation for fruit yield stability. Transcription factors and upstream signaling regulators, many of which have been characterized primarily through heterologous expression systems, should therefore be regarded as promising candidates whose breeding value will depend on strawberry-specific validation, careful regulatory tuning, and assessment of potential pleiotropic effects on growth, development, and fruit traits.

Despite progress in elucidating the molecular basis of salinity tolerance, translation of this knowledge into durable breeding outcomes remains uneven. Bridging this gap will require integrating mechanistic insights with long-term saline irrigation trials that extend to fruit yield and quality, together with genomic approaches that exploit natu-

 https://doi.org/10.3390/plants15030432

ral allelic variation rather than relying solely on transgenic overexpression [121]. Such coordinated strategies will be essential for enhancing strawberry resilience under saline cultivation conditions.

5. Future Perspectives

Although substantial progress has been made in elucidating physiological responses, identifying stress-responsive genes, and evaluating biological mitigation strategies, a clear gap persists between current knowledge and its effective deployment in breeding programs and commercial production systems.

A central priority is the functional validation of candidate genes in relation to reproductive performance rather than vegetative tolerance alone. Numerous genes involved in ion homeostasis, osmotic adjustment, antioxidant defense, and transcriptional regulation have been identified; however, most have been characterized at early developmental stages or in heterologous systems, and rarely under field conditions. Future studies should prioritize strawberry-specific validation under prolonged salinity, extending evaluation to flowering, fruit set, yield stability, and fruit quality under realistic field conditions. In this context, genes directly regulating Na^+ and Cl^- exclusion, compartmentation, and long-distance transport represent the most promising breeding targets, as they address the primary causes of yield loss while minimizing growth penalties.

The exploitation of natural allelic variation, rather than constitutive gene overexpression, should underpin future genetic improvement efforts. High-throughput phenotyping combined with genomic selection, marker-assisted breeding, and allele mining will be essential for identifying favorable variants that confer tolerance while preserving agronomic performance. CRISPR/Cas-based engineering may further accelerate the development of elite cultivars. Crop wild relatives, particularly salinity-tolerant accessions of *F. chiloensis*, remain an underutilized reservoir of adaptive traits. Their systematic integration through pre-breeding and advanced backcross populations will be critical for broadening the genetic base of cultivated strawberries, provided introgression is coupled with rigorous selection for agronomic and market traits.

In parallel, biological and agronomic interventions should be repositioned as complementary tools that enhance the expression of genetic tolerance rather than stand-alone mitigation measures. Biostimulants, small signaling molecules, and beneficial microorganisms consistently improve physiological resilience; however, evidence for their effects on yield and fruit quality remains limited. Future research should therefore focus on genotype × treatment interactions, enabling the co-development of tolerant cultivars and optimized management protocols. In this context, plant–microbe interactions represent a particularly promising yet underexploited research path. Arbuscular mycorrhizal fungi and plant growth–promoting rhizobacteria improve nutrient uptake, ion homeostasis, and root system functionality under saline conditions, and incorporating microbial responsiveness as a selection criterion could substantially enhance cultivar performance.

Additionally, advances in controlled-environment cultivation provide more opportunities to accelerate progress. Manipulation of light spectra can be used to refine stress exposure and improve salinity responses, thereby maintaining fruit quality and yield.

Overall, progress in this field will require a transition from stress tolerance screening to resilience-driven cultivar design. Successful strawberry improvement under salinity will depend on pyramiding ion-regulatory capacity, physiological plasticity, and yield stability, supported by targeted biological and agronomic interventions. The identification of cultivars capable of exploiting moderate salinity without compromising productivity, by achieving an optimal balance between mild salinity-induced improvements in fruit quality and yield penalties, could also become a specific breeding objective. By integrating

https://doi.org/10.3390/plants15030432

molecular knowledge with long-term field validation and breeding-oriented selection, it will be possible to develop strawberry cultivars that are not only salt-tolerant but also productive and economically viable under saline cultivation conditions.

6. Conclusions

This review provides a framework for breeding strategies and other mitigation solutions aimed at enhancing strawberry resilience to saline environments and offers practical insights to support growers managing production in salt-affected regions. In this context, it synthesizes current knowledge of strawberry responses to salinity stress, with emphasis on physiological, biochemical, and molecular mechanisms. However, available evidence indicates that most of these responses have been characterized primarily at vegetative or early developmental stages, and their relevance to fruit yield and quality remains limited or context dependent.

Significant genotypic variation shapes the degree of salinity tolerance among strawberry cultivars and CWRs, and many genes have been identified as stress-related, highlighting the complexity of tolerance mechanisms and the need for careful interpretation of stress-response traits, although often without a clear impact on yield. Taken together, the available evidence indicates that although several approaches exist to mitigate salinity stress in strawberries and can improve plant performance and survival under saline conditions, these strategies do not necessarily guarantee stable or enhanced fruit yield. Overall, this synthesis underscores the distinction between mechanistic indicators of salinity response and traits with demonstrated agronomic relevance in strawberry production under saline conditions.

Supplementary Materials: The following supporting information can be downloaded at: https://www.mdpi.com/article/10.3390/plants15030432/s1, Table S1: Responses of strawberry cultivars to salt stress at vegetative/no fruiting stage; Table S2: Studies referring to small signaling molecules and biostimulants involved in mitigating salinity stress, including the effects of supplemental lighting, microorganisms, and plant–microbe interactions under saline conditions; Table S3: Experimental settings, plant developmental stages, growth conditions, salinity stress protocols and parameters identified for all studies summarized in Table 2.

Author Contributions: Conceptualization, K.P., E.T. and V.P.; writing—original draft preparation, K.P., M.G. (Maria Goufa), M.G. (Maria Gerakari), E.T. and V.P.; writing—review and editing, K.P., M.G. (Maria Gerakari), E.T. and V.P.; visualization, M.G. (Maria Gerakari); supervision, E.T. and V.P. All authors have read and agreed to the published version of the manuscript.

Funding: This research received no external funding.

Data Availability Statement: No new data were created or analyzed in this study.

Conflicts of Interest: The authors declare no conflicts of interest.

References

1. Pisinaras, V.; Tsihrintzis, V.A.; Petalas, C.; Ouzounis, K. Soil Salinization in the Agricultural Lands of Rhodope District, Northeastern Greece. *Environ. Monit. Assess.* **2010**, *166*, 79–94. [CrossRef] [PubMed]
2. Hardie, M.; Doyle, R. Measuring Soil Salinity. In *Plant Salt Tolerance*; Shabala, S., Cuin, T.A., Eds.; Methods in Molecular Biology; Humana Press: Totowa, NJ, USA, 2012; Volume 913, pp. 415–425.
3. Tóth, T. Classification and Mitigation of Soil Salinization. In *Oxford Research Encyclopedia of Environmental Science*; Oxford University Press: Oxford, UK, 2017.
4. FAO. *Global Map of Salt-Affected Soils 2021*; Food and Agriculture Organization: Rome, Italy.
5. Singh, M.; Nara, U.; Kumar, A.; Choudhary, A.; Singh, H.; Thapa, S. Salinity Tolerance Mechanisms and Their Breeding Implications. *J. Genet. Eng. Biotechnol.* **2021**, *19*, 173. [CrossRef] [PubMed]

6. Pessarakli, M.; Szabolcs, I. Soil Salinity and Sodicity as Particular Plant/Crop Stress Factors. In *Handbook of Plant and Crop Stress, Fourth Edition*; CRC Press: Boca Raton, FL, USA, 2019; pp. 3–21.
7. Petalas, C.; Pisinaras, V.; Gemitzi, A.; Tsihrintzis, V.A.; Ouzounis, K. Current Conditions of Saltwater Intrusion in the Coastal Rhodope Aquifer System, Northeastern Greece. *Desalination* **2009**, *237*, 22–41. [CrossRef]
8. Karamanos, A.; Aggelides, S.; Londra, P. Water Use Efficiency and Water Productivity in Greece. *Options Méditerr.* **2005**, *57*, 92–100.
9. Daliakopoulos, I.N.; Tsanis, I.K.; Koutroulis, A.; Kourgialas, N.N.; Varouchakis, A.E.; Karatzas, G.P.; Ritsema, C.J. The Threat of Soil Salinity: A European Scale Review. *Sci. Total Environ.* **2016**, *573*, 727–739. [CrossRef]
10. Yuan, X.; Li, S.; Chen, J.; Yu, H.; Yang, T.; Wang, C.; Huang, S.; Chen, H.; Ao, X. Impacts of Global Climate Change on Agricultural Production: A Comprehensive Review. *Agronomy* **2024**, *14*, 1360. [CrossRef]
11. Terán, F.; Vives-Peris, V.; Gómez-Cadenas, A.; Pérez-Clemente, R.M. Facing Climate Change: Plant Stress Mitigation Strategies in Agriculture. *Physiol. Plant.* **2024**, *176*, e14484. [CrossRef]
12. Zaman, M.; Shahid, S.A.; Heng, L. *Guideline for Salinity Assessment, Mitigation and Adaptation Using Nuclear and Related Techniques*; Springer Nature: London, UK, 2018.
13. Negrão, S.; Schmöckel, S.M.; Tester, M. Evaluating Physiological Responses of Plants to Salinity Stress. *Ann. Bot.* **2017**, *119*, 1–11. [CrossRef]
14. Xiong, W.; Reynolds, M.; Xu, Y. Climate Change Challenges Plant Breeding. *Curr. Opin. Plant Biol.* **2022**, *70*, 102308. [CrossRef]
15. Shokat, S.; Großkinsky, D.K. Tackling Salinity in Sustainable Agriculture—What Developing Countries May Learn from Approaches of the Developed World. *Sustainability* **2019**, *11*, 4558. [CrossRef]
16. Hancock, J.F. *Temperate Fruit Crop Breeding: Germplasm to Genomics*; Springer Science & Business Media: Berlin/Heidelberg, Germany, 2008.
17. Ferreira, J.F.; Liu, X.; Suarez, D.L. Fruit Yield and Survival of Five Commercial Strawberry Cultivars under Field Cultivation and Salinity Stress. *Sci. Hortic.* **2019**, *243*, 401–410. [CrossRef]
18. Mazzoni, L.; Perez-Lopez, P.; Giampieri, F.; Alvarez-Suarez, J.M.; Gasparrini, M.; Forbes-Hernandez, T.Y.; Quiles, J.L.; Mezzetti, B.; Battino, M. The Genetic Aspects of Berries: From Field to Health. *J. Sci. Food Agric.* **2016**, *96*, 365–371. [CrossRef] [PubMed]
19. Giampieri, F.; Alvarez-Suarez, J.M.; Battino, M. Strawberry and Human Health: Effects beyond Antioxidant Activity. *J. Agric. Food Chem.* **2014**, *62*, 3867–3876. [CrossRef] [PubMed]
20. Galli, V.; da Silva Messias, R.; Perin, E.C.; Borowski, J.M.; Bamberg, A.L.; Rombaldi, C.V. Mild Salt Stress Improves Strawberry Fruit Quality. *Lwt* **2016**, *73*, 693–699. [CrossRef]
21. Grieve, C.M.; Grattan, S.R.; Maas, E.V. Plant Salt Tolerance. *ASCE Man. Rep. Eng. Pract.* **2012**, *71*, 405–459.
22. Denaxa, N.-K.; Nomikou, A.; Malamos, N.; Liveri, E.; Roussos, P.A.; Papasotiropoulos, V. Salinity Effect on Plant Growth Parameters and Fruit Bioactive Compounds of Two Strawberry Cultivars, Coupled with Environmental Conditions Monitoring. *Agronomy* **2022**, *12*, 2279. [CrossRef]
23. Ghaderi, N.; Hatami, M.R.; Mozafari, A.; Siosehmardeh, A. Change in Antioxidant Enzymes Activity and Some Morpho-Physiological Characteristics of Strawberry under Long-Term Salt Stress. *Physiol. Mol. Biol. Plants* **2018**, *24*, 833–843. [CrossRef]
24. Mozafari, A.; Ghaderi, N.; Havas, F.; Dedejani, S. Comparative Investigation of Structural Relationships among Morpho-Physiological and Biochemical Properties of Strawberry (*Fragaria* × *ananassa* Duch.) under Drought and Salinity Stresses: A Study Based on in Vitro Culture. *Sci. Hortic.* **2019**, *256*, 108601. [CrossRef]
25. Hussein, A.E.; El-Kerdany, A.Y.; Afifi, K.M. Effect of Drought and Salinity Stresses on Two Strawberry Cultivars during Their Regeneration in Vitro. *Int. J. Innov. Sci. Eng. Technol.* **2016**, *4*, 83–93.
26. Akbari, M.; Mahna, N.; Ramesh, K.; Bandehagh, A.; Mazzuca, S. Ion Homeostasis, Osmoregulation, and Physiological Changes in the Roots and Leaves of Pistachio Rootstocks in Response to Salinity. *Protoplasma* **2018**, *255*, 1349–1362. [CrossRef]
27. Sikder, R.K.; Wang, X.; Zhang, H.; Gui, H.; Dong, Q.; Jin, D.; Song, M. Nitrogen Enhances Salt Tolerance by Modulating the Antioxidant Defense System and Osmoregulation Substance Content in *Gossypium hirsutum*. *Plants* **2020**, *9*, 450. [CrossRef]
28. Asghar, R.; Biglarifard, A.; Mirdehghan, H.; Borghei, S.F. Influence of NaCl Salinity on Growth Analysis of Strawberry Cv. Camarosa. *J. Stress. Physiol. Biochem.* **2011**, *7*, 145–156.
29. Wu, Y.; Li, L.; Li, M.; Zhang, M.; Sun, H.; Sigrimis, N. Optimal Fertigation for High Yield and Fruit Quality of Greenhouse Strawberry. *PLoS ONE* **2020**, *15*, e0224588. [CrossRef] [PubMed]
30. Keutgen, A.; Pawelzik, E. Modifications of Taste-Relevant Compounds in Strawberry Fruit under NaCl Salinity. *Food Chem.* **2007**, *105*, 1487–1494. [CrossRef]
31. Endo, T.; Yamamoto, S.; Larrinaga, J.A.; Fujiyama, H.; Honna, T. Status and Causes of Soil Salinization of Irrigated Agricultural Lands in Southern Baja California, Mexico. *Appl. Environ. Soil Sci.* **2011**, *2011*, 1–12. [CrossRef]
32. Singh, A. Soil Salinization and Waterlogging: A Threat to Environment and Agricultural Sustainability. *Ecol. Indic.* **2015**, *57*, 128–130. [CrossRef]

https://doi.org/10.3390/plants15030432

33. Keutgen, A.J.; Pawelzik, E. Quality and Nutritional Value of Strawberry Fruit under Long Term Salt Stress. *Food Chem.* **2008**, *107*, 1413–1420. [CrossRef]
34. Keutgen, A.J.; Pawelzik, E. Impacts of NaCl Stress on Plant Growth and Mineral Nutrient Assimilation in Two Cultivars of Strawberry. *Environ. Exp. Bot.* **2009**, *65*, 170–176. [CrossRef]
35. Hu, Y.; Feng, C.; Yang, L.; Edger, P.P.; Kang, M. Genomic Population Structure and Local Adaptation of the Wild Strawberry *Fragaria* Nilgerrensis. *Hortic. Res.* **2022**, *9*, uhab059. [CrossRef]
36. Bringhurst, R.S.; Hancock, J.F.; Voth, V. The Beach Strawberry, an Important Natural Resource. *Calif. Agric.* **1977**, *31*, 10.
37. VanDerZanden, A.M.; Cameron, J.S. Effect of Water Deficit Stress on 11 Native *Fragaria chiloensis* Clones Selected as Ornamental Groundcovers. *Sci. Hortic.* **1996**, *66*, 241–253. [CrossRef]
38. Carrasco, B.; Garcés, M.; Rojas, P.; Saud, G.; Herrera, R.; Retamales, J.B.; Caligari, P.D. The Chilean Strawberry [*Fragaria chiloensis* (L.) Duch.]: Genetic Diversity and Structure. *J. Am. Soc. Hortic. Sci.* **2007**, *132*, 501–506. [CrossRef]
39. Nikoloudi, A. *Salinity Tolerance in Strawberry (Fragaria spp.) as Influenced by Genotype*; Michigan State University: East Lansing, MI, USA, 2003.
40. Garriga, M.; Muñoz, C.A.; Caligari, P.D.; Retamales, J.B. Effect of Salt Stress on Genotypes of Commercial (*Fragaria* × *ananassa*) and Chilean Strawberry (*F. chiloensis*). *Sci. Hortic.* **2015**, *195*, 37–47. [CrossRef]
41. Kortekamp, T.; Chen, T.-W.; Wagner, H.; Olbricht, K. Evaluation of Salinity Tolerance in a *Fragaria* F2 Population. In *Proceedings of the XVI EUCARPIA Symposium on Fruit Breeding and Genetics 1412*; International Society for Horticultural Science: Leuven (Korbeek-Lo), Belgium, 2023; pp. 111–116.
42. Demiral, M.A. Effect of Salt Stress on Concentration of Nitrogen and Phosphorus in Root and Leaf of Strawberry Plant. *Eurasian J. Soil Sci.* **2017**, *6*, 357–364. [CrossRef]
43. Orsini, F.; Alnayef, M.; Bona, S.; Maggio, A.; Gianquinto, G. Low Stomatal Density and Reduced Transpiration Facilitate Strawberry Adaptation to Salinity. *Environ. Exp. Bot.* **2012**, *81*, 1–10. [CrossRef]
44. Turhan, E.; Eris, A. Changes of Growth, Amino Acids, and Ionic Composition in Strawberry Plants under Salt Stress Conditions. *Commun. Soil Sci. Plant Anal.* **2009**, *40*, 3308–3322. [CrossRef]
45. Turhan, E.; Eris, A. Changes of Micronutrients, Dry Weight, and Chlorophyll Contents in Strawberry Plants Under Salt Stress Conditions. *Commun. Soil Sci. Plant Anal.* **2005**, *36*, 1021–1028. [CrossRef]
46. Turhan, E.; Eris, A. Effects of Sodium Chloride Applications and Different Growth Media on Ionic Composition in Strawberry Plant. *J. Plant Nutr.* **2005**, *27*, 1653–1665. [CrossRef]
47. Gulen, H.; Turhan, E.; Eris, A. Changes in Peroxidase Activities and Soluble Proteins in Strawberry Varieties under Salt-Stress. *Acta Physiol. Plant.* **2006**, *28*, 109–116. [CrossRef]
48. Alnayef, M.; Hartley, J.; Orsini, F.; Di Silvestro, R.; Sanoubar, R.; Marotti, I.; Gianquinto, G.; Dinelli, G.; Puniran-Hartley, N. Using Salinity to Improve Nutritional and Market Value of Strawberries. *Aust. J. Crop Sci.* **2022**, *16*, 7–17. [CrossRef]
49. Pirlak, L.; Eşitken, A. Salinity Effects on Growth, Proline and Ion Accumulation in Strawberry Plants. *Acta Agric. Scand. Sect. B—Soil Plant Sci.* **2004**, *54*, 189–192. [CrossRef]
50. Rahimi, A.; Biglarifard, A. Influence of NaCl Salinity and Different Substracts on Plant Growth, Mineral Nutrient Assimilation and Fruit Yield of Strawberry. *Not. Bot. Horti Agrobot. Cluj-Napoca* **2011**, *39*, 219–226. [CrossRef]
51. Ondrašek, G.; Romić, D.; Romić, M.; Duralija, B.; Mustač, I. Strawberry Growth and Fruit Yield in a Saline Environment. *Agric. Conspec. Sci.* **2006**, *71*, 155–158.
52. Alhamza Juameer, R.A.; Assi Obaid, A.; Ayed Yousif, S. Improved Micropropagation and Salinity Tolerance of Strawberry (*Fragaria* × *ananassa* L.) Cv. Albion. *Rev. Bionatura* **2022**, *7*, 34. [CrossRef]
53. Yİldİz, K.; Uzal, Ö.; Yİlmaz, H. Consequences of NaCl Salinity on Growth and Ion Accumulation in Selected Strawberry Cultivars. *Eur. J. Hortic. Sci.* **2008**, *73*, 69–72. [CrossRef]
54. Sun, Y.; Niu, G.; Wallace, R.; Masabni, J.; Gu, M. Relative Salt Tolerance of Seven Strawberry Cultivars. *Horticulturae* **2015**, *1*, 27–43. [CrossRef]
55. Chan, Z.; Shi, H. Improved Abiotic Stress Tolerance of Bermudagrass by Exogenous Small Molecules. *Plant Signal. Behav.* **2015**, *10*, e991577. [CrossRef]
56. Averina, N.G.; Gritskevich, E.R.; Vershilovskaya, I.V.; Usatov, A.V.; Yaronskaya, E.B. Mechanisms of Salt Stress Tolerance Development in Barley Plants under the Influence of 5-Aminolevulinic Acid. *Russ. J. Plant Physiol.* **2010**, *57*, 792–798. [CrossRef]
57. Wu, Y.; Liao, W.; Dawuda, M.M.; Hu, L.; Yu, J. 5-Aminolevulinic Acid (ALA) Biosynthetic and Metabolic Pathways and Its Role in Higher Plants: A Review. *Plant Growth Regul.* **2019**, *87*, 357–374. [CrossRef]
58. Wu, Y.; Jin, X.; Liao, W.; Hu, L.; Dawuda, M.M.; Zhao, X.; Tang, Z.; Gong, T.; Yu, J. 5-Aminolevulinic Acid (ALA) Alleviated Salinity Stress in Cucumber Seedlings by Enhancing Chlorophyll Synthesis Pathway. *Front. Plant Sci.* **2018**, *9*, 635. [CrossRef]
59. Lu, W.; Fan, H.; Zhang, Y.; Cai, B.; Wang, X.; Xue, Z.; Li, Q. Low-Concentration NaCl Foliar Spraying Enhances Photosynthesis, Mineral Concentration, and Fruit Quality of Strawberry during Greenhouse High-Temperature Periods. *BMC Plant Biol.* **2025**, *25*, 487. [CrossRef]

 https://doi.org/10.3390/plants15030432

60. Wei, W.; Li, Q.-T.; Chu, Y.-N.; Reiter, R.J.; Yu, X.-M.; Zhu, D.-H.; Zhang, W.-K.; Ma, B.; Lin, Q.; Zhang, J.-S. Melatonin Enhances Plant Growth and Abiotic Stress Tolerance in Soybean Plants. *J. Exp. Bot.* **2015**, *66*, 695–707. [CrossRef] [PubMed]
61. Rahman, M.M.; Keya, S.S.; Sahu, A.; Gupta, A.; Dhingra, A.; Tran, L.-S.P.; Mostofa, M.G. Acetic Acid: A Cheap but Chief Metabolic Regulator for Abiotic Stress Tolerance in Plants. *Stress. Biol.* **2024**, *4*, 34. [CrossRef] [PubMed]
62. Sun, Y.P.; Liu, J.; Cao, R.X.; Huang, Y.J.; Hall, A.M.; Guo, C.B.; Wang, L.J. Effects of 5-Aminolevulinic Acid Treatment on Photosynthesis of Strawberry. *Photosynthetica* **2017**, *55*, 276–284. [CrossRef]
63. Wu, W.; He, S.; An, Y.; Cao, R.; Sun, Y.; Tang, Q.; Wang, L. Hydrogen Peroxide as a Mediator of 5-aminolevulinic Acid-induced Na^+ Retention in Roots for Improving Salt Tolerance of Strawberries. *Physiol. Plant.* **2019**, *167*, 5–20. [CrossRef]
64. Zhong, Y.; Wei, X.; Zhang, J.; Wang, L. Transcriptome Sequencing Reveals Jasmonate Playing a Key Role in ALA-Induced Osmotic Stress Tolerance in Strawberry. *BMC Plant Biol.* **2025**, *25*, 41. [CrossRef]
65. Li, J.; Liu, J.; Zhu, T.; Zhao, C.; Li, L.; Chen, M. The Role of Melatonin in Salt Stress Responses. *Int. J. Mol. Sci.* **2019**, *20*, 1735. [CrossRef]
66. Pardo-Hernández, M.; López-Delacalle, M.; Rivero, R.M. ROS and NO Regulation by Melatonin under Abiotic Stress in Plants. *Antioxidants* **2020**, *9*, 1078. [CrossRef]
67. Georgiadou, E.C.; García, C.J.; Taliadorou, A.M.; Gedeon, S.; Valanides, N.; Varaldo, A.; Gohari, G.; Balsells-Llauradó, M.; Alcázar, R.; Hertog, M.L. Pre-Harvest Application of Sodium Alginate Functionalized with Melatonin Enhances Secondary Metabolism in Strawberry Fruit. *Curr. Plant Biol.* **2025**, *43*, 100515. [CrossRef]
68. Zahedi, S.M.; Hosseini, M.S.; Abadía, J.; Marjani, M. Melatonin Foliar Sprays Elicit Salinity Stress Tolerance and Enhance Fruit Yield and Quality in Strawberry (*Fragaria* × *ananassa* Duch.). *Plant Physiol. Biochem.* **2020**, *149*, 313–323. [CrossRef]
69. Bandurska, H. Salicylic Acid: An Update on Biosynthesis and Action in Plant Response to Water Deficit and Performance Under Drought. In *Salicylic Acid*; Hayat, S., Ahmad, A., Alyemeni, M.N., Eds.; Springer: Dordrecht, The Netherlands, 2013; pp. 1–14.
70. Karlidag, H.; Yildirim, E.; Turan, M. Melhora Do Efeito de Estresse Salino Em Morango Por Adição de Ácido Salicílico. *Sci. Agric.* **2009**, *66*, 180–187. [CrossRef]
71. Jamali, B.; Eshghi, S. Salicylic Acid–Induced Salinity Redressal in Hydroponically Grown Strawberry. *Commun. Soil Sci. Plant Anal.* **2015**, *46*, 1482–1493. [CrossRef]
72. Samadi, S.; Habibi, G.; Vaziri, A. Effects of Exogenous Salicylic Acid on Antioxidative Responses, Phenolic Metabolism and Photochemical Activity of Strawberry under Salt Stress. *Iran. J. Plant Physiol.* **2019**, *9*, 2685–2694.
73. Roshdy, A.E.-D.; Alebidi, A.; Almutairi, K.; Al-Obeed, R.; Elsabagh, A. The Effect of Salicylic Acid on the Performances of Salt Stressed Strawberry Plants, Enzymes Activity, and Salt Tolerance Index. *Agronomy* **2021**, *11*, 775. [CrossRef]
74. Faghih, S.; Ghobadi, C.; Zarei, A. Response of Strawberry Plant cv. Camarosa to Salicylic Acid and Methyl Jasmonate Application under Salt Stress Condition. *J. Plant Growth Regul.* **2017**, *36*, 651–659. [CrossRef]
75. Zhang, Y.; Deng, M.; Lin, B.; Tian, S.; Chen, Y.; Huang, S.; Lin, Y.; Li, M.; He, W.; Wang, Y.; et al. Physiological and Transcriptomic Evidence Revealed the Role of Exogenous GABA in Enhancing Salt Tolerance in Strawberry Seedlings. *BMC Genom.* **2025**, *26*, 196. [CrossRef] [PubMed]
76. Golnari, S.; Vafaee, Y.; Nazari, F.; Ghaderi, N. Gamma-Aminobutyric Acid (GABA) and Salinity Impacts Antioxidative Response and Expression of Stress-Related Genes in Strawberry Cv. Aromas. *Braz. J. Bot.* **2021**, *44*, 639–651. [CrossRef]
77. Nabi, F.; Sarfaraz, A.; Kama, R.; Kanwal, R.; Li, H. Structure-Based Function of Humic Acid in Abiotic Stress Alleviation in Plants: A Review. *Plants* **2025**, *14*, 1916. [CrossRef]
78. Canellas, L.P.; Da Silva, R.M.; Busato, J.G.; Olivares, F.L. Humic Substances and Plant Abiotic Stress Adaptation. *Chem. Biol. Technol. Agric.* **2024**, *11*, 66. [CrossRef]
79. Mora, V.; Olaetxea, M.; Bacaicoa, E.; Baigorri, R.; Fuentes, M.; Zamarreño, A.M.; Garcia-Mina, J.M. Abiotic Stress Tolerance in Plants: Exploring the Role of Nitric Oxide and Humic Substances. In *Nitric Oxide in Plants: Metabolism and Role in Stress Physiology*; Khan, M.N., Mobin, M., Mohammad, F., Corpas, F.J., Eds.; Springer International Publishing: Cham, Switzerland, 2014; pp. 243–264.
80. Saidimoradi, D.; Ghaderi, N.; Javadi, T. Salinity Stress Mitigation by Humic Acid Application in Strawberry (*Fragaria* × *ananassa* Duch.). *Sci. Hortic.* **2019**, *256*, 108594. [CrossRef]
81. Mirfattahi, Z.; Eshghi, S.; Gharaghani, A.; Etemadi, M.; Moghadam, A. Acetic Acid Application Timing on Strawberry: An Alleviator for Salinity Adverse Effect. *Acta Physiol. Plant* **2022**, *44*, 139. [CrossRef]
82. Moradi, P.; Vafaee, Y.; Mozafari, A.A.; Tahir, N.A. Silicon Nanoparticles and Methyl Jasmonate Improve Physiological Response and Increase Expression of Stress-Related Genes in Strawberry cv. Paros under Salinity Stress. *Silicon* **2022**, *14*, 10559–10569. [CrossRef]
83. Zeid, I.M.A.; Mohamed, F.H.; Metwali, E.M. Responses of Two Strawberry Cultivars to NaCl-Induced Salt Stress under the Influence of ZnO Nanoparticles. *Saudi J. Biol. Sci.* **2023**, *30*, 103623. [CrossRef]
84. Evelin, H.; Kapoor, R.; Giri, B. Arbuscular Mycorrhizal Fungi in Alleviation of Salt Stress: A Review. *Ann. Bot.* **2009**, *104*, 1263–1280. [CrossRef]

85. Boorboori, M.R.; Lackóová, L. Arbuscular Mycorrhizal Fungi and Salinity Stress Mitigation in Plants. *Front. Plant Sci.* **2025**, *15*, 1504970. [CrossRef] [PubMed]
86. Haque, S.I.; Matsubara, Y. Salinity Tolerance and Sodium Localization in Mycorrhizal Strawberry Plants. *Commun. Soil Sci. Plant Anal.* **2018**, *49*, 2782–2792. [CrossRef]
87. Karlidag, H.; Yildirim, E.; Turan, M.; Pehluvan, M.; Donmez, F. Plant Growth-Promoting Rhizobacteria Mitigate Deleterious Effects of Salt Stress on Strawberry Plants (*Fragaria* × *ananassa*). *HortScience* **2013**, *48*, 563–567. [CrossRef]
88. Yavuz, A.; Erdogan, U.; Turan, M.; Argın, S.; Kocaman, A. Synergistic Strategies for Overcoming Salt Stress in Strawberry Farming: The Use of Organic Fertilizers and Plant Growth Promoting Rhizobacteria (PGPR). *Appl. Fruit. Sci.* **2024**, *66*, 1787–1797. [CrossRef]
89. Redondo-Gómez, S.; Mesa-Marín, J.; Pérez-Romero, J.A.; López-Jurado, J.; García-López, J.V.; Mariscal, V.; Molina-Heredia, F.P.; Pajuelo, E.; Rodríguez-Llorente, I.D.; Flowers, T.J. Consortia of Plant-Growth-Promoting Rhizobacteria Isolated from Halophytes Improve Response of Eight Crops to Soil Salinization and Climate Change Conditions. *Agronomy* **2021**, *11*, 1609. [CrossRef]
90. Sangiorgio, D.; Cellini, A.; Spinelli, F.; Donati, I. Promoting Strawberry (*Fragaria* × *ananassa*) Stress Resistance, Growth, and Yield Using Native Bacterial Biostimulants. *Agronomy* **2023**, *13*, 529. [CrossRef]
91. Esmaeilizadeh, M.; Malekzadeh Shamsabad, M.R.; Roosta, H.R.; Dąbrowski, P.; Rapacz, M.; Zieliński, A.; Wróbel, J.; Kalaji, H.M. Manipulation of Light Spectrum Can Improve the Performance of Photosynthetic Apparatus of Strawberry Plants Growing under Salt and Alkalinity Stress. *PLoS ONE* **2021**, *16*, e0261585. [CrossRef]
92. Malekzadeh Shamsabad, M.R.; Esmaeilizadeh, M.; Roosta, H.R.; Dąbrowski, P.; Telesiński, A.; Kalaji, H.M. Supplemental Light Application Can Improve the Growth and Development of Strawberry Plants under Salinity and Alkalinity Stress Conditions. *Sci. Rep.* **2022**, *12*, 9272. [CrossRef] [PubMed]
93. Malekzadeh, M.R.; Roosta, H.R.; Kalaji, H.M. Enhancing Strawberry Resilience to Saline, Alkaline, and Combined Stresses with Light Spectra: Impacts on Growth, Enzymatic Activity, Nutrient Uptake, and Osmotic Regulation. *BMC Plant Biol.* **2024**, *24*, 1038. [CrossRef] [PubMed]
94. Wang, R.; Chu, S.; Zhang, D.; Hayat, K.; Zhang, X.; Chi, Y.; Ma, X.; Chen, X.; Yang, H.; Ding, W.; et al. Alleviation of Salt Stress in Strawberries by Hydrogen-rich Water: Physiological, Transcriptomic and Metabolomic Responses. *Physiol. Plant.* **2025**, *177*, e70151. [CrossRef] [PubMed]
95. Roychowdhury, R.; Das, S.P.; Gupta, A.; Parihar, P.; Chandrasekhar, K.; Sarker, U.; Kumar, A.; Ramrao, D.P.; Sudhakar, C. Multi-Omics Pipeline and Omics-Integration Approach to Decipher Plant's Abiotic Stress Tolerance Responses. *Genes* **2023**, *14*, 1281. [CrossRef]
96. Zhang, J.-L.; Shi, H. Physiological and Molecular Mechanisms of Plant Salt Tolerance. *Photosynth. Res.* **2013**, *115*, 1–22. [CrossRef]
97. Chen, H.; He, X.; Qing, L.; Wu, Y.; Ren, C.; Sheriff, R.E.; Zhu, C. Real-World Single Image Super-Resolution: A Brief Review. *Inf. Fusion* **2022**, *79*, 124–145. [CrossRef]
98. Sandhu, D.; Pudussery, M.V.; Ferreira, J.F.; Liu, X.; Pallete, A.; Grover, K.K.; Hummer, K. Variable Salinity Responses and Comparative Gene Expression in Woodland Strawberry Genotypes. *Sci. Hortic.* **2019**, *254*, 61–69. [CrossRef]
99. Wang, S.; Shi, M.; Zhang, Y.; Xie, X.; Sun, P.; Fang, C.; Zhao, J. FvMYB24, a Strawberry R2R3-MYB Transcription Factor, Improved Salt Stress Tolerance in Transgenic Arabidopsis. *Biochem. Biophys. Res. Commun.* **2021**, *569*, 93–99. [CrossRef]
100. Chu, C.-Y.; Lin, L.-F.; Lai, S.-C.; Yang, J.-H.; Chou, M.-L. FaTEDT1L of Octoploid Cultivated Strawberry Functions as a Transcriptional Activator and Enhances Abiotic Stress Tolerance in Transgenic Arabidopsis. *Int. J. Mol. Sci.* **2024**, *25*, 10091. [CrossRef]
101. Singh, N.K.; Nelson, D.E.; Kuhn, D.; Hasegawa, P.M.; Bressan, R.A. Molecular Cloning of Osmotin and Regulation of Its Expression by ABA and Adaptation to Low Water Potential. *Plant Physiol.* **1989**, *90*, 1096–1101. [CrossRef]
102. Husaini, A.M.; Abdin, M.Z. Development of Transgenic Strawberry (*Fragaria* × *ananassa* Duch.) Plants Tolerant to Salt Stress. *Plant Sci.* **2008**, *174*, 446–455. [CrossRef]
103. Liang, J.; Zheng, J.; Wu, Z.; Wang, H. Strawberry FaNAC2 Enhances Tolerance to Abiotic Stress by Regulating Proline Metabolism. *Plants* **2020**, *9*, 1417. [CrossRef] [PubMed]
104. Li, W.; Li, H.; Wei, Y.; Han, J.; Wang, Y.; Li, X.; Zhang, L.; Han, D. Overexpression of a *Fragaria* Vesca NAM, ATAF, and CUC (NAC) Transcription Factor Gene (FvNAC29) Increases Salt and Cold Tolerance in *Arabidopsis thaliana*. *Int. J. Mol. Sci.* **2024**, *25*, 4088. [CrossRef] [PubMed]
105. Li, W.; Zhang, W.; Li, H.; Yao, A.; Ma, Z.; Kang, R.; Guo, Y.; Li, X.; Yu, W.; Han, D. Overexpression of a *Fragaria* × *ananassa* AP2/ERF Transcription Factor Gene (FaTINY2) Increases Cold and Salt Tolerance in *Arabidopsis thaliana*. *Int. J. Mol. Sci.* **2025**, *26*, 2109. [CrossRef]
106. Zhang, Y.; Chen, Y.; Gao, W.; Tian, S.; Lin, B.; Gu, X.; Lin, Y.; Li, M.; Wang, Y.; He, W. Characterization of Superoxide Dismutase (SOD) Gene Family and Their Responses to Salinity Stress and Fruit Development in Octoploid Strawberry. *Agronomy* **2024**, *14*, 2514. [CrossRef]
107. Lei, D.; Lin, Y.; Luo, M.; Zhao, B.; Tang, H.; Zhou, X.; Yao, W.; Zhang, Y.; Wang, Y.; Li, M. Genome-Wide Investigation of G6PDH Gene in Strawberry: Evolution and Expression Analysis during Development and Stress. *Int. J. Mol. Sci.* **2022**, *23*, 4728. [CrossRef]
108. Li, H.; Yue, M.; Jiang, L.; Liu, Y.; Zhang, N.; Liu, X.; Ye, Y.; Lin, X.; Zhang, Y.; Lin, Y. Genome-Wide Identification of Strawberry C2H2-ZFP C1-2i Subclass and the Potential Function of FaZAT10 in Abiotic Stress. *Int. J. Mol. Sci.* **2022**, *23*, 13079. [CrossRef]

 https://doi.org/10.3390/plants15030432

109. Kula, N.; Öztürk Erdem, S.; Ünal, D. Salt Tolerance Mechanisms in Strawberry Cultivars: Comparing the Physiological and Molecular Responses of Petaluma and Cabrillo. *Appl. Fruit. Sci.* **2025**, *67*, 379. [CrossRef]
110. Zhang, H.; Kang, H.; Su, C.; Qi, Y.; Liu, X.; Pu, J. Genome-Wide Identification and Expression Profile Analysis of the NAC Transcription Factor Family during Abiotic and Biotic Stress in Woodland Strawberry. *PLoS ONE* **2018**, *13*, e0197892. [CrossRef]
111. Li, Z.; Xie, Q.; Yan, J.; Chen, J.; Chen, Q. Genome-Wide Identification and Characterization of the Abiotic-Stress-Responsive GRF Gene Family in Diploid Woodland Strawberry (*Fragaria vesca*). *Plants* **2021**, *10*, 1916. [CrossRef] [PubMed]
112. Li, W.; Zhong, J.; Zhang, L.; Wang, Y.; Song, P.; Liu, W.; Li, X.; Han, D. Overexpression of a *Fragaria vesca* MYB Transcription Factor Gene (FvMYB82) Increases Salt and Cold Tolerance in *Arabidopsis thaliana*. *Int. J. Mol. Sci.* **2022**, *23*, 10538. [CrossRef] [PubMed]
113. Crizel, R.L.; Perin, E.C.; Vighi, I.L.; Woloski, R.; Seixas, A.; da Silva Pinto, L.; Rombaldi, C.V.; Galli, V. Genome-Wide Identification, and Characterization of the CDPK Gene Family Reveal Their Involvement in Abiotic Stress Response in *Fragaria* × *ananassa*. *Sci. Rep.* **2020**, *10*, 11040. [CrossRef] [PubMed]
114. Jiang, L.; Lin, Y.; Wang, L.; Peng, Y.; Yang, M.; Jiang, Y.; Hou, G.; Liu, X.; Li, M.; Zhang, Y. Genome-Wide Identification and Expression Profiling Reveal the Regulatory Role of U-Box E3 Ubiquitin Ligase Genes in Strawberry Fruit Ripening and Abiotic Stresses Resistance. *Front. Plant Sci.* **2023**, *14*, 1171056. [CrossRef]
115. Kong, J.; Xiong, R.; Qiu, K.; Lin, X.; Li, D.; Lu, L.; Zhou, J.; Zhu, S.; Liu, M.; Sun, Q. Genome-Wide Identification and Characterization of the Laccase Gene Family in *Fragaria vesca* and Its Potential Roles in Response to Salt and Drought Stresses. *Plants* **2024**, *13*, 3366. [CrossRef]
116. Li, S.; Chang, L.; Sun, R.; Dong, J.; Zhong, C.; Gao, Y.; Zhang, H.; Wei, L.; Wei, Y.; Zhang, Y. Combined Transcriptomic and Metabolomic Analysis Reveals a Role for Adenosine Triphosphate-Binding Cassette Transporters and Cell Wall Remodeling in Response to Salt Stress in Strawberry. *Front. Plant Sci.* **2022**, *13*, 996765. [CrossRef]
117. López, M.-E.; Roquis, D.; Becker, C.; Denoyes, B.; Bucher, E. DNA methylation dynamics during stress response in woodland strawberry (*Fragaria vesca*). *Hortic. Res.* **2022**, *9*, uhac174. [CrossRef]
118. Fernando, V.D. *Major Transcription Factor Families Involved in Salinity Stress Tolerance in Plants. Transcription Factors for Abiotic Stress Tolerance in Plants*; Academic Press: Cambridge, MA, USA, 2020; pp. 99–109.
119. Shao, H.; Wang, H.; Tang, X. NAC Transcription Factors in Plant Multiple Abiotic Stress Responses: Progress and Prospects. *Front. Plant Sci.* **2015**, *6*, 902. [CrossRef]
120. Licausi, F.; Ohme-Takagi, M.; Perata, P. APETALA 2/Ethylene Responsive Factor (AP 2/ERF) Transcription Factors: Mediators of Stress Responses and Developmental Programs. *New Phytol.* **2013**, *199*, 639–649. [CrossRef]
121. Tibbs Cortes, L.; Zhang, Z.; Yu, J. Status and Prospects of Genome-wide Association Studies in Plants. *Plant Genome* **2021**, *14*, e20077. [CrossRef]

https://doi.org/10.3390/plants15030432

plants

MDPI

Article

Exogenous Melatonin Reinforces Photosynthesis, Antioxidant Defense and Gene Expression to Ameliorate Na_2CO_3 Stress in Maize

Guoxiang Qi, Xiaoqiang Zhao *, Fuqiang He, Siqi Sun, Zhenzhen Shi and Yining Niu

State Key Laboratory of Aridland Crop Science, College of Agronomy, Gansu Agricultural University, Lanzhou 730070, China; qigx1321@163.com (G.Q.); hefq6125@163.com (F.H.); 15045240973@163.com (S.S.); shizz@gsau.edu.cn (Z.S.); niuyn@gsau.edu.cn (Y.N.)

* Correspondence: zhaoxiaoq@gsau.edu.cn; Tel.: +86-183-9415-8662

Abstract: Salt stress can seriously affect the growth and development of maize (*Zea mays* L.), resulting in a great yield loss. Melatonin (MT), an indole hormone, is a potential enhancer of plant tolerance against salt stress. However, the complex mechanisms of MT application in enhancing maize salt tolerance are still unclear. Herein, three-leaf seedlings of salt-susceptible P138 and its salt-resistant ethyl methane sulfonate (EMS)-104 mutant were cultured with or without 150 μM MT application under 0 and 100 mM Na_2CO_3 treatments for seven days, to systematically explore the response mechanisms of exogenous MT in improving the salt tolerance of maize. The results showed that salt stress triggered an escalation in reactive oxygen species production, enhanced multiple antioxidant enzymes' activities, impaired cellular membrane permeability, inhibited photosynthetic pigment accumulation, and ultimately undermined the vigor and photosynthetic prowess of the seedlings. While suitable MT application counteracted the detrimental impacts of Na_2CO_3 on seedlings' growth and photosynthetic capacity, the seedling length and net photosynthetic rate of P138 and EMS-104 were increased by 5.5% and 18.7%, and 12.7% and 54.5%, respectively. Quantitative real-time PCR (qRT-PCR) analysis further showed that MT application activated the expression levels of antioxidant enzyme-related genes (*Zm00001d025106*, *Zm00001d031908*, *Zm00001d027511*, and *Zm00001d040364*) and pigment biosynthesis-related genes (*Zm00001d011819* and *Zm00001d017766*) in both maize seedlings under Na_2CO_3 stress; they then formed a complex interaction network of gene expression, multiple physiological metabolisms, and phenotype changes to influence the salt tolerance of maize seedlings under MT or Na_2CO_3 stress. To sum up, these observations underscore that 150 μM MT can alleviate salt injury of maize seedlings, which may provide new insights for further investigating MT regulation mechanisms to enhance maize seedlings' salt resistance.

Keywords: maize seedlings; Na_2CO_3 stress; melatonin; ethyl methane sulfonate (EMS) mutant; antioxidant system; photosynthesis; quantitative real-time PCR

1. Introduction

Salt stress represents a significant constraint on crop yields worldwide, jeopardizing global food security [1]. Globally, saline–alkali soil covers about 9.54×10^8 ha, with China alone possessing 9.91×10^7 ha [2–4]. By 2050, it is expected that about 50% of agricultural land worldwide will be affected by salinity [4]. The salinization problem is caused by a variety of factors, including little rainfall, high evaporation rates, improper irrigation, and excessive fertilizer use [5,6]. Due to high salt levels severely hindering the absorption of essential nutrients and water in different crop species, their growth and development are inhibited and toxic ion accumulation is aggravated [7].

Maize (*Zea mays* L.), an important cereal crop, plays a pivotal role in providing food, feed, and bioenergy. Generally, maize is extremely sensitive to salt stress during its whole growth period, especially the seedling stage [8,9]. Salt stress influences the activities of multiple photosynthetic enzymes in maize, such as adenosine triphosphate synthetase (ATPase), phosphoenol-pyruvate carboxylase (PEPcase), and ribulose-1,5-bisphosphate carboxylase (RuBPase), which further inhibits CO_2 fixation and photosynthesis efficiency [10–12], ultimately contributing to yield losses of 20–50% [13,14]. At the same time, excess free radicals and reactive oxygen species (ROS) elevate lipid peroxidation and lipoxygenase acidity, which further decomposes phospholipids of cellular membranes in maize seedlings under salt stress [15]. In addition, stressed maize seedlings cannot efficiently absorb and assimilate nutrients from the soil, leading to nutritional imbalances [16].

Melatonin (MT), an indole compound, shares structural similarities with indole-3-acetic acid (IAA) and exhibits a characteristic "low concentration promotion, high concentration inhibition" behavior [17]. Previous studies have reported that MT could benefit growth, biomass accumulation, flowering, and grain formation in different plants [18–20]. Meanwhile, MT has been shown to alleviate salt stress in various crop species, including rice (*Oryza sativa* L.), soybean (*Glycine max* L.), cucumber (*Cucumissativus* L.), and sunflower (*Helianthus annuus* L.) [21–24]. In particular, MT effectively inhibited chlorophyll degradation while ensuring the normal production of photosynthetic pigments under salt stress [25]. The application of 100 μM MT enhances the net photosynthetic rate (Pn) in tomato (*Solanum lycopersicum* L.) seedlings under salt stress, which exhibit better growth [26]. Likewise, 100 μM MT application increases stomatal conductance (Gs) and intercellular CO_2 accumulation (Ci) of soybean leaves under drought stress, thereby enhancing the drought resistance of soybean plants [17]. In addition, MT also serves as an effective inducer and direct eliminator of excessive ROS inside cells, thus modulating oxidative damage resulting from environmental stresses [27]. However, its role in alleviating Na_2CO_3 stress in maize seedlings is not well known.

Considering the crucial importance of MT in boosting plant resistance, it is of great significance to reveal the mechanism of exogenous MT application in the mitigation of salt stress in maize seedlings. In this investigation, the salt-susceptible inbred maize line P138 and its salt-resistant EMS-104 (from a library of ethyl methane sulfonate (EMS)-induced mutants) mutant were used as materials. Their seedlings at the three-leaf stage were treated with or without 150 μM MT application under 0 and 100 mM Na_2CO_3 treatments for seven days to analyze seedling growth status, photosynthetic pigment accumulation, photosynthetic ability, ROS levels, and antioxidant enzyme activities; meanwhile, the expression levels of six candidate genes responsible for antioxidant enzymes and photosynthesis in both genotypes seedlings under all treatments were also quantitatively analyzed by quantitative real-time PCR (qRT-PCR). Thereby, this study will help us determine the mechanism by which MT alleviates the impact of salt stress on maize seedlings, and gather insights into the mechanisms of salt tolerance in maize.

2. Results

2.1. Combined Analysis of All Traits

Multiple factorial analysis of variance (ANOVA) proved significant for the tested variables (including maize genotype (G), MT (M) application, and Na_2CO_3 (N) stress) and their physiological and molecular interactions with twenty-five tested traits including four growth parameters, eight photosynthetic performances, one membrane characteristic, one ROS level, four antioxidant enzymes' activities, one cell osmotic pressure, and six stomatal morphology traits in two maize genotype seedlings under different treatments (Figures 1–6; Table S1). The findings demonstrated that resistance to salt stress was controlled by maize's own genetic constitution, exogenous MT application, Na_2CO_3 stress, and their interaction effects. Our results also showed that the genotypes, MT application, and Na_2CO_3 stress imposed higher influences on these traits than their interactions.

2.2. Effect of MT on Changes in Growth Phenotypes of Maize Seedlings under Na_2CO_3 Stress

The growth phenotypes can directly reflect the changes in plants under different treatments. In this study, under N (100 mM Na_2CO_3 + 0 μM MT) stress, P138 and EMS-104 seedlings showed significant decreases in seedling weight (SW), root weight (RW), seedling length (SL), and leaf area (LA) compared to CK (0 mM Na_2CO_3 + 0 μM MT) treatment, with 29.0% and 27.2%, 32.8% and 29.9%, 26.5% and 12.8%, and 26.4% and 24.9% decreases, respectively ($p < 0.05$) (Figure 1A–D). This phenomenon suggested that 100 mM Na_2CO_3 stress can significantly inhibit maize seedling growth. In the M (0 mM Na_2CO_3 + 150 μM MT) treatment, only P138 seedlings had higher seedling weight and root weight, which increased by 11.7% and 6.6%, respectively (Figure 1A,B). When both maize genotypes were treated with MN (100 mM Na_2CO_3 + 150 μM MT), P138 root weight, seedling length, and leaf area increased by 27.9%, 18.8%, and 26.8%, respectively (Figure 1B–D). The EMS-104 seedlings showed significantly increased seedling weight and leaf area by 13.8% and 18.0% under MN treatment (Figure 1A,D). These results showed that 150 μM MT application clearly alleviates the inhibition of maize seedling growth when they are growing in a Na_2CO_3 environment. Overall, the growth of mutant EMS-104 seedlings was better than the wild P138 seedlings under all treatments (Figure 1).

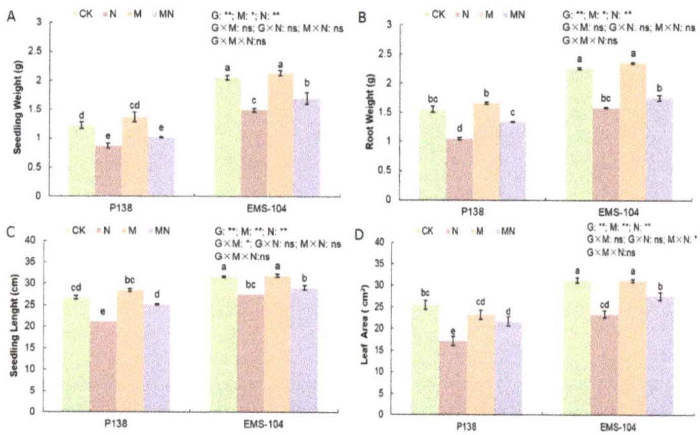

Figure 1. Changes in seedling weight (**A**), root weight (**B**), seedling length (**C**), and leaf area (**D**) of P138 and EMS-104 seedlings under different treatments, including 0 mM Na_2CO_3 + 0 μM melatonin (MT) treatment (CK); 100 mM Na_2CO_3 + 0 μM MT treatment (N); 0 mM Na_2CO_3 + 150 μM MT treatment (M); and 100 mM Na_2CO_3 + 150 μM MT treatment (MN). Different lowercase letters indicate significant differences at the $p < 0.05$ level. ns: multiple factorial analysis of variance (ANOVA) was nonsignificant at the $p > 0.05$ level; *: multiple factorial ANOVA was significant at the $p \leq 0.05$ level; **: multiple factorial ANOVA was significant at the $p \leq 0.01$ level.

2.3. Effects of MT on H_2O_2 Content and Antioxidant Enzyme Activities of Maize Seedlings under Na_2CO_3 Stress

The maintenance of reactive oxygen species (ROS) homeostasis is pivotal for normal growth, development, and stress tolerance in plants [28]. Regarding changes in the content of hydrogen peroxide (H_2O_2), an important ROS, it was significantly increased by 70.4% and 79.5% in P138 and EMS-104 seedlings under N stress compared to the control ($p < 0.05$) (Figure 2E). The findings suggested that the presence of high Na_2CO_3 concentration stimulates high H_2O_2 production in maize seedlings. Under M treatment, the H_2O_2 content in P138 and EMS-104 seedlings decreased by 14.7% and 11.9%, respectively ($p < 0.05$) (Figure 2E). Compared to N stress, the H_2O_2 content in P138 and EMS-104 seedlings decreased further by 32.2% and 30.5% under MN treatment, respectively (Figure 2E). The findings indicated that MT possibly inhibits the generation of H_2O_2 or scavenges its deposition (Figure 2E).

The antioxidant enzyme system plays an important role in protecting plant cells from stress [29]. In this study, compared with the control, N stress significantly increased the activities of superoxide dismutase (SOD), peroxidase (POD), catalase (CAT), and ascorbate peroxidase (APX) in P138 and EMS-104 seedlings ($p < 0.05$), they increased by 76.5% and 94.9%, 17.5% and 25.7%, 187.8% and 216.9%, and 144.2% and 84.8%, respectively (Figure 2A–D). The results showed that Na_2CO_3 stress enhances antioxidant enzyme activities to scavenge excess ROS in maize. Moreover, the SOD activity of P138 seedlings significantly increased by 24.9%, while the APX activity clearly decreased by 19.2% under M treatment; similarly, SOD activity increased significantly by 55.8%; however, CAT activity decreased clearly by 84.9% in EMS-104 seedlings under M treatment (Figure 2A,D). Additionally, when both seedlings of P138 and EMS-104 were exposed to MN treatment, their four enzymes' activities increased by 7.9% and 3.4%, 8.9% and 6.9%, 22.1% and 10.7%, and 30.7% and 27.3% compared to N stress, respectively (Figure 2A–D).

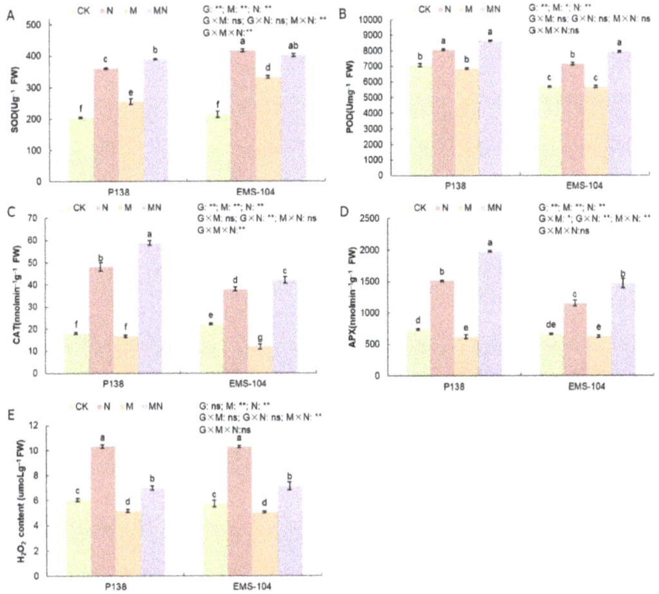

Figure 2. Changes in superoxide dismutase (SOD) activity (**A**), peroxidase (POD) activity (**B**), catalase (CAT) activity (**C**), ascorbate peroxidase (APX) activity (**D**), and H_2O_2 content (**E**) of P138 and EMS-104 seedlings under different treatments, including 0 mM Na_2CO_3 + 0 μM melatonin (MT) treatment (CK); 100 mM Na_2CO_3 + 0 μM MT treatment (N); 0 mM Na_2CO_3 + 150 μM MT treatment (M); and 100 mM Na_2CO_3 + 150 μM MT treatment (MN). Different lowercase letters indicate significant differences at the $p < 0.05$ level. ns: multiple factorial analysis of variance (ANOVA) was nonsignificant at the $p > 0.05$ level; *: multiple factorial ANOVA was significant at the $p \leq 0.05$ level; **: multiple factorial ANOVA was significant at the $p \leq 0.01$ level.

2.4. *Effect of MT on Relative Electrical Conductivity and Relative Water Content of Maize Seedlings under Na_2CO_3 Stress*

To evaluate the cell status of maize seedlings under Na_2CO_3 stress, the relative electrical conductivity (REC) and relative water content (RWC) in the leaves of two maize genotypes under different treatments were measured. Compared to CK, both P138 and EMS-104 leaves had significantly increased REC, with 33.3% and 20.5%, and had a significant decrease in RWC, with 4.2% and 1.8% under N treatment (Figure 3A,B). The results showed that Na_2CO_3 stress could aggravate cell membrane peroxidation to cause a significant decrease in RWC in maize. In the MN treatment, P138 leaves showed a lower REC than that under N treatment (9.4%; $p < 0.05$), while the RWC in P138 leaves increased by 2.5% (Figure 3A,B).

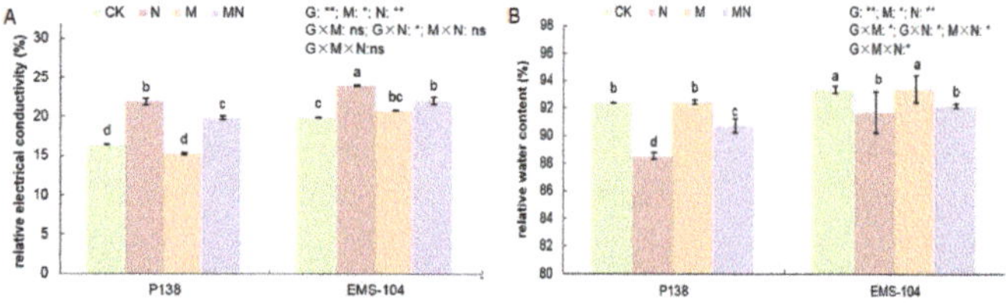

Figure 3. Changes in relative electrical conductivity (REC (**A**)), and relative water content (RWC (**B**)) of P138 and EMS-104 seedlings under different treatments, including 0 mM Na_2CO_3 + 0 μM melatonin (MT) treatment (CK); 100 mM Na_2CO_3 + 0 μM MT treatment (N); 0 mM Na_2CO_3 + 150 μM MT treatment (M); and 100 mM Na_2CO_3 + 150 μM MT treatment (MN). Different lowercase letters indicate significant differences at the $p < 0.05$ level. ns: multiple factorial analysis of variance (ANOVA) was nonsignificant at the $p > 0.05$ level; *: multiple factorial ANOVA was significant at the $p \leq 0.05$ level; **: multiple factorial ANOVA was significant at the $p \leq 0.01$ level.

2.5. *Effects of MT on Photosynthetic Characteristics of Maize Seedlings under Na_2CO_3 Stress*

Compared to CK treatment, N stress led to significant decreases in Pn, Gs, and transpiration rate (Tr) of both P138 and EMS-104 genotypes, with 57.6% and 31.9%, 56.6% and 30.1%, and 72.3% and 56.8%, respectively, while the Ci increased significantly by 79.3% and 142.7% ($p < 0.05$) (Figure 4A–D). The M treatment had lower values of Pn, Gs, and Ci for P138 material, while its Tr increased by 11.3%; EMS-104 seedlings showed increased Gs, Ci, and Tr by 4.6%, 23.1%, and 1.4% under M treatment (Figure 4A–D). Under MN treatment, P138 and EMS-104 seedlings exhibited increases in Pn, Gs, and Tr by 54.6% and 12.8%, 75.1% and 29.5%, and 146.9% and 60.1%, respectively, while their Ci decreased by 22.0% and 23.5%, respectively (Figure 4A–D).

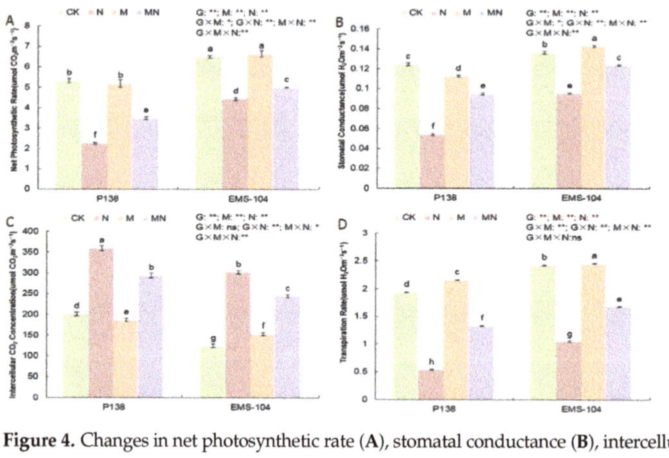

Figure 4. Changes in net photosynthetic rate (**A**), stomatal conductance (**B**), intercellular CO_2 concentration (**C**), and transpiration rate (**D**) of P138 and EMS-104 seedlings under different treatments, including 0 mM Na_2CO_3 + 0 μM melatonin (MT) treatment (CK); 100 mM Na_2CO_3 + 0 μM MT treatment (N); 0 mM Na_2CO_3 + 150 μM MT treatment (M); and 100 mM Na_2CO_3 + 150 μM MT treatment (MN). Different lowercase letters indicate significant differences at the $p < 0.05$ level. ns: multiple factorial analysis of variance (ANOVA) was nonsignificant at the $p > 0.05$ level; *: multiple factorial ANOVA was significant at the $p \leq 0.05$ level; **: multiple factorial ANOVA was significant at the $p \leq 0.01$ level.

2.6. Effect of MT on Photosynthetic Pigments in Maize Seedlings under Na_2CO_3 Stress

The contents of multiple photosynthetic pigments of two maize genotypes also showed the same changes under different treatments. In comparison to CK, the N treatment significantly decreased chlorophyll a (Chl a) content, chlorophyll b (Chl b) content, carotenoid (Car) content, and the chlorophyll a/b ratio (Chl a/b) in all maize materials, by 29.4% and 28.7%, 14.1% and 18.1%, 53.7% and 64.1%, and 23.9% and 25.6%, respectively ($p < 0.05$) (Figure 5A–D). The results showed that Na_2CO_3 stress could damage chloroplasts' structure and inhibit photosynthetic pigment synthesis in maize, and the influence of Chl a content was larger than that of Chl b content. Conversely, the M treatment led to significant increases in these parameters for P138 and EMS-104 seedlings, with 22.0% and 5.0%, 5.9% and 6.1%, 15.5% and 3.8%, and 16.3% and 5.4%, respectively ($p < 0.05$) (Figure 5A–D). Under MN treatment, P138 and EMS-104 seedlings had increases in these traits by 29.1% and 17.4%, 9.5% and 16.5%, 60.4% and 46.7%, and 21.2% and 17.1%, respectively ($p < 0.05$) (Figure 5A–D). The results thus showed that 150 μM MT application might protect chloroplasts' structure and increase multiple photosynthetic pigments' accumulation in maize seedlings under Na_2CO_3 stress.

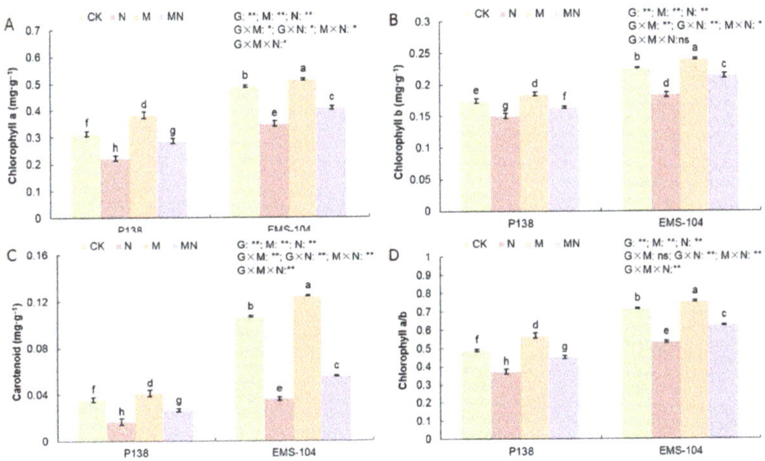

Figure 5. Changes in chlorophyll a (Chl a (**A**)), chlorophyll b (Chl b (**B**)), carotenoids (Car; (**C**)), and chlorophyll a/b (Chl a/b (**D**)) of P138 and EMS-104 seedlings under different treatments, including 0 mM Na_2CO_3 + 0 μM melatonin (MT) treatment (CK); 100 mM Na_2CO_3 + 0 μM MT treatment (N); 0 mM Na_2CO_3 + 150 μM MT treatment (M); and 100 mM Na_2CO_3 + 150 μM MT treatment (MN). Different lowercase letters indicate significant differences at the $p < 0.05$ level. ns: multiple factorial analysis of variance (ANOVA) was nonsignificant at the $p > 0.05$ level; *: multiple factorial ANOVA was significant at the $p \leq 0.05$ level; **: multiple factorial ANOVA was significant at the $p \leq 0.01$ level.

2.7. Effects of MT on Stomata of Maize Seedlings under Na_2CO_3 Stress

To verify the cause of Pn decrease after Na_2CO_3 stress, the stomatal morphology of the leaves in maize seedlings was observed. Under N stress, the paraxial surface stomatal length (PSL), paraxial surface stomatal width (PSW), and paraxial surface stomatal area (PSA) of P138 and EMS-104 experienced a significant decrease by 21.0% and 9.0%, 13.7% and 11.6%, and 28.0% and 16.6%, respectively ($p < 0.05$) (Figure 6A,C,E). The same phenomenon was also observed in dorsal surface stomatal length (DSL), dorsal surface stomatal width (DSW), and dorsal surface stomatal area (DSA) in these maize materials, which decreased by 28.9% and 6.9%, 29.2% and 27.2%, and 30.6% and 27.6%, respectively ($p < 0.05$) (Figure 6B,D,F). We thus speculated that the decrease in Pn is correlated with the changes in paraxial/dorsal surface stomata in maize seedling leaves under Na_2CO_3 stress.

Under M treatment, there were no clear changes in the paraxial surface stomatal morphology of P138 leaves; meanwhile, the paraxial surface stomatal morphology parameters, including PSL, PSW, and PSA, in EMS-104 leaves were increased by 11.0%, 4.2%, and 8.9%, respectively ($p < 0.05$) (Figure 6A,C,E). Interestingly, for the dorsal surface stomata, the DSL, DSW, and DSA in P138 leaves showed significant increases under M treatment by 22.9%, 8.4%, and 12.5%, respectively; except for DSW, other dorsal surface stomatal morphology parameters in EMS-104 leaves had significant increases ($p < 0.05$) (Figure 6B,D,F). When P138 and EMS-104 seedlings were cultured in MN treatment, their three paraxial surface stomatal morphology parameters were increased by 10.7% and 3.4%, 12.2% and 9.9%, and 17.3% and 8.3%, respectively; similarly, their three dorsal surface stomatal morphology parameters were also increased by 43.3% and 10.4%, 15.4% and 24.8%, and 25.6% and 31.0%, respectively ($p < 0.05$) (Figure 6A–F). Therefore, these results showed that MT application could improve multiple paraxial/dorsal surface stomatal morphology in maize leaves under Na_2CO_3 stress.

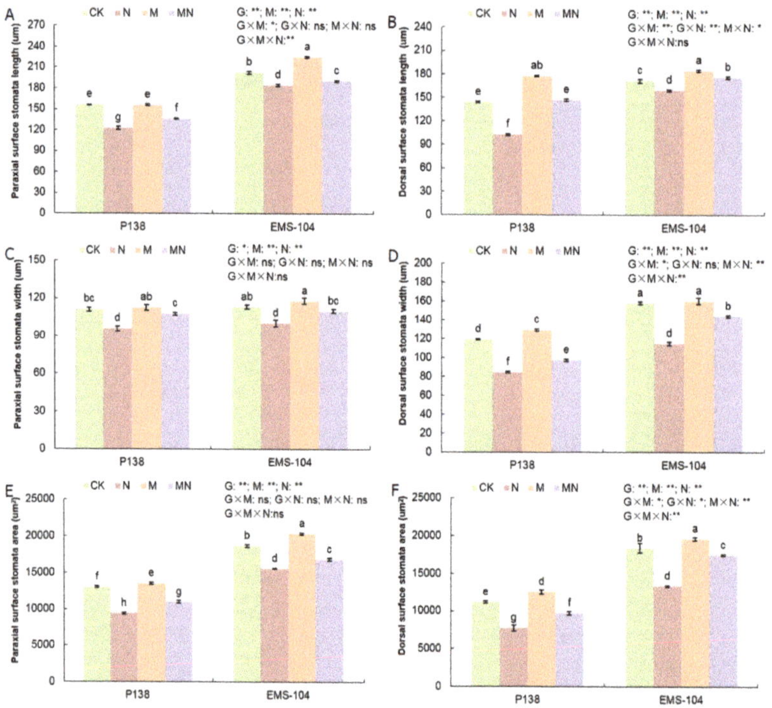

Figure 6. Changes in paraxial surface stomatal length (**A**), dorsal surface stomatal length (**B**), paraxial surface stomatal width (**C**), dorsal surface stomatal width (**D**), paraxial surface stomatal area (**E**), and dorsal surface stomatal area (**F**) of P138 and EMS-104 seedlings under different treatments, including 0 mM Na_2CO_3 + 0 μM melatonin (MT) treatment (CK); 100 mM Na_2CO_3 + 0 μM MT treatment (N); 0 mM Na_2CO_3 + 150 μM MT treatment (M); and 100 mM Na_2CO_3 + 150 μM MT treatment (MN). Different lowercase letters indicate significant differences at the $p < 0.05$ level. ns: multiple factorial analysis of variance (ANOVA) was nonsignificant at the $p > 0.05$ level; *: multiple factorial ANOVA was significant at the $p \leq 0.05$ level; **: multiple factorial ANOVA was significant at the $p \leq 0.01$ level.

2.8. Correlation Relationships among 25 Traits in Two Maize Genotypes under All Treatments

In this study, a Pearson correlation analysis among 25 tested traits in two maize genotypes under all treatments was further conducted to reveal the internal relationships among all traits under Na_2CO_3 stress and MT stimulation. The results showed that the REC was significantly negatively correlated with Tr, PSW, and SOD activity; in contrast, it was significantly positively correlated with Ci, CAT activity, APX activity, and H_2O_2 levels. The Ci was negatively correlated with Tr, PSL, DSL, PSW, DSW, PSA, DSA, Chl a content, Chl b content, Car content, Chl a/b, and SOD activity; however, it was significantly positively correlated with POD activity, CAT activity, APX activity, and H_2O_2 content. The Pn was significantly positively correlated with Gs, Tr, PSL, DSL, PSW, DSW, PSA, DSA, Chl a content, Chl b content, Car content, Chl a/b, and SOD activity; it was significantly negatively correlated with POD activity, CAT activity, APX activity, and H_2O_2 content. The H_2O_2 content was significantly negatively correlated with SW, RW, SL, LA, RWC, Pn, Gs, Tr, PSL, DSL, PSW, DSW, PSA, DSA, Chl a content, Chl b content, Car content, Chl a/b, and SOD activity. Conversely, there was a significant positive correlation with REC, Ci, SOD activity, CAT activity, and APX activity (Figure 7). These results suggested that multiple traits synergistically influence salt tolerance in maize under salt stress or MT treatment.

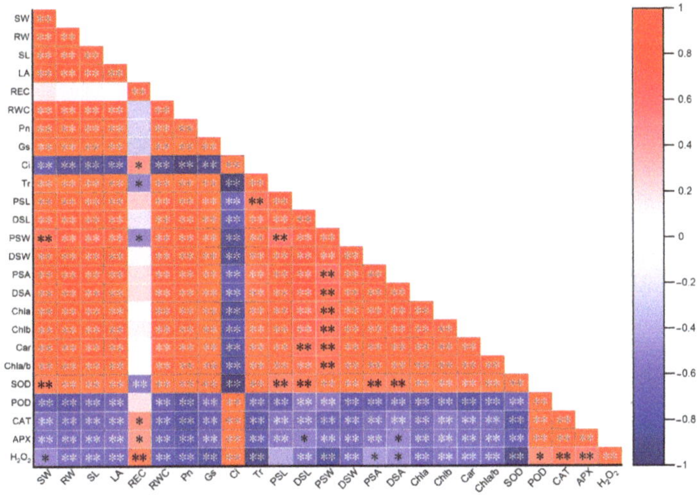

Figure 7. The correlation analysis of 25 tested traits of maize seedlings, including seedling length (SL), leaf area (LA), fresh seedling weight (SW), fresh root weight (RW), relative electrical conductivity (REC), and relative water content (RWC), net photosynthetic rate (Pn), stomatal conductance (Gs), intercellular CO_2 concentration (Ci), transpiration rate (Tr), paraxial surface stomatal length (PSL), paraxial surface stomatal width (PSW), paraxial surface stomatal area (PSA), dorsal surface stomatal length (DSL), dorsal surface stomatal width (DSW), dorsal surface stomatal area (DSA), chlorophyll a content (Chl a), chlorophyll b content (Chl a/b), carotenoid content (Car), chlorophyll a/b ratio (Chl a/b), superoxide dismutase (SOD), peroxidase (POD), catalase (CAT), ascorbate peroxidase (APX), and H_2O_2 content (H_2O_2). ** indicates a significant correlation at the $p < 0.01$ level and * indicates a significant association at the $p < 0.05$ level. The red squares represent a positive correlation between the two metrics, and the blue squares represent a negative correlation. The strength of the correlation is indicated by the color of the blocks.

2.9. Expression Analysis of Genes Involved in Photosynthesis and Antioxidant Homeostasis under Na_2CO_3 Stress

To further verify the regulatory mechanism of Na_2CO_3 stress in maize seedlings, the differential expression levels of six candidate genes involved in photosynthesis and antioxidant homeostasis were analyzed. Compared to CK treatment, N stress significantly increased the corresponding gene expression levels for *superoxide dismutase* (*SOD*; *Zm00001d025106* and *Zm00001d031908*), *catalase 2* (*CAT2*; *Zm00001d027511*), and *peroxidase 72* (*POD72*; *Zm00001d040364*) in both P138 and EMS-104 leaves by 98.8% and 137.4%, 39.6%

and 26.4%, 63.8% and 88.2%, and 15.9% and 11.4%, respectively ($p < 0.05$) (Figure 8A–D). The expression levels of two photosynthesis-related genes, i.e., *nine-cis-epoxycarotenoid dioxygenase8* (*Zm00001d017766*) and *chlorophyllide a oxygenase chloroplastic* (*Zm00001d011819*), were significantly decreased by 53.6% and 25.5%, and 49.2% and 9.8%, respectively ($p < 0.05$) (Figure 8E,F). Under M treatment, the expressions of *Zm00001d025106*, *Zm00001d011819*, and *Zm00001d017766* in P138 and EMS-104 increased by 36.8% and 53.7%, 28.3% and 45.6%, and 57.7% and 69.1%, compared to CK, and *Zm00001d031908* and *Zm00001d040364* in P138 were up-regulated by 24.7% and 11.4%; notably, the expression pattern of *Zm00001d027511* decreased by 12.1% and 46.1% ($p < 0.05$) (Figure 8A–F). Moreover, in the MN treatment, *Zm00001d025106*, *Zm00001d031908*, *Zm00001d040364*, *Zm00001d011819*, and *Zm00001d017766* in P138 and EMS-104 were significantly up-regulated by 59.5% and 33.9%, 53.0% and 36.3%, 15.2% and 16.8%, 137.4% and 69.2%, and 194.6% and 48.8% compared to N stress, respectively ($p < 0.05$) (Figure 8A–F).

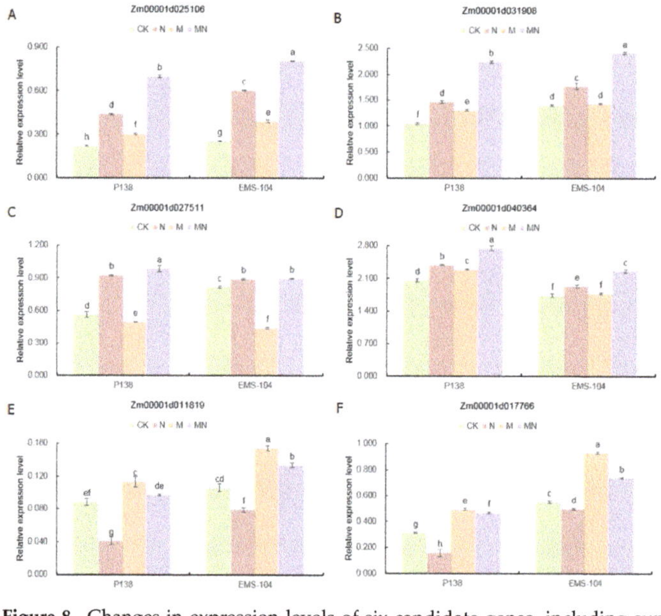

Figure 8. Changes in expression levels of six candidate genes, including superoxide dismutase (*SOD; Zm00001d025106* (**A**), *Zm00001d031908* (**B**), catalase 2 (*CAT2; Zm00001d027511*) (**C**), peroxidase 72 (*POD72; Zm00001d040364*) (**D**), chlorophyllide a oxygenase chloroplastic (*Zm00001d011819*) (**E**), and nine-cis-epoxycarotenoid dioxygenase8 (*Zm00001d017766*) (**F**) in leaves of P138 and EMS-104 seedlings under different treatments: 0 mM Na_2CO_3 + 0 μM melatonin (MT) treatment (CK); 100 mM Na_2CO_3 + 0 μM MT treatment (N); 0 mM Na_2CO_3 + 150 μM MT treatment (M); and 100 mM Na_2CO_3 + 150 μM MT treatment (MN). Different lowercase letters indicate significant differences in $p < 0.05$ level.

3. Discussion

It is well known that salt stress is the major environmental factor in the soil that inhibits and negatively affects plant growth, development, and yield formation [22,30]. Fortunately, different strategies have been developed in some plant species to resist various environmental stresses. Previous studies have shown that exogenous MT application can enhance salt tolerance in rice, soybean, cucumber, and sunflower [21–24]. However, our knowledge about the mechanisms involved in MT-mediated tolerance to Na_2CO_3 stress in salt-susceptible P138 and its salt-resistant EMS-104 mutant seedlings still remains unclear.

Many studies clearly observed that the phenotypes of plants under salt stress exhibited unfavorable effects, especially in biomass and plant height [31–33]. Surprisingly, 150 μM

MT increased the leaf area and biomass of cotton (*Gossypium hirsutum* L.) plants under salt stress [34]. A similar phenomenon was also found in our study, i.e., under the stress of 100 mM Na_2CO_3, the seedling length, leaf area, seedling weight, and root weight of P138 and theEMS-104 mutant were significantly reduced (Figure 1A–D), indicating that salt stress severely inhibits the survival or healthy growth of maize. We also found that the growth status of theEMS-104 mutant was better than its wild-type P138 (Figure 1A–D), which provided valuable genetic resources for salt tolerance breeding in future. After applying 150 μM MT, these growth phenotypes were improved significantly (Figure 1A–D). From this, we infer that MT application may improve multiple physiological processes of maize seedlings to resist Na_2CO_3 stress injury.

The RWC can directly reflect the water status of plants [35], and the REC is generally considered to show membrane permeability and membrane integrity under stress [36]. Zhao et al. [37] found that drought stress reduced the RWC of maize, and Wang et al. [38] found that salt stress combined with drought stress increased the REC of boxwood seedlings. Interestingly, early studies have shown that the application of MT was beneficial for maintaining intracellular turgor pressure and improving plasma ion homeostasis [38,39]. In this study, we found that the RWC of maize P138 and EMS-104 genotypes decreased significantly, while their REC was significantly increased under the stress of 100 mM Na_2CO_3, which was consistent with the results of previous studies [37,38]. After applying 150 μM MT, the RWC of P138 and EMS-104 seedlings was significantly increased, but the REC was significantly decreased (Figure 3A,B), indicating that 150 μM MT treatment may maintain relatively high osmotic pressure to protect cell membrane integrity in maize seedlings under 100 mM Na_2CO_3 stress.

Photosynthetic pigments are closely related to photosynthesis and stress resistance in plants [40,41]. Salt stress accelerated the degradation of chlorophyll and inhibited the biosynthesis of chlorophyll in maize [42]. Under drought stress, Ahmad et al. [43] discovered that chlorophyll levels were clearly increased in maize after 100 μM MT application. Regarding other exogenous hormones' applications, similar results have also confirmed that exogenous salicylic acid (SA) increased the chlorophyll content of maize under salt stress [44]. Our study showed that 100 mM Na_2CO_3 stress significantly decreased the contents of Chl a, Chl b, and Car in P138 and EMS-104 seedling leaves. In contrast, 150 μM MT treatment significantly increased these photosynthetic pigments' contents (Figure 5A–D). This is consistent with the results of previous studies [43,44]. In addition, our gene expression analysis also showed that *Zm00001d017766* and *Zm00001d011819*, two photosynthetic pigment-related genes had positive expression levels under MT application in both maize genotypes (Figure 8E,F). Therefore, these findings indicated that 150 μM MT could promote photosynthetic pigments' gene expression to increase photosynthetic pigment accumulation under 100 mM Na_2CO_3 stress. In addition, some studies have shown that appropriate ROS levels are beneficial for chloroplast development and function [45]. We also found that levels of ROS, such as H_2O_2, were significantly decreased in P138 and EMS-104 seedlings when they were treated with 150 μM MT application under 100 mM Na_2CO_3 stress (Figure 2E). Thereby, lower ROS levels may benefit photosynthetic pigment production and maintain high photosynthesis rates in maize under various environmental stimulations.

The changes in stomatal morphology can positively reflect the physiological changes in plants under different environmental conditions [46]. Some studies have shown that salt stress led to stomatal closure, while exogenous silicon (Si) treatment improved the Gs of alfalfa (*Medicago sativa* L.) seedlings under salt stress [47]. Similarly, exogenous SA application improved the Gs of barley under salt stress [48]. We also found that the application of 150 μM MT improved the paraxial/dorsal surface stomatal morphology under 100 mM Na_2CO_3 stress in the current study (Figure 6A–F).

Photosynthesis is the fundament for plants' survival, growth, reproduction, and resilience to environmental stresses [49,50]. Salt stress reduced the light energy absorption and conversion efficiency of leaves [51]. Previous studies have shown that exogenous MT protected the structure of chloroplasts and photosynthetic organs, induced stomatal reopening, sustained high levels of chlorophyll, and enhanced leaves' ability to capture light energy; this action thus alleviated the inhibition of photosynthesis caused by NaCl stress in alfalfa leaves [52–55]. Similar studies also found that MT stimulation significantly enhanced the photosynthetic efficiency in maize under drought stress [56]. Like drought stress in maize, when the P138 and EMS-104 seedlings were treated with 150 μM MT application under 100 mM Na_2CO_3 stress, their Pn, Gs, and Tr increased significantly, while Ci decreased significantly (Figure 4A–D).

During the evolution of plants, they have developed powerful antioxidant enzyme systems, such as SOD, CAT, POD, and APX, which synergistically enhance plant stress resistance and trigger protective response against oxygen toxicity in cells and throughout the plant [57]. Many studies indicated that MT treatment under abiotic stress boosted the activities of these antioxidant enzymes [58–60]. Consistently, our study also showed that the activities of SOD, POD, CAT, and APX in the leaves of P138 and EMS-104 seedlings were significantly increased under 100 mM Na_2CO_3 stress and 150 μM MT treatment (Figure 2A–D). At the same time, the expression levels of the four antioxidant enzyme genes, i.e., *Zm00001d025106*, *Zm00001d031908*, *Zm00001d027511*, and *Zm00001d040364*, were up-regulated in P138 and EMS-104 seedlings under MT stimulation and Na_2CO_3 stress (Figure 8A–D). Hence, this suggested that 150 μM MT application improved the salt tolerance of maize seedlings by enhancing antioxidant enzyme activities.

In conclusion, under 100 mM Na_2CO_3 stress, 150 μM MT application elevated RWC, Pn, Gs, Tr, and Chl content, while reducing the REC, Ci, and H_2O_2 content of leaves in maize seedlings. Meanwhile, the six candidate genes related to photosynthetic pigments and antioxidant homeostasis were up-regulated. They then formed complex interaction relationships to enhance salt tolerance in maize seedlings. Briefly, 150 μM MT application maintains the integrity of photosynthetic organs, scavenges excess ROS, and improves photosynthetic characteristics, subsequently alleviating damage and promoting the growth of maize seedlings under a Na_2CO_3 stress environment (Figure 9).

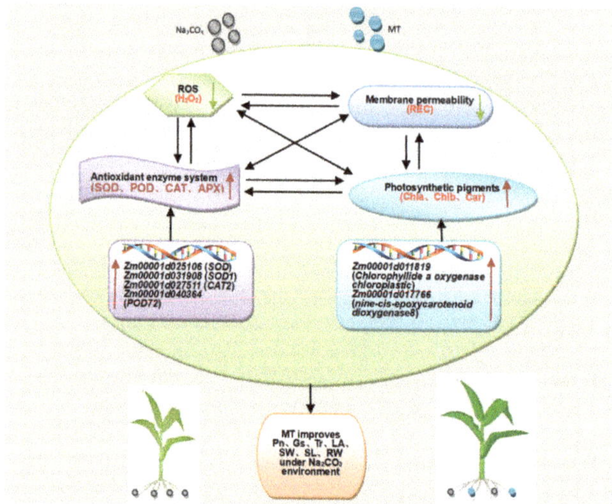

Figure 9. The response mechanism of melatonin (MT) application in maize seedlings under Na_2CO_3 stress. Chlorophyll a content (Chl a); chlorophyll b content (Chl b); chlorophyll a:b ratio (Chl a/b); photosynthetic performance, including net photosynthetic rate (Pn), stomatal conductance (Gs), and transpiration rate (Tr); antioxidant enzyme system, namely superoxide dismutase activity (SOD), peroxidase activity (POD), catalase activity (CAT), and ascorbate peroxidase activity (APX); reactive oxygen species (ROS) levels; H_2O_2 content (H_2O_2); membrane permeability (MP); relative electrical conductivity (REC). The red and green arrows indicate increases and decreases in the corresponding traits, respectively.

4. Materials and Methods

4.1. Materials

Previously, a total of 1041 M_3 maize mutant lines were constructed with pollen from the elite inbred line P138, which was treated with 0.5 mg·L^{-1} ethyl methane sulfonate solution (EMS) by our team. P138 is a representative inbred line, derived from the P group, with weak salt tolerance [61]. The mutant EMS-104 is more salt tolerant than the wild-type P138. Therefore, we conducted a follow-up study on P138 and its mutant EMS-104.

4.2. Plant Growth Conditions

The seeds of P138 and EMS-104 were sterilized with 0.5% (*v*/*v*) sodium hypochlorite solution for 15 min, rinsed five times with ddH_2O, and then soaked in ddH_2O for 24 h at 22 ± 0.5 °C. Subsequently, the ten soaked seeds were planted in sterilized vermiculite plastic boxes (13 cm diameter × 11 cm high), and then cultured in a greenhouse (25 ± 0.5 °C temperature; 300 μM·m^{-2} s^{-1} light intensity; 65% relative humidity; 12/12 h light/dark) for 15 d. During this period, the seedlings were washed by 50 mL ddH_2O every two days. The three-leaf seedlings received four treatments for 7 d, i.e., 0 mM Na_2CO_3 + 0 μM MT (ddH_2O, CK), 100 mM Na_2CO_3 + 0 μM MT(N), 0 mM Na_2CO_3 + 150 μM MT(M), and 100 mM Na_2CO_3 + 150 μM MT(MN). During all the treatments, the seedlings were washed by 50 mL of the above corresponding mixed solution every two days. Each treatment was repeated three times.

4.3. Growth Phenotype Measurements

For the seedlings of the two maize materials under four treatments, their seedling length (SL), seedling weight (SW), and root weight (RW) were measured according to Zhao et al.'s [62] method. The seedling leaf area (LA) was calculated as follows:

$$LA = \text{Leaf length} \times \text{Leaf width} \times 1.5.$$

4.4. Stomatal Morphology in Seedlings

The 3rd fresh leaves of P138 and EMS-104 seedlings under all treatments were cut in the middle of the leaf (1 cm × 1 cm), and were stuck on a slide coated with glue for 20 min. The leaf sections were then delicately detached by tweezers, leaving the temporary stomatal slides for collection. Stomatal morphology was examined utilizing an integrated forward and inverted fluorescence microscope (Revolve RVL-100-G, ECHO, San Diego, CA, USA). For each treatment, five random fields of view were selected from the paraxial/dorsal surface. Stomatal length, width, and area were measured and statistically recorded under a 4× objective (image size 3226 × 3024).

4.5. Photosynthetic Parameter Determination

The four photosynthetic parameters, including net photosynthetic rate (Pn; μM CO_2·m^{-2}·s^{-1}), intercellular CO_2 (Ci; μM CO_2·mol^{-1}), stomatal conductance (Gs; M H_2O·m^{-2}·s^{-1}), and transpiration rate (Tr; mM H_2O·m^{-2}·s^{-1}), of the 3rd leaf of P138 and EMS-104 seedlings under all treatments were measured by an LI-6400XT (LI-COR; Biosciences, Inc. Lincoln, NE, USA) portable photosynthetic measurement system. The illuminance inside the leaf chamber was set to 1000 μM/(m^2·s), a CO_2 concentration of 400 μM·M^{-1}, and a temperature of 25 °C.

4.6. Relative Moisture Content Assay

We measured 1.0 g of fresh leaves (FW) from P138 and EMS-104 seedlings under all treatments, which were then completely soaked in 50 mL ddH_2O until a constant weight (TW) was reached. The soaked leaves were then dried in an oven to a constant weight (DW). The relative water content (RWC) was calculated as follows:

$$[(FW - DW)/(TW - DW)] \times 100\%.$$

4.7. Relative Electrical Conductivity Measurement

We placed 0.1 g fresh leaves of P138 and EMS-104 seedlings under all treatments in a test tube of 10 mL ddH_2O, and left them at room temperature for 12 h to determine the electrical conductivity (R_1). Then, the same set of samples was stored in a 100 °C water bath for 15 min and the electrical conductivity (R_2) was recorded. The conductivity (R_1, R_2) was measured by a DDSJ-308F conductivity meter (Rex Electric Chemical, Shanghai, China). The REC was estimated as follows [62]:

$$REC = (R_1/R_2) \times 100\%.$$

4.8. Chlorophyll Content Measurement

We cut 1.0 g of fresh leaves from P138 and EMS-104 seedlings under all treatments and soaked them in 10 mL of 95% alcohol for 48 h; they were then centrifuged at 12,000 rpm (Centrifuge 5425/5425 R; Eppendorf, Hamburg, Germany) for 10 min. The absorbance of the supernatant was measured by utilizing a multifunctional enzyme marker model (SynergyHTX; BioTek Instruments, Inc., Winooski, VT, USA), with readings taken at wavelengths of 665 nm, 649 nm, and 470 nm. The concentrations of chlorophyll a, chlorophyll b, and carotenoids were calculated by the following formulas: $Cac = 13.95 \times A665 - 6.88 \times A649$; $Cbc = 24.96 \times A649 - 7.32 \times A665$; and $Ccc = 1000 \times A470 - 2.05 \times Ca - 114 \times Cb$. The pigment content was calculated by the following equations: Chl a content (mg g^{-1}) = $Cac \times Vt \times n/FW \times 1000$; Chl b content (mg g^{-1}) = $Cbc \times Vt \times n/FW \times 1000$; and Car content (mg/g) = $Ccc \times Vt \times n/FW \times 1000$. Cac, Cbc, and Ccc represent the concentrations of Chl a, Chl b, and Car, respectively. FW is the fresh weight (g), Vt is the total volume of extract (mL), and n is the dilution factor.

4.9. Antioxidant Enzyme Activity Determination

Referring to Zhao et al.'s [62] method, the SOD activity, POD activity, CAT activity, APX activity, and H_2O_2 content were determined using corresponding Solarbio kits (Beijing Solarbio Science and Technology Co., Ltd., Beijing, China) and a multifunctional microplate reader (SynergyHTX; BioTek Instruments, Inc., USA), following the manufacturers' instructions.

4.10. qRT-PCR Analysis

The total RNA was extracted from eight samples of the 3rd leaf of salt-sensitive P138 and salt-resistant EMS-104 seedlings (three biological replicates for each sample) under the above four treatments with TRIZOL reagent (Invitrogen, Carlsbad, CA, USA), which was reverse-transcribed into cDNA using a SuperScript III First strand Kit (Invitrogen, Gibco, Grand Island, NE, USA). qRT-PCR was conducted using TransStart Tip Green qPCR SuperMix (Tran, Beijing, China). Primers (Table 1) for six candidate genes involved in antioxidant enzyme activities and photosynthetic pigment biosynthesis were designed via Primer3web v.4.1.0 (https://primer3.ut.ee/; accessed on 13 May 2024). The positions of these genes were mapped in the *Zea mays* B73_V4 reference genome (https://www.maizegdb.org/; accessed on 13 May 2024), and their functional annotations were performed using the tool AgBase v2.00 (https://agbase.arizona.edu/; accessed on 13 May 2024). Relative gene expression levels were calculated by the $2^{-\Delta\Delta Ct}$ method, with *Zm00001d010159* as an internal reference gene.

Table 1. Sequences of primers used in reverse-transcription quantitative PCR (qRT-PCR) and functional annotation of six candidate genes.

Gene ID (Encoded Protein)	Gene Position	Primer Sequence (5′ to 3′)	Gene Functional Annotation
Zm00001d025106 (*superoxide dismutase*)	Chromosome 10 (104541524_104545523 bp)	F: TTGAACTTCACTGGGGTAAGC R: ACAAAAGACTCTGCACG-CATC	Superoxide dismutase activity (GO:0004784); removal of superoxide radicals (GO:0019430); superoxide metabolic process (GO:0006801)

Table 1. *Cont.*

Gene ID (Encoded Protein)	Gene Position	Primer Sequence (5′ to 3′)	Gene Functional Annotation
Zm00001d031908 (*superoxide dismutase (SOD1a)*)	Chromosome 1 (206405747_206408926 bp)	F: TTCGCCGCTCCCTATTCC R: GTCCTGTCGATATGCACCCA	Superoxide dismutase activity (GO:0004784); oxidoreductase activity (GO:0016491); antioxidant activity (GO:0016209)
Zm00001d027511 (*catalase 2*)	Chromosome 1 (7144389_7146665 bp)	F: CCCCAACTACCTGCTGCTAC R: TGGTTATGAACCGCTCTTGC	Oxidoreductase activity (GO:0016491); catalase activity (GO:0004096); peroxidase activity (GO:0004601); response to abiotic stimulus (GO:0009628); response to hormone (GO:0009725); response to reactive oxygen species (GO:0000302); cellular oxidant detoxification (GO:00098869); hydrogen peroxide catabolic process (GO:0042744); response to oxidative stress (GO:0006979); response to hydrogen peroxide (GO:0042542); peroxisome (GO:0005777)
Zm00001d040364 (*peroxidase 72*)	Chromosome 3 (40090724_40092823 bp)	F: GGATGTATCCTACGCCGCAA R: TTGTCAAACTTGGCAGGGGT	Oxidoreductase activity (GO:0016491); peroxidase activity (GO:0004601); cellular oxidant detoxification (GO:0098869); response to oxidative stress (GO:0006979); hydrogen peroxide catabolic process (GO:0042744)
Zm00001d011819 (*chlorophyllide a oxygenase chloroplastic*)	Chromosome 8 (162062135_162065501)	F: CCATCAA-GAAGGGCAAGTTCC R: TCTTTCCTCAAGGTCCCGAT	Oxidoreductase activity (GO:0016491); chlorophyllide a oxygenase [overall] activity (GO:0010277); chlorophyll biosynthetic process (GO:0015995); plastid (GO:0009536); chloroplast (GO:0009507); cytoplasm (GO:0005737)
Zm00001d017766 (*nine–cis–epoxycarotenoid dioxygenase8*)	Chromosome 5 (206198899_206201211 bp)	F: CCACGCACACCAGAGTTACA R: GCTGGGCGCCTTTCTACTAA	Carotene catabolic process (GO:0016121); Obsolete oxidation–reduction process (GO:0055114); Chloroplast (GO:0009507); Chloroplast stroma (GO:0009570); Carotenoid dioxygenase activity (GO:0010436); oxidoreductase activity, acting on single donors with incorporation of molecular oxygen, incorporation of two atoms of oxygen (GO:0016702); 9–cis–epoxycarotenoid dioxygenase activity (GO:0045549); metal ion binding (GO:0046872); dioxygenase activity (GO:0051213)
Zm00001d010159 (*actin 1*)	Chromosome 8 (102413768_102417536 bp)	F: CGATTGAGCATGGCATTGTCA R: CCCACTAGCGTACAACGAA	ATP binding (GO:0005524); nucleotide binding (GO:0000166)

4.11. Statistical Analysis

For all tested traits of the two maize genotypes under all treatments, three-way ANOVAs were performed using IBM-SPSS Statistics v.20.0 software (SPSS, Chicago, IL, USA; https://www.Ibm.com/products/spss-statistics; accessed on June 16 2024). Pearson correlation among all traits was analyzed using Origin 2021 (v. 21.0, OriginPro, Northampton, MA, USA; https://www.genescloud.cn; accessed on 16 June 2024).

Supplementary Materials: The following supporting information can be downloaded at https://www.mdpi.com/article/10.3390/plants13202844/s1: Table S1: Multiple factorial analysis of variance (ANOVA) of the three examined variables (maize genotypes, melatonin (MT) application, and Na_2CO_3 stress), and their interactions on 25 tested traits in 2 maize genotype seedlings under different treatments.

Author Contributions: Conceptualization, X.Z.; methodology, X.Z.; analysis, G.Q.; investigation, G.Q., F.H., S.S., Z.S., and Y.N.; resources, X.Z.; writing—original draft preparation, G.Q. and X.Z.; writing—review and editing, X.Z.; supervision, X.Z.; acquisition, X.Z.; project administration, X.Z.; funding acquisition, X.Z. All authors have read and agreed to the published version of the manuscript.

Funding: This research was supported by the University Research and Innovation Platform Project of Gansu, China (2024CXPT-01), the National Natural Science Foundation of China (32360478, 32060486), the Fuxi Yong Talent Foundation of Gansu Agricultural University, China (Gaufx-05Y08), the Youth mentor Support Fund of Gansu Agricultural University, China (GAU-QDFC-2024-19), the horizontal project of Gansu Agricultural University, China (GSAU-JSXK-2024-001), and the State Key Laboratory of Aridland Crop Science, Gansu Agricultural University, China (GSCS-2023-08; GSCS-2023-Z01).

Data Availability Statement: Data are contained within the article and Supplementary Materials.

Acknowledgments: We would like to thank the staff at the Longxi experimental stations, Gansu, China, for providing maize materials.

Conflicts of Interest: The authors declare no conflicts of interest.

References

1. Chen, D.Q.; Yin, L.N.; Deng, X.P.; Wang, S.W. Silicon increases salt tolerance by influencing the two-phase growth response to salinity in wheat (*Triticum aestivum* L.). *Acta Physiol. Plant* **2014**, *36*, 2531–2535. [CrossRef]
2. Jiang, C.Q.; Cui, Q.R.; Feng, K.; Xu, D.F.; Li, C.F.; Zheng, Q.S. Melatonin improves antioxidant capacity and ion homeostasis and enhances salt tolerance in maize seedlings. *Acta Physiol. Plant* **2016**, *38*, 82. [CrossRef]
3. Shrivastava, P.; Kumar, R. Soil salinity: A serious environmental issue and plant growth promoting bacteria as one of the tools for its alleviation. *Saudi J. Biol. Sci.* **2015**, *22*, 123–131. [CrossRef]
4. Guo, S.J.; Lv, L.J.; Zhao, Y.X.; Wang, J.L.; Lu, X.J.; Zhang, M.G.; Wang, R.H.; Zhang, Y.; Guo, X.Y. Using high-throughput phenotyping analysis to decipher the phenotypic components and genetic architecture of maize seedling salt tolerance. *Genes* **2023**, *14*, 1771. [CrossRef] [PubMed]
5. Tanjimul, I.A.T.M.; Thawatchai, K.; Hayat, U.; Rujira, T.; Sujin, J.; Suriyan, C.; Avishek, D. Salt tolerance of hybrid baby corn genotypes in relation to growth, yield, physiological, and biochemical characters. *S. Afr. J. Bot.* **2022**, *147*, 808–819.
6. Wu, J.X.; Muhammad, N.; Lakshman, G.; Raymond, T.; Mumtaz, C. Effects of chilling stress on morphological, physiological, and biochemical attributes of silage corn genotypes during seedling establishment. *Plants* **2022**, *11*, 1217. [CrossRef]
7. Rohman, M.M.; Islam, M.R.; Mahmuda, B.M.; Begum, S.; Fakir, O.A.; Amiruzzaman, M. Higher K^+/Na^+ and lower reactive oxygen species and lipid peroxidation are related to higher yield in maize under saline condition. *Afr. J. Agric. Resour. E.* **2018**, *13*, 239–247. [CrossRef]
8. Ahmed, S.; Ashraf, S.; Yasin, N.A.; Sardar, R.; Ashkar, I.A.; Abdelhamid, M.T.; Sabagh, A.E. Exogenously applied nano-zinc oxide mitigates cadmium stress in *Zea mays* L. through modulation of physiochemical activities and nutrients homeostasis. *Int. J. Phytoremediation* **2024**, 1–16. [CrossRef] [PubMed]
9. Swati, V.; Prabha, N.N.; Shalini, P.; Gaurav, M.; Deepak, K. Auxin response factors in plant adaptation to drought and salinity stress. *Physiol. Plantarum* **2022**, *174*, e13714.
10. Maia, C.F.; da Silva, B.R.S.; Batista, B.L.; Bajguz, A.; Lobato, A.K.D.S. 24-Epibrassinolide simultaneously stimulates photosynthetic machinery and biomass accumulation in tomato plants under lead stress: Essential contributions connected to the antioxidant system and anatomical structures. *Agronomy* **2022**, *12*, 1985. [CrossRef]
11. Hosseinpour, M.; Ebadi, A.; Habibi, H.; Nabizadeh, E.; Jahanbakhsh, S. Enhancing enzymatic and nonenzymatic response of echinacea purpurea by exogenous 24-epibrassinolide under drought stress. *Ind. Crops Prod.* **2020**, *146*, 112045. [CrossRef]
12. Faizan, M.; Bhat, J.A.; Noureldeen, A.; Ahmad, P.; Yu, F. Zinc oxide nanoparticles and 24-epibrassinolide alleviates Cu toxicity in tomato by regulating ROS scavenging, stomatal movement and photosynthesis. *Ecotoxicol. Environ. Saf.* **2021**, *218*, 112293. [CrossRef]
13. Xu, S.H.; Wang, S.T.; Wang, Z.C.; Lu, Y.; Tao, T.Y.; Huang, Q.F.; Lu, Z.; Wang, H.Y.; Su, Y.; Ahmed, G.; et al. Integrative analyses of transcriptome, microRNA-seq and metabolome reveal insights into exogenous melatonin-mediated salt tolerance during seed germination of maize. *J. Plant Growth Regul.* **2024**, *103*, 689–704. [CrossRef]
14. Chen, Z.P.; Gu, Q.; Yu, X.L.; Huang, L.Q.; Xu, S.; Wang, R.; Shen, W.; Shen, W.B. Hydrogen peroxide acts downstream of melatonin to induce lateral root formation. *Ann. Bot.* **2018**, *121*, 1127–1136. [CrossRef] [PubMed]

15. Wang, L.Y.; Liu, J.L.; Wang, W.X.; Sun, Y. Exogenous melatonin improves growth and photosynthetic capacity of cucumber under salinity-induced stress. *Photosynthetica* **2016**, *54*, 19–27. [CrossRef]
16. Mosaad, I.S.M.; Serag, A.H.I.; Moustafa-Farag, M.; Seadh, A.K. Effect of exogenous proline application on maize yield and the optimum rate of mineral nitrogen under salinity stress. *J. Plant Nutr.* **2020**, *43*, 354–370. [CrossRef]
17. Wei, W.; Li, Q.T.; Chu, Y.N.; Reiter, R.J.; Yu, X.M.; Zhu, D.H.; Zhang, W.K.; Ma, B.; Lin, Q.; Zhang, J.S.; et al. Melatonin enhances plant growth and abiotic stress tolerance in soybean plants. *J. Exp. Bot.* **2015**, *66*, 695–707. [CrossRef]
18. Barman, D.; Kumar, R.; Ghimire, O.P.; Ramesh, R.; Gupta, S.; Nagar, S.; Pal, M.; Dalal, M.; Chinnusamy, V.; Arora, A. Melatonin induces acclimation to heat stress and pollen viability by enhancing antioxidative defense in rice (*Oryza sativa* L.). *Environ. Exp. Bot.* **2024**, *220*, 105693. [CrossRef]
19. Yan, F.Y.; Zhang, G.L.; Zhao, H.L.; Huang, Z.W.; Niu, Y.; Zhu, M.C. Foliar application of melatonin improve the number of secondary branches and secondary branch grains quality of rice. *PLoS ONE* **2024**, *19*, e0307368. [CrossRef] [PubMed]
20. Ye, F.; Jiang, M.; Zhang, P.; Liu, L.; Liu, S.Q.; Zhao, C.S.; Li, X.N. Exogenous melatonin reprograms the rhizosphere microbial community to modulate the responses of barley to drought stress. *Int. J. Mol. Sci.* **2022**, *23*, 9665. [CrossRef] [PubMed]
21. Martinez, V.; Nieves-Cordones, M.; Lopez-Delacalle, M.; Rodenas, R.; Mestre, T.C.; Garcia-Sanchez, F.; Rubio, F.; Nortes, P.A.; Mittler, R.; Rivero, R.M. Tolerance to stress combination in tomato plants: New insights in the protective role of melatonin. *Molecules* **2018**, *23*, 535. [CrossRef]
22. Muhammad, I.; Yang, L.; Ahmad, S.; Mosaad, I.S.M.; Al-Ghamdi, A.A.; Abbasi, A.M.; Zhou, X.-B. Melatonin application alleviates stress-induced photosynthetic inhibition and oxidative damage by regulating antioxidant defense system of maize: A meta-analysis. *Antioxidants* **2022**, *11*, 512. [CrossRef] [PubMed]
23. Hassan, M.U.; Mahmood, A.; Awan, M.I.; Maqbool, R.; Aamer, M.; Alhaithloul, H.A.S.; Huang, G.; Skalicky, M.; Brestic, M.; Pandey, S.; et al. Melatonin-induced protection against plant abiotic stress: Mechanisms and prospects. *Front. Plant Sci.* **2022**, *13*, 902694. [CrossRef] [PubMed]
24. Arslan, H.; Kiremit, M.S.; Güngör, A. Impacts of different water salinity levels on salt tolerance, water use, yield, and growth of chives (*Allium schoenoprasum* L.). *Commun. Soil Sci. Plant Anal.* **2018**, *49*, 2614–2625. [CrossRef]
25. Li, P.C.; Yang, X.Y.; Wang, H.M.; Pan, T.; Yang, J.Y.; Wang, Y.Y.; Xu, Y.; Yang, Z.F.; Xu, C.W. Metabolic responses to combined water deficit and salt stress in maize primary roots. *J. Integr. Agric.* **2021**, *20*, 109–119. [CrossRef]
26. Zhang, H.J.; Zhang, N.; Yang, R.C.; Wang, L.; Sun, Q.Q.; Li, D.B.; Cao, Y.Y.; Sarah, W.; Zhao, B.; Ren, S.X.; et al. Melatonin promotes seed germination under high salinity by regulating antioxidant systems, ABA and GA_4 interaction in cucumber (*Cucumis sativus* L.). *J. Pineal Res.* **2014**, *57*, 269–279. [CrossRef] [PubMed]
27. Gopalakrishnan, T.; Kumar, L. Linking long-term changes in soil salinity to paddy land abandonment in Jaffna Peninsula, Sri Lanka. *Agriculture* **2021**, *11*, 211. [CrossRef]
28. Tharani, G.; Lalit, K.; Thushyanthy, M. Assessment of spatial and temporal trend of groundwater salinity in Jaffna Peninsula and its link to paddy land abandonment. *Sustainability* **2020**, *12*, 3681. [CrossRef]
29. Zhao, C.Z.; Zhang, H.; Song, C.P.; Zhu, J.K.; Sergey, S. Mechanisms of plant responses and adaptation to soil salinity. *Innovation* **2020**, *1*, 100017. [CrossRef] [PubMed]
30. Shakeel, A.; Cui, W.W.; Muhammad, K.; Irshad, A.; Meng, X.P.; Wu, X.R.; Su, W.N.; Tehseen, J.; Hamed, A.E.S.; Jia, Z.K.; et al. Exogenous application of melatonin induces tolerance to salt stress by improving the photosynthetic efficiency and antioxidant defense system of maize seedling. *J. Plant Growth Regul.* **2020**, *40*, 1270–1283.
31. AbdElgawad, H.; Zinta, G.; Hegab, M.M.; Pandey, R.; Asard, H.; Abuelsoud, W. High salinity induces different oxidative stress and antioxidant responses in maize seedlings organs. *Front. Plant Sci.* **2016**, *7*, 276. [CrossRef]
32. Liu, Y.; Huang, J.; Ma, Y.R.; Qi, T.; Feng, Y.Z.; Meng, A.J.; Wang, X.Y. Effects induced by inputting biochar into the saliferous gray desert soil on the soil moisture movement. *Xinjiang Agric. Sci.* **2017**, *54*, 343–351.
33. Wang, L.; Lin, G.; Li, Y.; Qu, W.; Wang, Y.; Lin, Y.; Huang, Y.; Li, J.; Qian, C.; Yang, G.; et al. Phenotype, biomass, carbon and nitrogen assimilation, and antioxidant response of rapeseed under salt stress. *Plants* **2024**, *13*, 1488. [CrossRef] [PubMed]
34. Fu, Y.Y.; Xin, L.; Abdoul, K.M.H.; Sun, W.H.; Wang, H.B.; Abubakar, S.A.; Wang, X.P.; Qin, A.Z.; Gao, Y. Foliar application of melatonin positively affects the physio-biochemical characteristics of cotton (*Gossypium hirsutum* L.) under the combined effects of low temperature and salinity stress. *Plants* **2023**, *12*, 3730. [CrossRef] [PubMed]
35. Alagoz, S.M.; Hadi, H.; Toorchi, M.; Pawowski, T.A.; Lajayer, B.A.; Price, G.W.; Farooq, M.; Astatkie, T. Morpho-physiological responses and growth indices of triticale to drought and salt stresses. *Sci. Rep.* **2023**, *13*, 8896.
36. Zhang, W.; Bao, G.; Tang, W.; Dai, G.; Xiao, J.; Liu, J.; Wang, Z.; Xi, J. Physiological response of barley seedlings to salinity and artemisinin combined stresses under freeze-thaw environment. *Environ. Sci. Pollut. Res.* **2022**, *29*, 70552–70563. [CrossRef]
37. Zhao, C.; Yang, M.; Wu, X.; Wang, Y.; Zhang, R. Physiological and transcriptomic analyses of the effects of exogenous melatonin on drought tolerance in maize (*Zea mays* L.). *Plant Physiol. Biochem.* **2021**, *168*, 128–142. [CrossRef]
38. Wang, N.N.; Luo, X.M.; Wang, Z.; Liu, J.G. Mitigating effect of exogenous melatonin on salt and drought stress in *Cyperus esculentus* L. during the tilling stage. *Agronomy* **2024**, *14*, 1009. [CrossRef]
39. Wicke, B.; Smeets, E.; Dornburg, V.; Vashey, B.; Gaiser, T.; Turkenburg, W.; Faaij, A. The global technical and economic potential of bioenergy from salt-affected soils. *Energy Environ. Sci.* **2011**, *4*, 2669–2681. [CrossRef]
40. Yemets, O.; Gauslaa, Y.; Solhaug, K.A. Monitoring with lichens-conductivity methods assess salt and heavy metal damage more efficiently than chlorophyll fluorescence. *Ecol. Indic.* **2015**, *55*, 59–64. [CrossRef]

41. Kostopoulou, Z.; Therios, I.; Roumeliotis, E.; Kanellis, A.K.; Molassiotis, A. Melatonin combined with ascorbic acid provides salt adaptation in *Citrus aurantium* L. seedlings. *Plant Physiol. Biochem.* **2015**, *86*, 155–165. [CrossRef] [PubMed]
42. Navdeep, K.; Kamal, K.; Shivani, S.; Isha, S.; Pascal, G.; Kumar, P.P. Reactive oxygen species generating system and brassinosteroids are linked to salt stress adaptation mechanisms in rice. *Plant Signal. Behav.* **2016**, *11*, e1247136.
43. Shakeel, A.; Wang, G.Y.; Ihsan, M.; Muhammad, Z.; Zhou, X.B. Melatonin and KNO_3 application improves growth, physiological and biochemical characteristics of maize seedlings under waterlogging stress conditions. *Biology* **2022**, *11*, 99. [CrossRef] [PubMed]
44. Ullah, M.A.; Sagar, A.; Mia, M.A.; Tania, S.S.; Jannat-E-Tajkia; Kabir, M.H.; Hossain, A.K.M.Z.; Rauf, F.; Rhaman, M.S. Mitigation of salinity-induced growth inhibition of maize by seed priming and exogenous application of salicylic acid. *J. Agric. Ecol. Res. Int.* **2023**, *24*, 100–109. [CrossRef]
45. Li, C.H.; Li, J.X.; Du, X.H.; Zhang, J.X.; Zou, Y.R.; Liu, Y.D.; Li, Y.; Lin, H.Y.; Li, H.; Liu, D.; et al. Chloroplast thylakoidal ascorbate peroxidase, ptotAPX, has enhanced resistance to oxidative stress in populous tormentors. *Int. J. Mol. Sci.* **2022**, *23*, 3340. [CrossRef] [PubMed]
46. Meng, Y.; Yin, Q.; Yan, Z.; Wang, Y.; Niu, J.; Zhang, J.; Fan, K. Exogenous silicon enhanced salt resistance by maintaining K^+/Na^+ homeostasis and antioxidant performance in Alfalfa leaves. *Front. Plant Sci.* **2020**, *11*, 1183. [CrossRef]
47. Nidal, O. Molecular and biochemical responses of barley (*Hordeum vulgare* L.) to NaCl salinity stress and salicylic acid. *Res. Crop.* **2018**, *19*, 101–106.
48. Mahmoud, F.S.; Awais, A.; Bushra, A.A.; ElKamil, T. Exogenous application of zinc oxide nanoparticles improved antioxidants, photosynthetic, and yield traits in salt-stressed maize. *Agronomy* **2023**, *13*, 2645. [CrossRef]
49. Li, L.J.; Gu, W.R.; Zhang, L.G.; Li, C.F.; Chen, X.C.; Qian, C.R.; Wang, Z.H.; Li, W.H.; Zuo, S.Y.; Wei, S. Exogenous 2-(3,4-Dichlorophenoxy) triethylamine alleviates salinity stress in maize by enhancing photosynthetic capacity, improving water status and maintaining K^+/Na^+ homeostasis. *BMC Plant Biol.* **2020**, *20*, 348. [CrossRef] [PubMed]
50. Talaat, N.B.; Shawky, B.T. Synergistic effects of salicylic acid and melatonin on modulating ion homeostasis in salt-stressed wheat (*Triticum aestivum* L.) plants by enhancing root H^+-pump activity. *Plants* **2022**, *11*, 416. [CrossRef]
51. Maghsoudi, K.; Emam, Y.; Pessarakli, M. Effect of silicon on photosynthetic gas exchange, photosynthetic pigments, cell membrane stability and relative water content of different wheat cultivars under drought stress conditions. *J. Plant Nutr.* **2016**, *39*, 1001–1015. [CrossRef]
52. Chen, Y.E.; Mao, J.J.; Sun, L.Q.; Huang, B.; Ding, C.B.; Gu, Y.; Liao, J.Q.; Hu, C.; Zhang, Z.W.; Yuan, S.; et al. Exogenous melatonin enhances salt stress tolerance in maize seedlings by improving antioxidant and photosynthetic capacity. *Physiol. Plant.* **2018**, *164*, 349–363. [CrossRef] [PubMed]
53. Yang, Y.R.; Cao, Y.P.; Li, Z.X.; Anastasiia, Z.; Yang, S.T.; Wang, J.L.; Tang, Z.H.; Cao, Y.H.; Zhang, Y.F.; Wang, D.L. Interactive effects of exogenous melatonin and rhizophagies intraradical on saline-alkaline stress tolerance in lemus chiesas. *Mycorrhiza* **2020**, *30*, 357–371. [CrossRef]
54. Liang, W.J.; Ma, X.L.; Wan, P.; Liu, L.Y. Plant salt-tolerance mechanism: A review. *Biochem. Biophys. Res. Commun.* **2018**, *495*, 286–291. [CrossRef] [PubMed]
55. Li, J.; Liu, J.; Zhu, T.; Zhao, C.; Li, L.; Chen, M. The role of melatonin in salt stress responses. *Int. J. Mol. Sci.* **2019**, *20*, 1735. [CrossRef]
56. Shakeel, A.; Yun, W.G.; Ihsan, M.; Saqib, F.; Muhammad, K.; Irshad, A.; Muhammad, Z.; Tehseen, J.; Saif, U.; Hua, H.J.; et al. Application of melatonin-mediated modulation of drought tolerance by regulating photosynthetic efficiency, chloroplast ultrastructure, and endogenous hormones in maize. *Chem. Biol. Technol. Ag.* **2022**, *9*, 5.
57. Zhan, X. A Specific Protective Mechanism against chloroplast photo-reactive oxygen species in phosphate-starved rice plants. *Adv. Biol.* **2023**, *7*, e2300106.
58. Anjum, N.A.; Sharma, P.; Gill, S.S.; Hasanuzzaman, M.; Khan, E.A.; Kachhap, K.; Mohamed, A.A.; Thangavel, P.; Devi, G.D.; Vasudhevan, P.; et al. Catalase and ascorbate peroxidase-representative H_2O_2-detoxifying heme enzymes in plants. *Environ. Sci. Pollut. Res.* **2016**, *23*, 19002–19029. [CrossRef] [PubMed]
59. Li, C.; Wang, P.; Wei, Z.W.; Liang, D.; Liu, C.H.; Yin, L.H.; Jia, D.F.; Fu, M.Y.; Ma, F.W. The mitigation effects of exogenous melatonin on salinity-induced stress in *Malus hupehensis*. *J. Pineal Res.* **2012**, *53*, 298–306. [CrossRef] [PubMed]
60. Zeng, L.; Cai, J.S.; Li, J.J.; Lu, G.Y.; Li, C.S.; Fu, G.P.; Zhang, X.K.; Ma, H.Q.; Liu, Q.Y.; Zou, X.L.; et al. Exogenous application of a low concentration of melatonin enhances salt tolerance in rapeseed (*Brassica napus* L.) seedlings. *J. Integr. Agric.* **2016**, *17*, 328–335. [CrossRef]
61. Niu, Y.N.; Zhao, X.Q.; Chao, W.; Lu, P.N.; Bai, X.D.; Mao, T.T. Genetic variation, DIMBOA accumulation, and candidate gene identification in maize multiple insect-resistance. *Int. J. Mol. Sci.* **2023**, *24*, 2138. [CrossRef] [PubMed]
62. Zhao, X.Q.; Zhao, C.; Niu, Y.N.; Chao, W.; He, W.; Wang, Y.F.; Bai, X.D. Understanding and comprehensive evaluation of cold resistance in the seedlings of multiple maize-genotypes. *Plants* **2022**, *11*, 1881. [CrossRef] [PubMed]

plants

MDPI

Article

Low NaCl Concentrations Increase Cotyledon Growth in Chinese White Radish (*Raphanus sativus* L. var. *longipinnatus* Bailey) Seedlings via Aquaporin-Mediated Water Transport

Nutkamol Masepan [1,2], Sitthisak Intarasit [1], Aussara Panya [1] and Jarunee Jungklang [1,*]

1 Department of Biology, Faculty of Science, Chiang Mai University, Chiang Mai 50200, Thailand; nutkamol_m@cmu.ac.th (N.M.); sitthisak.inta@cmu.ac.th (S.I.)
2 Ph.D.'s Degree Program in Biology (International Program), Faculty of Science, Chiang Mai University, Chiang Mai 50200, Thailand
* Correspondence: jarunee.j@cmu.ac.th

Abstract: This study aimed to elucidate the specific role of low NaCl concentrations, particularly 10 and 20 mM, in stimulating cotyledon growth in Chinese white radish (*Raphanus sativus* L. var. *longipinnatus* Bailey) seedlings. Chinese white radish seeds were cultivated in sand culture and subjected to daily watering with solutions containing 0, 10, 20, 50, or 100 mM NaCl. Growth, water status, aquaporin gene expression, ion contents, and physiology-related parameters were assessed 4 days after sowing. Applying 10 and 20 mM NaCl significantly promoted the growth of 4-day-old seedlings. Notably, the cotyledons exhibited the most significant growth, achieving a rate of 177% compared with the 125–138% growth observed in the hypocotyl and root parts. This substantial enhancement in cotyledon growth, including biomass, cotyledon area, cotyledon thickness, and mesophyll cell size, was induced by an optimal concentration of 10 mM NaCl. This induction correlated with the increased water content, degree of succulence, and expression of aquaporin genes, specifically within *PIP1-1*, *PIP1-2*, *PIP2-1*, *PIP2-2*, and *TIP1-1*, in addition to the maintenance of the Hill reaction, heightened free radical scavenging, and the elevated accumulation of Na^+, Cl^-, K^+, proline, total N, and C. These findings suggest a beneficial role of low NaCl levels in optimising early-stage seedling growth.

Keywords: *Raphanus sativus*; low NaCl concentration; growth; water status; aquaporin genes; ion contents

1. Introduction

In general, elevated NaCl concentrations in soil can lead to ionic imbalance, osmotic stress, and oxidative stress, resulting in reduced plant growth and development [1]. Conversely, lower NaCl concentrations (<50 mM) have been shown to have positive effects on the growth and development of certain plant species, especially halophytes. However, research on the effects of low NaCl concentrations on plants has been less extensive compared to high NaCl levels. The application of low NaCl concentrations can produce suitable levels of sodium ions (Na^+) and chloride ions (Cl^-), which can act as beneficial elements and osmotica, thereby enhancing plant growth through various developmental processes observed in numerous plant species. For example, *Arabidopsis thaliana* exhibited increased shoot and root biomass when exposed to a low concentration of 5 mM NaCl, accompanied by elevated sulfur (S), carbon (C), zinc (Zn), and copper (Cu) contents [2]. Supplying Na^+ to highly salt-tolerant sugar beets increased their leaf area, stomatal density, and shoot fresh weight [3]. Cl^- has been shown to stimulate fresh weight growth by enhancing cell expansion in leaves and cell elongation in tobacco leaves and roots, which is attributed to its role as an osmoregulator that enhances water use efficiency and maintains turgor pressure within the cell [4]. Similarly, Cl^- accumulation in the shoots of *Pugionium cornutum* induces turgor pressure, thereby enhancing the plant's osmotic adjustment ability under saline and drought conditions [5]. These plants employ various strategies to enhance water use efficiency, including the utilisation of osmoregulators, maintenance of ion homeostasis, and regulation of water flux [6].

Aquaporin (AQP) proteins play a crucial role in regulating water influx and efflux in plant cells. AQPs are membrane channel proteins that play a key role in transporting water, hydrogen peroxide, and other small molecules, such as CO_2, NO, glycerol, ammonia, urea, and boron. They are categorised into five major subfamilies within higher plants: plasma membrane intrinsic proteins (PIPs), tonoplast intrinsic proteins (TIPs), nodulin 26-like intrinsic proteins (NIPs), small basic intrinsic proteins (SIPs), and uncategorised intrinsic proteins (XIPs). In plants, PIPs and TIPs facilitate water translocation across cellular membranes and subcellular compartments [7–9]. AQPs are involved in drought and salt stress tolerance in several plant species. For instance, *MaPIP1-1*, known as an AQP gene in banana, exhibited overexpression in transgenic *Arabidopsis* plants under 50–150 mM NaCl conditions, leading to enhanced primary root length, increased root hair development, reduced ion leakage and malondialdehyde (MDA) content, decreased cytosolic K^+ and Na^+ accumulation, and an improved K^+/Na^+ ratio compared to wild-type seedlings [10]. The upregulation of *SsPIP* expression in the halophytic plant *Suaeda salsa* L. under 100 mM NaCl treatment correlated with increased shoot fresh weight, osmotic potential, root hydraulic conductance, and succulence degree [11]. However, the impact of low NaCl concentrations on AQP expression in plant cells has received less attention compared to high NaCl concentrations.

The seedling or sprout stage of Chinese white radish (*Raphanus sativus* L. var. *longipinnatus* Bailey) is rich in vitamins (B1, B2, B3, B5, B6, B9, and C) and minerals (Cl, K, Zn, Fe, Mg, Mn, and P), and contains bioactive compounds such as polyphenols and glucosinolates. These seedlings have been recommended for consumption due to their high nutritional value and the presence of anti-cancer compounds that contribute to human health [12,13]. To date, this sprout has been distributed on the shelves of numerous supermarkets across Thailand. Chinese white radish has been identified as a halophyte exhibiting moderate tolerance to salinity stress, as it is capable of surviving and completing its life cycle under high-salinity conditions (0.61–1.2 dS/m EC 1:5), which corresponds to approximately 0.4–0.8% soil salinity [14,15]. Seven-day-old seedlings of this species can also grow under saline conditions (0.4–0.8 dS/m EC 1:5 in sand culture) [16]. Yuan et al. [17] reported that treatment with 10 mM NaCl increased the yields of 5- and 7-day-old radish seedlings by 50% and 64%, respectively, while treatment with 50 mM NaCl resulted in yield increases of 74% and 100%, respectively.

Despite these findings, the effects of low NaCl concentrations on the growth of Chinese white radish seedlings remain relatively unexplored. This study aims to elucidate the positive effects of a low NaCl concentration on increasing growth, particularly in the cotyledon, in 4-day-old Chinese white radish seedlings. We hypothesised that low NaCl concentrations would promote seedling growth by enhancing water influx into cotyledon cells, thereby increasing water status, maintaining ion homeostasis, and ultimately promoting cotyledon growth. This research contributes to the understanding of how reduced NaCl levels stimulate the growth of Chinese white radish seedlings and may offer insights into enhancing seedling viability, particularly under environmental stresses such as salinity and water deficit.

2. Results

2.1. Effects of Low to High NaCl Concentrations on Growth of Chinese White Radish Seedlings

All the tested NaCl concentrations (10, 20, 50, and 100 mM) did not significantly inhibit seed germination or seedling growth in 4-day-old Chinese white radish. However, seedling growth rates were observed to be slower and unevenly germinated under moderate (50 mM) and high (100 mM) NaCl concentrations compared to the control (Supplementary Figure S1). Specifically, the 50 and 100 mM NaCl treatments led to reductions in shoot height, root length, and hypocotyl fresh weight, as well as cotyledon, hypocotyl, and root dry weight compared to those of the control (Figures 1 and 2). In contrast, lower NaCl concentrations (10 and 20 mM) did not negatively affect these growth parameters. Instead, they resulted in a significant increase in cotyledon fresh weight, with enhancements of up to 177% and 178%, respectively, compared to those of the control (Figure 2D).

Figure 1. Effects of 0, 10, 20, 50, and 100 mM NaCl on growth of 4-day-old Chinese white radish (*Raphanus sativus* L. var. *longipinnatus* Bailey) seedlings.

2.2. Effects of Low NaCl Concentrations on Structure and Function of Chinese White Radish Cotyledons

2.2.1. Growth Parameters

Low NaCl concentrations (10 and 20 mM) affected cotyledon growth. The cotyledon area, cotyledon thickness, palisade cell width, and spongy cell length were significantly increased by these NaCl concentrations compared to the control. In addition, only the palisade mesophyll cell length of the cotyledons treated with 10 mM NaCl increased significantly (Figure 3).

2.2.2. Water Status and Aquaporin Gene Expression

Concentrations of 10 and 20 mM NaCl had different effects on the water status of 4-day-old Chinese white radish seedlings. Although they significantly decreased the RWC of the cotyledons (Figure 4A), they conversely increased the water content and degree of succulence. Notably, this increase exceeded 150% compared to the control (Figure 4B,C).

Moreover, these low NaCl concentrations increased the expression of AQP genes, which are ubiquitous membrane channel proteins that play key roles in water transport across membranes. Treatment with 10 mM NaCl significantly increased the expression of the *PIP1-1*, *PIP1-2*, *PIP2-1*, *PIP2-2*, and *TIP1-1* genes, while the 20 mM NaCl treatment significantly increased the expression of the *PIP1-1*, *PIP1-2*, and *TIP1-1* genes. However, the most remarkable increase in expression was observed in the *TIP1-1* gene, which rose by 1084% and 483%, respectively, compared to the control (Figure 4D–H).

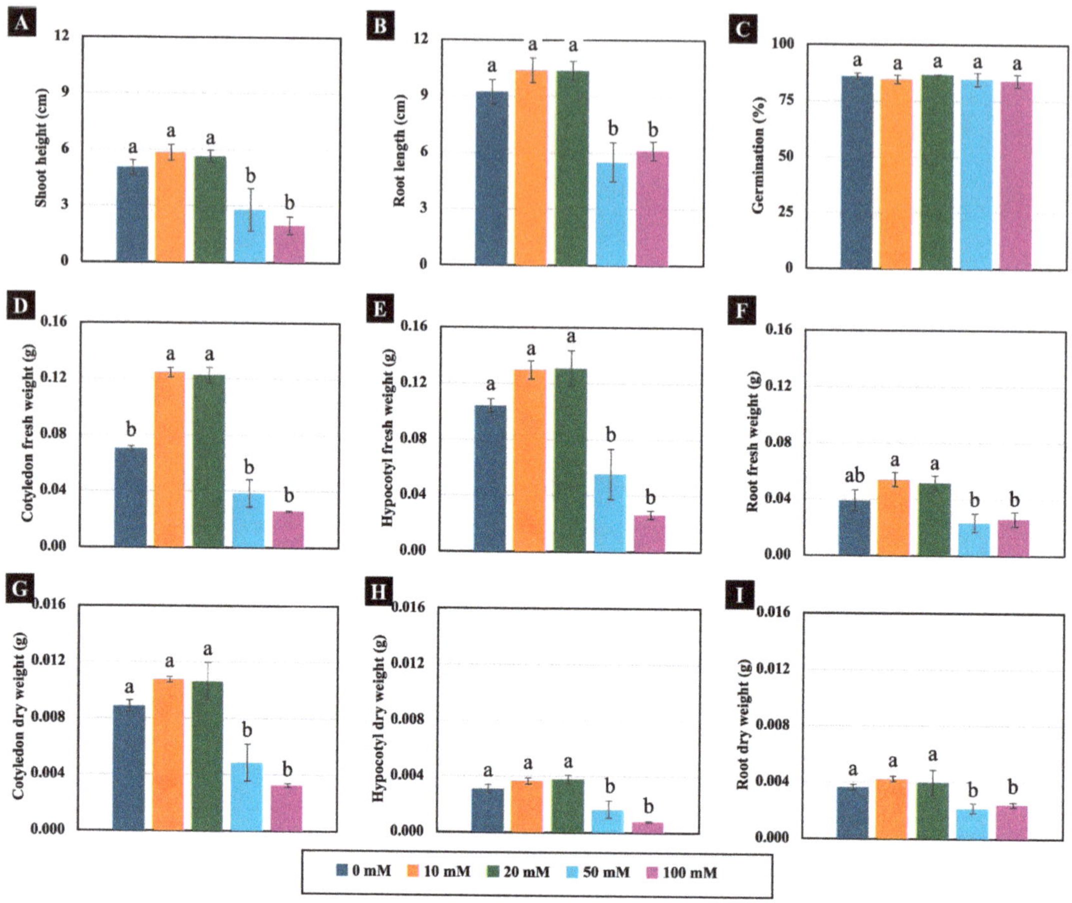

Figure 2. The effects of 0, 10, 20, 50, and 100 mM NaCl on the shoot height (**A**), root length (**B**), germination (**C**), fresh weight (**D–F**), and dry weight (**G–I**) of 4-day-old Chinese white radish seedlings. The results are the means of the three replications ± standard deviation (SD). Different letters (a and b) for the same parameter indicate significant differences at $p < 0.05$, determined by Duncan's multiple range test.

2.2.3. Ion Contents

Low NaCl concentrations (10 and 20 mM) significantly increased the accumulation of Na^+ and Cl^- in the cotyledons of 4-day-old Chinese white radish seedlings compared the control. In the cotyledons of the seedlings treated with 10 and 20 mM NaCl, Na^+ accumulation was more than 1000% higher, whereas Cl^- accumulation was more than 450% greater compared to the control (Figure 5A,B). Additionally, these low concentrations exhibited distinct trends in the macronutrient content of the cotyledons of 4-day-old Chinese white radish seedlings. The levels of C, total N, and K^+ increased significantly in cotyledons treated with 10 and 20 mM NaCl. The K^+ content increased by over 120% compared to the control (Figure 5C,D,F). Conversely, the contents of PO_4^{3-}, Ca^{2+}, and extract S decreased in the cotyledons treated with low NaCl. Specifically, both the Ca^{2+} and extract S contents were reduced by more than 18% compared to the control (Figure 5E,G,I). However, the Mg^{2+} content was not significantly different between the treated and untreated seedlings (Figure 5H).

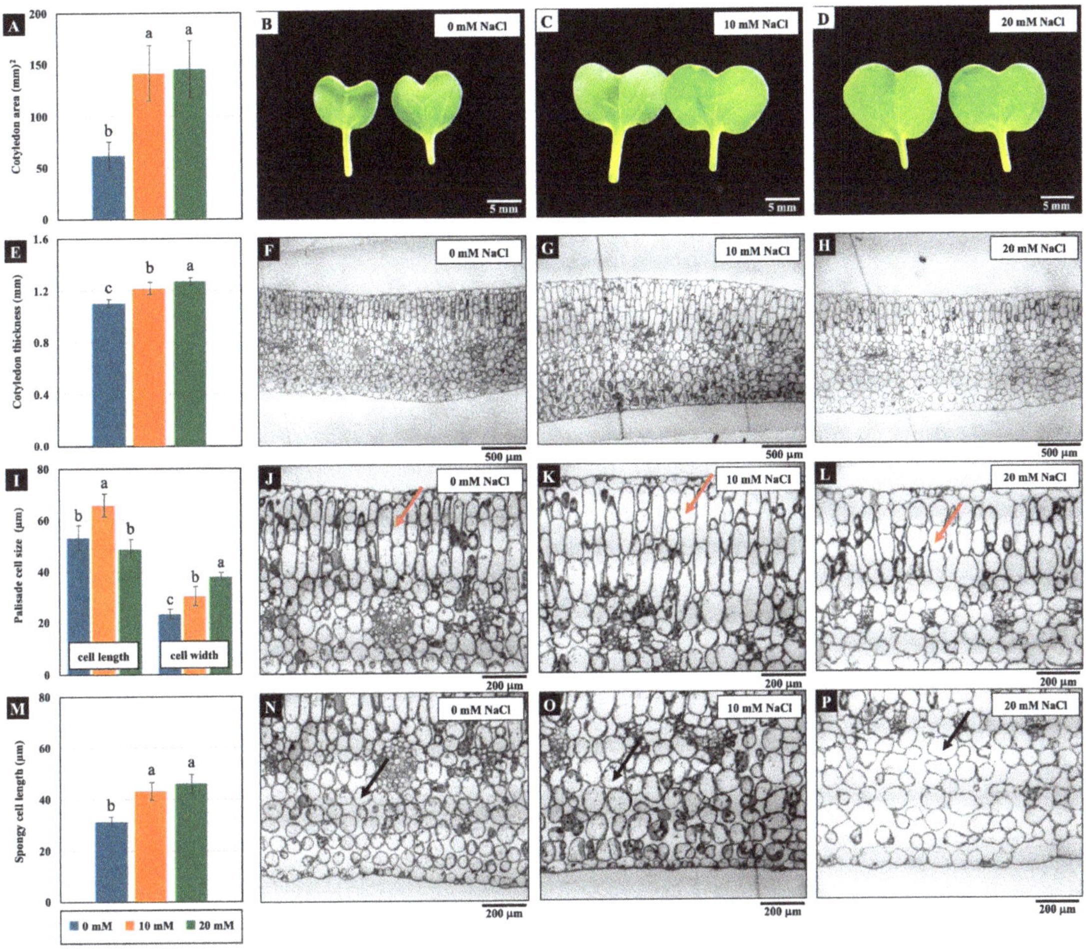

Figure 3. The effects of 0, 10, and 20 mM NaCl on the cotyledon area (**A–D**), cotyledon thickness (**E–H**), palisade (red arrow) mesophyll cell size (**I–L**), and spongy (black arrow) mesophyll cell width (**M–P**), respectively, of the cotyledons of 4-day-old Chinese white radish seedlings. The results are the means of three replications ± standard deviation (SD). Different letters (a–c) for the same parameter indicate significant differences at $p < 0.05$, determined by Duncan's multiple range tests.

2.2.4. Physiology-Related Parameters

Low NaCl concentrations (10 and 20 mM) had varying effects on physiology-related parameters. The Hill reaction, which represents the activity of the electron transport system, was not significantly different in the cotyledons treated with 10 mM NaCl for 4 days. However, this function was reduced in the cotyledons treated with 20 mM NaCl (Figure 6A). Additionally, 10 and 20 mM NaCl slightly increased the free radical scavenging activity, assessed using the DPPH assay, measuring 101% compared to the control in the cotyledons of 4-day-old Chinese white radish seedlings (Figure 6B). In contrast, proline content significantly increased with increasing NaCl concentration, showing levels of up to 133% of that of the control in the cotyledons (Figure 6C).

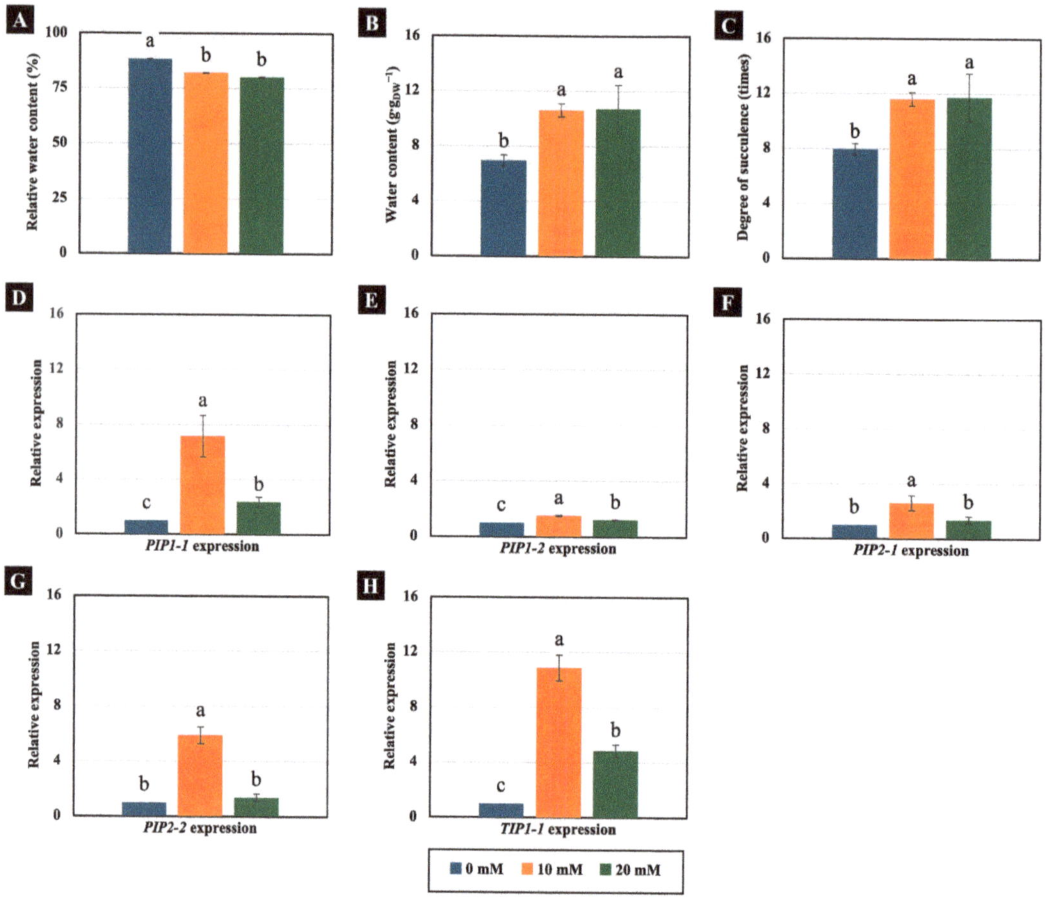

Figure 4. The effects of 0, 10, and 20 mM NaCl on the relative water content (**A**), water content (**B**), degree of succulence (**C**), and expression of the *PIP1-1* (**D**), *PIP1-2* (**E**), *PIP2-1* (**F**), *PIP2-2* (**G**), and *TIP1-1* (**H**) genes, respectively, in the cotyledons of 4-day-old Chinese white radish seedlings. The results are the means of three replications ± standard deviation (SD). Different letters (a–c) for the same parameter indicate significant differences at $p < 0.05$, determined by Duncan's multiple range test.

2.2.5. Correlation Coefficients Among Cotyledon Parameters Under Low NaCl Concentrations

The correlation coefficients among various parameters in the cotyledons of 4-day-old Chinese white radish exhibited considerable variability. Fresh weight demonstrated a positive correlation with dry weight, water content, degree of succulence, sodium, chloride, potassium, nitrogen, the expression of aquaporin genes (*PIP1-1*, *PIP1-2*, *PIP2-1*, and *TIP1-1*), area, thickness, palisade cell width, spongy cell length, free radical scavenging activity measured by DPPH assay, and proline content in the cotyledons. Conversely, fresh weight showed a negative correlation with relative water content (RWC), carbon, calcium, phosphorus, and sulfur levels. Furthermore, sodium exhibited a positive correlation with potassium and the expression of aquaporin genes, while both sodium and chloride displayed negative correlations with calcium, phosphorus, and sulfur levels (Figure 7).

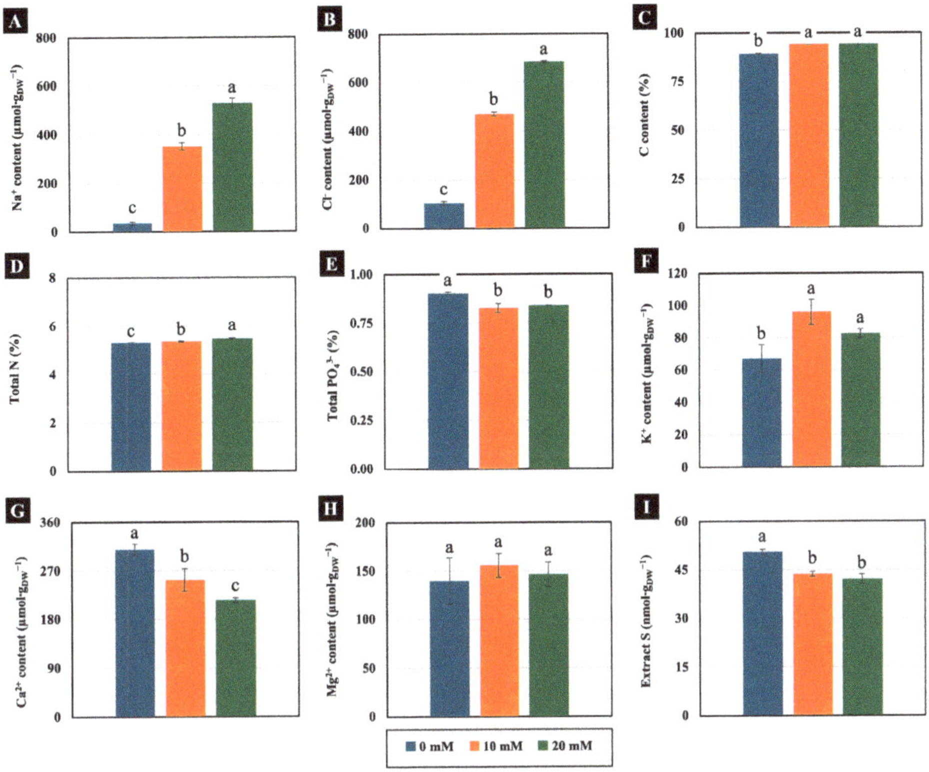

Figure 5. The effects of 0, 10, and 20 mM NaCl on the Na^+ content (**A**), Cl^- content (**B**), C content (**C**), total N (**D**), total PO_4^{3-} (**E**), K^+ content (**F**), Ca^{2+} content (**G**), Mg^{2+} content (**H**), and extract S (**I**), respectively, of the cotyledons of 4-day-old Chinese white radish seedlings. The results are the means of three replications ± standard deviation (SD). Different letters (a–c) for the same parameter indicate significant difference at $p < 0.05$, determined by Duncan's multiple range test.

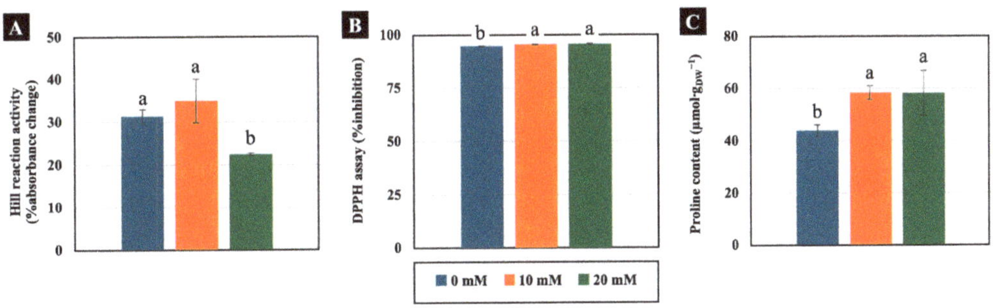

Figure 6. The effects of 0, 10, and 20 mM NaCl on Hill reaction activity (**A**), free radical scavenging activity by DPPH assay (**B**), and proline content (**C**), respectively, in the cotyledons of 4-day-old Chinese white radish seedlings. The results are the means of three replications ± standard deviation (SD). Different letters (a and b) for the same parameter indicate significant differences at $p < 0.05$, determined by Duncan's multiple range test.

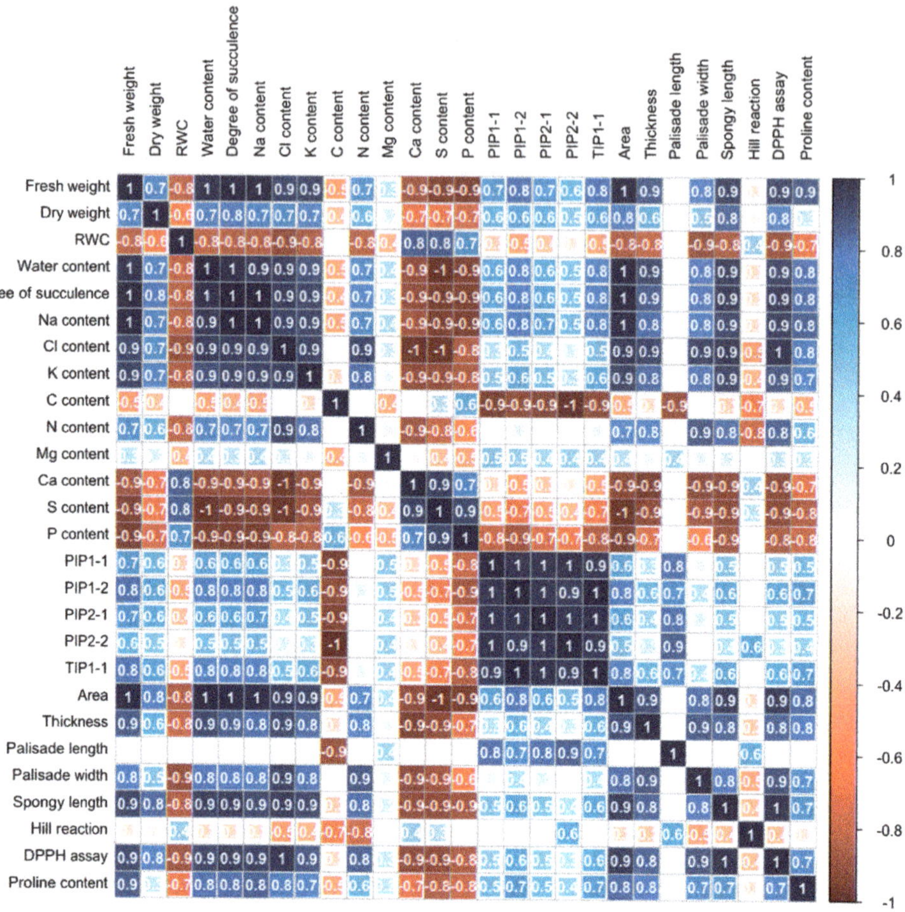

Figure 7. Correlation coefficients among cotyledon parameters under low NaCl concentrations in 4-day-old Chinese white radish seedlings. Pearson's correlation coefficient (r) significant at $p < 0.05$.

3. Discussion

This study employed moderate (50 mM) and high (100 mM) concentrations of NaCl, which reduced growth in 4-day-old Chinese white radish seedlings by decreasing shoot height, root length, fresh weight, and dry weight, but low concentrations of 10 and 20 mM NaCl resulted in significant enhancements in growth, as measured by cotyledon fresh weight (Figures 1 and 2). These observed effects can be attributed to the slight accumulation of soluble salts in sand culture due to the low levels of NaCl concentration, thereby creating optimal conditions for seedling development. This assertion is supported by the preliminary experiment (Supplementary Figure S2), where the electrical conductivity (EC) in the sand culture watered with 10 and 20 mM NaCl solutions ranged from 0.08 to 0.15 dS/m over a period of 0–4 days. These values indicate a slight salinity in the sand culture [14,15,18], providing an optimal supply of Na^+ and Cl^- ions in the sand culture, which serve as essential nutrients for these seedlings, while 50 and 100 mM NaCl, which had the higher EC values in watered sand culture, ranged from 0.21 to 0.35 (moderate salinity) and 0.40 to 0.65 dS/m (strong salinity), respectively. These concentrations may represent higher doses that are not optimal for promoting growth.

Previous studies have indicated that Na^+ is a beneficial element for certain halophytic species, while Chinese white radish is a moderately salt-tolerant plant [19]. This ion shares chemical and structural similarities with K^+ and can partially substitute K^+ in various functions, facilitating the long-distance transport of anions and enhancing photosynthetic activity in these species [3,20,21]. Additionally, Cl^- is a plant micronutrient required in small quantities for various physiological processes. Cl^- plays a pivotal role in stabilising the photosynthetic reactions involved in oxygen evolution and cell division in both leaves and roots. Moreover, it is a regulator of specific enzyme activities, such as vacuolar proton-pumping ATPase activity [22,23]. Our findings align with Yuan et al. [17], who showed that radish seedlings aged 3–7 days and exposed to 10 mM NaCl exhibited increased fresh weight without reducing germination. Similarly, Hongqiao et al. [2] observed that 5 mM NaCl increased shoot and root biomass in 11-day-old *Arabidopsis thaliana* seedlings.

However, the increase in cotyledon fresh weight was remarkable among the three plant parts, with a significant increase of 177–178% compared to the control (Figure 2). This augmentation in fresh weight observed in the cotyledons of Chinese white radish under low NaCl concentrations can be ascribed to heightened cell expansion, signifying changes in mesophyll cell width and length. Consequently, this led to an expansion in the cotyledon area and thickness (Figures 3 and 7). This phenomenon was closely associated with the enhancement of water status, as indicated by the plants' water content, degree of succulence, and elevated levels of Na^+, Cl^-, and K^+ ions and proline content (Figures 5, 6C and 7). In radish, spraying with proline treatment mitigated salinity stress by increasing the transpiration rate, stomatal frequency, leaf area, dry matter, water content, and photosynthetic pigments [24]. Proline is recognised as an osmotic solute pivotal in regulating the water balance between plant cells and the environment under osmotic and salinity stress [25–27]; Na^+, Cl^-, and K^+ ions may serve as osmoregulators, mitigating the osmotic potential of seedling tissues. The role of Na^+, Cl^-, and K^+ as osmotic agents facilitating osmotic adjustment in plants under salinity stress has been corroborated by prior investigations [5,21].

The exceptionally high accumulation of Na^+ and Cl^- in the cotyledons of Chinese white radish may serve as a primary mechanism for facilitating water uptake into the cells (Figure 5A,B), consequently aiding in various metabolic processes, such as the hydrolysis of adenosine triphosphate (ATP), cell division, and photosynthesis, while also mitigating the toxicity of excess Na^+ and Cl^- ions by diluting tissue concentrations [28,29]. Moreover, Cl^- benefits the growth of chloride-tolerant species, such as *Pugionium cornutum*, by enhancing leaf hydration, photosynthetic activity, and biomass [5]. Assessing RWC is crucial to gauge water insufficiency, as it reflects plant tissue hydration in a fully turgid state [26]. The RWC, which indicates the free energy of water within the cell (typically above 80%), signifies adequate hydration. Although our study revealed a 4–6% decrease in cotyledon RWC with low NaCl concentrations (Figure 4), it did not imply water stress, remaining within an acceptable range (above 80% RWC) for Chinese white radish seedlings and having no detrimental effect on their growth [30,31].

Aquaporin (AQP) genes play a crucial role in regulating water transport across cellular and organellar membranes, thereby influencing essential physiological processes in plants. These genes encode water channel proteins that facilitate the movement of water molecules, contributing to cellular hydration and turgor pressure, which are both vital for plant growth and development. Previous studies have identified *RsPIP1*, *RsPIP2*, and *RsTIP1* as the predominant AQPs expressed in radish seedling tissues. The *PIP1* and *PIP2* genes encode plasma membrane intrinsic proteins (PIPs) that mediate water transport across the plasma membrane, whereas *TIP* genes encode tonoplast intrinsic proteins (TIPs), which function at the vacuolar membrane [7]. Among these, specific members of the RsPIP2 group are upregulated in response to osmotic stress, while the RsPIP1 and RsTIP1 groups act as constitutive aquaporins [32]. Fricke and Knipfer [33] emphasised that proteins of the PIP and TIP families do more than facilitate water entry into individual cells; they play a pivotal role in enabling transcellular water flow, particularly in rapidly expanding tissues. This process is essential for maintaining turgor pressure, which is a key driver of continuous cell expansion and development. Supporting this, the constitutive overexpression of the *PtoPIP1-1* gene in *Arabidopsis* enhanced plant water status and turgor pressure, promoting cell expansion in both roots and leaves, ultimately accelerating bolting and flowering [34]. Furthermore, under prolonged high-salinity stress (100 mM NaCl for 10 days), *PIP1* expression in melon roots remained unchanged, while *TIP1-1* was significantly upregulated, suggesting its prominent role in maintaining turgor pressure and facilitating Na^+ sequestration into vacuoles [35].

Our results revealed a notable upregulation of the *PIP1-1, PIP1-2, PIP2-1, PIP2-2,* and *TIP1-1* genes in cotyledons treated with 10 mM NaCl, whereas the 20 mM NaCl treatment induced the expression only of *PIP1-1, PIP1-2,* and *TIP1-1* (Figures 4 and 7). These findings suggest that 10 mM NaCl more effectively stimulates the expression of specific AQPs, particularly those associated with the plasma and vacuolar membranes. According to previous reports, PIP1 proteins exhibit high water channel activity when coexpressed with PIP2 proteins, forming PIP1–PIP2 heterotetramers. This complex not only facilitates the trafficking of PIP1 proteins to the plasma membrane but also enhances PIP2 protein activity by increasing membrane water permeability [36]. The enhanced expression of AQP genes under 10 mM NaCl conditions appears to be closely associated with increased water channel function, thereby promoting cell expansion. Microscopic analysis revealed greater cell expansion and reduced intercellular spaces in cotyledon cross-sections under 10 mM NaCl treatment compared to those exposed to 20 mM NaCl (Figure 3). This morphological change corresponds with the upregulation of *PIP1-1, PIP1-2, PIP2-1, PIP2-2,* and *TIP1-1* (Figure 4), indicating that the lower salt concentration more effectively enhances AQP-mediated water transport. This improved water permeability likely facilitated rapid water movement across both the plasma and vacuolar membranes, thereby supporting turgor-driven cell expansion. The resulting increase in cell volume would enhance the availability of water and solutes necessary for key physiological functions, including photosynthesis, ultimately contributing to improved cotyledon growth in 4-day-old Chinese white radish seedlings.

The high accumulation of sodium ion (Na^+) and chloride ion (Cl^-) disrupts ion homeostasis by either increasing or decreasing the uptake of essential macronutrients [6]. This disruption can significantly affect plant growth and development, as macronutrients play crucial roles in various physiological processes. Macronutrients such as nitrogen (N), phosphorus (P), potassium (K), calcium (Ca), magnesium (Mg), and sulfur (S) are vital for plant health. N is a key component of proteins, enzymes, and chlorophyll, directly influencing photosynthesis and energy transformation. P contributes to ATP production, nucleic acids (DNA and RNA), and cell membrane structure. K is essential for metabolite transport, water balance, and stomatal regulation. Ca maintains cell wall integrity and acts as a signalling molecule for growth and stress responses. Mg, as the central atom in chlorophyll, plays a major role in photosynthesis and enzyme activation. S is necessary for chlorophyll synthesis, protein formation, and amino acid production, particularly for methionine and cysteine [22,37]. In fact, excessive Na^+ accumulation primarily disrupts the uptake of cations such as Ca^{2+}, K^+, and Mg^{2+}, whereas high Cl^- levels interfere with the uptake of anions like nitrate (NO_3^-), sulfate (SO_4^{2-}), and phosphate (PO_4^{3-}) [2,38]. This imbalance alters nutrient availability and may impair plant growth. The disruption of ion homeostasis has been observed in Chinese white radish seedlings. Exposure to 10 and 20 mM NaCl led to an increase in K^+ and N content, while Ca^{2+}, S, and P levels decreased. Interestingly, Mg^{2+} content remained unchanged (Figure 5). This suggests that these low NaCl concentrations selectively influence macronutrient uptake. Additionally, 10 mM NaCl induced greater K^+ accumulation compared to 20 mM NaCl, correlating positively with water status (Figure 7). This finding implies that a low NaCl concentration (10 mM) may be more optimal for promoting cotyledon growth, particularly through water accumulation regulated by K^+. K^+ is known to enhance Na^+ sequestration within leaf vacuoles, a key mechanism for maintaining cytosolic K^+ retention, which is crucial for salt tolerance. Studies have shown that plants with higher cytosolic K^+ concentrations exhibit improved resilience to NaCl stress [39,40]. Similarly, Cl^- accumulation disrupts NO_3^-, SO_4^{2-}, and PO_4^{3-} uptake, leading to reductions in N, S, and P levels [2]. Despite the observed declines in S and P content in this study, N content increased under low NaCl concentrations (Figure 5). Previous research has reported similar findings, where NaCl treatments enhanced N uptake in various plants. For instance, upland rice treated with 8 mM NaCl showed increased N content [41], and *Arabidopsis* roots exposed to 5 and 10 mM NaCl exhibited elevated N levels [2]. The simultaneous increase in K^+, N, Na^+, and Cl^- content may have contributed to the enhanced cotyledon growth in Chinese white radish seedlings, possibly due to improved carbon (C) accumulation under low-NaCl conditions.

At 4 days old, Chinese white radish typically displays two heart-shaped cotyledons (Figure 1). These cotyledons may be crucial for photosynthetic processes, providing carbon sources for seedling growth and development. They have functions similar to those of mature leaves and play a vital role in photosynthesis. As shown in Figure 6, activities related to the Hill reaction, a marker of H_2O oxidation during photosynthesis, also occur in these cotyledons. According to findings in *Arabidopsis thaliana*, the initiation of the photosynthetic process occurs when the first cotyledon is exposed [42]. However, it is well documented that the high accumulation of Na^+ and Cl^- can lead to reduced levels of aminolevulinic acid (ALA), a precursor for chlorophyll synthesis [43]. Additionally, these conditions can enhance chlorophyllase activity, decreasing chlorophyll content and photosynthetic activity [44]. Our findings revealed the maintenance of Hill reaction ability in cotyledons under low NaCl concentrations, especially 10 mM (Figure 6A). Despite the large accumulation of Na^+ and Cl^- in the leaves, theses ions might be compartmentalised in vacuoles [45]. Moreover, the assessment of scavenging activities through DPPH assays revealed a slight increase in cotyledons treated with 10 and 20 mM NaCl (Figure 6B), suggesting that plant cells can regulate free radical accumulation. This suggests that the use of 10 mM NaCl is more suitable than 20 mM NaCl for inducing Na^+ and Cl^- accumulation in cotyledons, thereby maintaining the photosynthetic ability involved in oxygen evolution. These results are supported by the elevated expression levels of AQP–water gate proteins located on the cell membrane, specifically the *PIP1-1*, *PIP1-2*, *PIP2-1*, and *PIP2-2* genes (Figure 4), which facilitate the utilisation of free water as a substrate for this process.

4. Materials and Methods

Based on our preliminary experiment, the application of low concentrations of NaCl (10 and 20 mM) for 0–5 days promoted an increase in the fresh weight of the whole plant, including both the shoots and roots, of Chinese white radish (*Raphanus sativus* L. var. *longipinnatus* Bailey). In contrast, higher concentrations of 50 and 100 mM NaCl reduced these parameters by approximately 50% compared to the control (Figure 2). Moreover, the highest growth rate was observed in 4-day-old seedlings (see Supplementary Figure S1). The application of NaCl in the range of 0–100 mM also led to an increase in soil electrical conductivity (EC). EC values were measured using the 1:5 (soil/water) method and conductivity meter (Eutech, Vernon Hills, IL, USA), revealing a range of 0.08–0.15 dS/m in soils treated with 10 and 20 mM NaCl, which corresponds to slight salinity. In contrast,

50 mM NaCl resulted in EC values of 0.21–0.38 dS/m, indicating moderate salinity, while 100 mM NaCl led to EC values of 0.40–0.72 dS/m, also classified as high-salinity [18,46,47] (see Supplementary Figure S2). Therefore, NaCl concentrations of 10 and 20 mM were designated as low, while 50 and 100 mM were classified as moderate and high concentrations, respectively. Additionally, 4-day-old seedlings were selected as the optimal developmental stage for this study.

4.1. Plant Material and NaCl Treatments

The seedlings of 4-day-old Chinese white radish were prepared. The seeds were soaked in distilled water for 18 h, and 10 fully swollen seeds were selected for germination in a plastic container containing 200 g of fine sand. The sand cultures were initially watered with 40 mL of 5 NaCl solutions: 0 (control), 10, 20, 50, or 100 mM. Subsequently, 10 mL of the same concentration was applied daily to each treatment group. The cultivation of the seedlings was conducted in a greenhouse under a 12/12 h light/dark photoperiod at 30 ± 5 °C, a relative humidity of $70 \pm 5\%$, and a photosynthetic photon flux density of 3000 ± 100 lux. Four-day-old seedlings were chosen for the analysis of growth, water status, AQP gene expression, ion contents, and physiology-related parameters.

4.2. Measurement of Growth Parameters

The percentage of germination was calculated using the following formula: germination (%) = (number of germinated seeds/number of sown seeds) $\times$ 100. Shoot heights and root lengths were measured using a ruler and recorded in centimetre intervals. The fresh and dry weights, which were obtained after drying the fresh plant samples at 80 °C for 2 days in hot air oven (Binder, Camarillo, CA, USA), of the cotyledons, hypocotyls, and roots, were determined using a precision balance (Ohaus, Parsippany, NJ, USA) and recorded in g. Cotyledon areas were determined by capturing images and utilising ImageJ (version 1.54g), a Java-based image processing programme. The measurements were recorded in mm^2. Cotyledon thickness and palisade and spongy mesophyll cell sizes were determined under a light microscope (Olympus, Center Valley, PA, USA) using a cross-sectional technique and recorded in µm. Palisade mesophyll cells were measured in terms of width and length, whereas spongy mesophyll cells were measured only in terms of length or diameter.

4.3. Measurement of Water Status Parameters

The plant samples were first measured for fresh weight (FW). Subsequently, the samples were soaked in distilled water until they reached saturation or exhibited no change in fresh weight (approximately 4 h for the seedlings) to obtain their turgid weight (TW). Then, they were dried at 80 °C for 2 days to obtain the dry weight (DW). Following a method modified from Smart and Bingham [48], the relative water content (RWC) was calculated using the following formula: RWC (%) = [(FW − DW)/(TW − DW)] $\times$ 100. Following the modified method of Zeng et al. [49], the water content (WC) was calculated using the following formula: WC ($g \cdot g_{DW}^{-1}$) = (FW − DW)/DW. And the degree of succulence (DS) was calculated using the following formula: DS (times) = FW/DW.

4.4. Measurement of Aquaporin Gene Expression

The expression of AQP genes (*PIP1-1*, *PIP1-2*, *PIP2-1*, *PIP2-2*, and *TIP1-1*) was determined using real-time polymerase chain reaction (real-time PCR). The fresh plant samples were ground in liquid nitrogen, and the frozen powder was mixed with TRIzol reagent (Invitrogen, Waltham, MA, USA) for 5 min to extract the mRNA. Chloroform was added, and the mixture was shaken and left to separate into two layers. The mixture was then centrifuged at 12,000 rpm for 15 min at 4 °C using a microcentrifuge (Hercuvan Laboratory Systems, Subang Jaya, Malaysia). The supernatant was transferred to a new tube, isopropanol was added, and the mixture was inverted and incubated at room temperature (25–30 °C) for 10 min. After centrifuging again at 12,000 rpm for 10 min at 4 °C, the supernatant was discarded. The mRNA pellet was washed with 70% ethanol, centrifuged at 10,000 rpm for 5 min at 4 °C, and the supernatant was discarded. The pellet was air-dried, incubated with molecular-grade water at −20 °C overnight, and then converted to cDNA using the Tetro cDNA Synthesis Kit (Bioline, Essex, UK) and PCR technique at 45 °C in a thermal cycler (Thermo Fisher Scientific, Waltham, MA, USA). The primers for the AQPs and GAPDH used in this study (Supplementary Table S1) were designed based on previously published sequences by Suga et al. [32,50], Tsuchihira et al. [51], Higuchi et al. [52], and Xu et al. [53]. GAPDH was selected as the reference gene for normalisation, as it has been identified as the most stable internal control for gene expression analysis across various radish organs under multiple conditions, including salt stress [54]. The cDNA concentration was measured at 260 nm using a nanodrop spectrophotometer (Thermo Fisher Scientific, Waltham, MA, USA). The expression of AQPs was detected using a Luna qPCR dye (New England Biolabs, Ipswich, MA, USA) and a real-time PCR system (QIAGEN, Hilden, Germany).at 59 °C.

The relative expression levels of the target genes were calculated using the $2^{-\Delta\Delta Ct}$ method [55]. The calculation was performed in the following steps:

1. ΔCt (Delta Ct) was obtained by subtracting the Ct (Cycle threshold) value of the reference gene from the Ct value of the target gene in each sample: $\Delta Ct = Ct_{AQPs} - Ct_{GADPH}$
2. ΔΔCt (Delta Delta Ct) was calculated by subtracting the ΔCt of the control group from the ΔCt of the treatment group for each target gene: $\Delta\Delta Ct = \Delta Ct_{treatment} - \Delta Ct_{control}$
3. The relative expression level of each gene was determined by the following equation: relative expression = $2^{-\Delta\Delta Ct}$

4.5. Measurement of Ion Contents

The cation contents, Na^+, K^+, Ca^+, and Mg^+, were determined using a method modified from that described by Hoo et al. [56]. The dried plant samples were digested in 1 N HNO_3 until evaporation was complete. Subsequently, deionised water was added, and the resulting solution was filtered. The cation contents were quantified utilising a flame photometer (Cole-Parmer, Québec, QC, Canada), with the outcomes expressed in $\mu mol \cdot g_{DW}^{-1}$.

The quantification of P (total phosphate; PO_4^{3-}), total N, extract S, C, and Cl^- was conducted using a technique modified from that outlined by the Science Center of Land Development [57]. For the analysis of total PO_4^{3-}, the dried plant samples were digested and completely evaporated in a 2:1 (v/v) mixture of HNO_3 and $HClO_4$. Subsequently, the sample was appropriately diluted with deionised water and Barton's reagent (10.7 mM NH_4VO_3 and 21.5 mM $(NH_4)_6Mo_7O_{24}{\cdot}4H_2O$) was added during the digestion process to generate the yellow complex samples. The absorbance was measured at 420 nm using a spectrophotometer (BMG LABTECH, Ortenberg, Germany). The total PO_4^{3-} was determined using a standard curve and recorded as a percentage of dry weight (%DW).

For the total N analysis, the dried plant samples were digested in H_2SO_4 (conc.) and the samples were appropriately diluted with deionised water. A Kjeldahl digestion apparatus with a mixed catalyst was used until the resultant samples became colourless solutions. Subsequently, 40% NaOH was introduced into the resulting samples, leading to the generation of ammonia (NH_3). A mixing indicator and 3% boric acid (H_3BO_3) were used to facilitate the conversion of NH_3 into ammonium dihydrogen borate ($NH_4H_2BO_3$).

The total N was determined in a titration process using H_2SO_4, used to convert $NH_4H_2BO_3$ to ammonium hydrogen sulfate (NH_4HSO_4), and recorded as %DW.

Extract S analysis was conducted using the turbidimetric method. The dried samples were digested and evaporated using a 2:1 (*v*/*v*) mixture of HNO_3 and $HClO_4$. Subsequently, the sample was appropriately diluted with deionised water, and 2 M ammonium acetate (CH_3COONH_4) and 1g barium chloride ($BaCl_2{\cdot}2H_2O$) were added to induce turbidity. The resulting turbidity was quantified by measuring the absorbance at 420 nm using a spectrophotometer. Extract S was subsequently calculated using a standard curve and expressed in nanomoles per gram of dry weight ($nmol{\cdot}g_{DW}{}^{-1}$).

Carbon content was calculated using the organic matter by immersing a dried sample in a mixture of 1 N potassium dichromate ($K_2Cr_2O_7$) and H_2SO_4 (conc.), followed by shaking overnight to convert the organic carbon into ammonium iron (*II*) sulfate hexahydrate ($Fe(NH_4)_2(SO_4)_2{\cdot}6H_2O$). An indicator (0.025 M O-phenanthroline ferrous sulfate indicator) was added, and 0.5 N ferrous ammonium sulfate was used to titrate the solution, resulting in a brown-red solution. The organic matter was calculated using the formula outlined by Walkley and Black [58] and recorded as %DW.

For Cl^- content analysis, the dried plant samples were mixed with calcium oxide (0.4 g CaO:1 g sample) and distilled water in a porcelain crucible and burned at 550 °C for 90 min. The samples were then allowed to cool overnight. The dry ash was washed multiple times with hot water and filtered. The Cl^- content was then measured using ion chromatography (Thermo Fisher Scientific, Waltham, MA, USA), and the results were calculated in $\mu mol{\cdot}g_{DW}{}^{-1}$.

4.6. Measurement of Physiology-Related Parameters

The Hill reaction was performed using a method modified from Dean and Miskiewicz [59] and Latkowska et al. [60]. The fresh plant samples were ground in a grinding medium (0.04 M K_3PO_4, 0.008 M KCl, and 0.04 M sucrose at pH 6.8), filtered, and centrifuged at 4000 rpm for 5 min at 0 °C. The pellet was resuspended in the grinding medium. The chloroplast suspension was mixed with a 0.04 M phosphate buffer (pH 6.8), 0.1% 2,6-dichlorophenol indophenol (DCPIP), and distilled water. The absorbance was measured at 600 nm using a spectrophotometer as A_{600S} (using the chloroplast suspension without DCPIP as a blank). Subsequently, the chloroplast suspension was irritated with DCPIP under a 100 W light bulb at a 15 cm distance for 10 min, and the absorbance was measured again at 600 nm as A_{600E}. The results were expressed as changes in OD. The activity of the Hill reaction was calculated using the formula $[(A_{600S} - A_{600E})/A_{600S}] \times 100)$, and the outcomes were expressed as percentages of absorbance change (%absorbance change).

Free radical scavenging activity by the 2,2-diphenyl-1-picrylhydrazyl (DPPH) assay was determined using a method modified from Kumari et al. [61]. Plant extracts were prepared by grinding the fresh plant samples in 80% methanol (with 80% methanol used only as a blank) and filtering them through a filter paper. The plant extract was mixed with a 0.3 mM DPPH solution and incubated at room temperature for 10 min. The absorbance was measured at 517 nm using a spectrophotometer, and the results were calculated using the formula $[(A_{Blank} - A_{Sample})/A_{Blank}] \times 100$, as a percentage of inhibition (%inhibition).

The proline content was determined following the methods described by Bates et al. [62] and Ghoulam et al. [63]. The fresh plant samples were ground in 40% methanol, filtered, and treated with an acid ninhydrin reagent (0.25 g of ninhydrin dissolved in a 3:2 (*v*/*v*) mixture of glacial acetic acid and 85% orthophosphoric acid). The mixture was incubated in a water bath at 100 °C. The reaction was stopped by incubating the samples on ice, and 5 mL of toluene was added. The absorbance was measured at 528 nm using a spectrophotometer. The proline content was calculated using a standard curve and recorded in $\mu mol \cdot g_{DW}^{-1}$.

4.7. Statistical Analysis

The experiments were conducted using a completely randomised design (CRD). Data from each experiment were statistically analysed using RStudio (version 2024.04.0+735). Analysis of variance (ANOVA), followed by Duncan's multiple range test ($p < 0.05$), was performed using the 'agricolae' package (version 1.3-5). Pearson's correlation coefficients were calculated using the base 'cor()' function, and correlation matrices were visualised using the 'corrplot' package (version 0.92).

5. Conclusions

Low concentrations of NaCl (10 and 20 mM) have been shown to enhance the growth of 4-day-old Chinese white radish seedlings, particularly in the cotyledon region. Among these concentrations, 10 mM NaCl proved more effective at promoting cotyledon development, leading to the increased accumulation of Na^+, Cl^-, K^+, total nitrogen, carbon, and proline. This treatment also upregulated the expression of aquaporin (AQP) genes, notably *PIP1-1*, *PIP1-2*, *PIP2-1*, *PIP2-2*, and *TIP1-1*. Furthermore, low NaCl concentrations facilitated cell expansion, maintained Hill reaction activity, and enhanced free radical scavenging capacity. These physiological and biochemical changes collectively resulted in greater cotyledon area and thickness, as well as increased fresh and dry biomass.

These findings suggest that the application of 10 mM NaCl may serve as a simple, low-cost agronomic strategy to improve early seedling vigour and sprout biomass production in Chinese white radish. This approach could be particularly advantageous in controlled cultivation systems, such as hydroponics or greenhouses, where early growth is crucial for market value. Further research is recommended to elucidate the signalling mechanisms involved in AQP gene regulation under low-NaCl conditions. Moreover, understanding these responses could inform strategies to enhance seedling tolerance to abiotic stresses, such as water-deficit conditions.

Supplementary Materials: The following supporting information can be downloaded at: https://www.mdpi.com/article/10.3390/plants14111616/s1, Supplementary Figure S1: The effects of NaCl on the growth rate of fresh weight were evaluated after cultivating Chinese white radish seedlings for 5 days., Supplementary Figure S2: The electrical conductivity (EC) of the sand culture was measured using the 1:5 (soil/water) method after cultivating Chinese white radish seedlings for 5 days. Supplementary Table S1: Primers of AQP genes [32,50–52] and *GADPH* [53] used in this study.

Author Contributions: N.M. designed the research, performed data collection and analysis, and wrote and revised the manuscript. S.I. and A.P. revised the manuscript. J.J. conceptualised and reviewed the manuscript. All authors have read and agreed to the published version of the manuscript.

Funding: This work was supported by Thai Government (Science Achievement Scholarship of Thailand, SAST) and the Faculty of Science, Chiang Mai University (R000030192).

Data Availability Statement: Data will be made available on request.

Acknowledgments: The authors acknowledge the Science Achievement Scholarship of Thailand (SAST) and Faculty of Science, Chiang Mai University, for providing financial and facilities support. During the preparation of this work, the authors used ChatGPT-3.5 to enhance their English writing skills. After using this tool, the authors reviewed and edited the content as needed, and they take full responsibility for the content of the publication.

Conflicts of Interest: The authors declare no conflict of interest.

References

1. Ouerghi, Z.; Cornic, G.; Roudani, M.; Ayadi, A.; Brulfert, J. Effect of NaCI on photosynthesis of two wheat species *(Triticum durum* and *T. aestivum)* differing in their sensitivity to salt stress. *J. Plant Physiol.* **2000**, *156*, 335–340. [CrossRef]
2. Hongqiao, L.; Suyama, A.; Mitani-Ueno, N.; Hell, R.; Maruyama-Nakashita, A. A low level of NaCl stimulates plant growth by improving carbon and sulfur assimilation in *Arabidopsis thaliana*. *Plants* **2021**, *10*, 2138. [CrossRef] [PubMed]
3. Wakeel, A.; Steffens, D.; Schubert, S. Potassium substitution by sodium in sugar beet (*Beta vulgaris*) nutrition on K-fixing soils. *J. Plant Nutr. Soil Sci.* **2010**, *173*, 127–134. [CrossRef]
4. Franco-Navarro, J.D.; Brumós, J.; Rosales, M.A.; Cubero-Font, P.; Talón, M.; Colmenero-Flores, J.M. Chloride regulates leaf cell size and water relations in tobacco plants. *J. Exp. Bot.* **2016**, *67*, 873–891. [CrossRef]
5. Cui, Y.N.; Li, X.T.; Yuan, J.Z.; Wang, F.Z.; Guo, H.; Xia, Z.R.; Wang, S.M.; Ma, Q. Chloride is beneficial for growth of the xerophyte *Pugionium cornutum* by enhancing osmotic adjustment capacity under salt and drought stresses. *J. Exp. Bot.* **2020**, *71*, 4215–4231. [CrossRef]
6. Hasegawa, P.M.; Bressan, R.A.; Zhu, J.K.; Bohnert, H.J. Plant cellular and molecular responses to high salinity. *Annu. Rev. Plant Physiol.* **2000**, *51*, 463–499. [CrossRef]
7. Wang, Y.; Zhao, Z.; Liu, F.; Sun, L.; Hao, F. Versatile roles of aquaporins in plant growth and development. *Int. J. Mol. Sci.* **2020**, *21*, 9485. [CrossRef]
8. Wang, J.; Yang, L.; Chai, S.; Ren, Y.; Guan, M.; Ma, F.; Liu, J. An aquaporin gene MdPIP1;2 from *Malus domestica* confers salt tolerance in transgenic *Arabidopsis*. *J. Plant Physiol.* **2022**, *273*, 153711. [CrossRef]
9. Kong, W.; Huang, H.; Du, W.; Jiang, Z.; Luo, Y.; Yi, D.; Yang, G.; Pang, Y. Overexpression of *MsNIP2* improves salinity tolerance in *Medicago sativa*. *J. Plant Physiol.* **2024**, *295*, 154207. [CrossRef]
10. Xu, Y.; Hu, W.; Liu, J.; Zhang, J.; Jia, C.; Miao, H.; Xu, B.; Jin, Z. A banana aquaporin gene, *MaPIP1;1*, is involved in tolerance to drought and salt stresses. *BMC Plant Biol.* **2014**, *14*, 59. [CrossRef]
11. Qi, C.H.; Chen, M.; Song, J.; Wang, B.S. Increase in aquaporin activity is involved in leaf succulence of the euhalophyte *Suaeda salsa*, under salinity. *Plant Sci.* **2009**, *176*, 200–205. [CrossRef]
12. Banihani, S.A. Radish (*Raphanus sativus*) and diabetes. *Nutrients* **2017**, *9*, 1014. [CrossRef] [PubMed]
13. Manivannan, A.; Kim, J.; Kim, D.; Lee, E.; Lee, H. Deciphering the nutraceutical potential of *Raphanus sativus*—A comprehensive overview. *Nutrients* **2019**, *11*, 402. [CrossRef] [PubMed]
14. Arunin, S. *Saline Soil in Thailand*; Land Development Department: Bangkok, Thailand, 1996. (In Thai)
15. Hoffman, G.J.; Shannon, M.C. Salinity. In *Microirrigation for Crop Production*; Lamm, F.R., Ayars, J.E., Nakayama, F.S., Eds.; Elsevier: Amsterdam, The Netherlands, 2007; pp. 131–160. [CrossRef]
16. Jungklang, J. Effects of sodium chloride on germination, growth, relative water content, and chlorophyll, proline, malondialdehyde and vitamin C contents in Chinese white radish seedlings (*Raphanus sativus* L. var. *longipinnatus* Bailey). *Maejo Int. J. Sci. Technol.* **2018**, *12*, 89–100.
17. Yuan, G.; Wang, X.; Guo, R.; Wang, Q. Effect of salt stress on phenolic compounds, glucosinolates, myrosinase and antioxidant activity in radish sprouts. *Food Chem.* **2010**, *121*, 1014–1019. [CrossRef]
18. Omuto, C.T.; Vargas, R.R.; El Mobarak, A.M.; Mohamed, N.; Viatkin, K.; Yigini, Y. *Mapping of Salt-Affected Soils: Technical Manual;* FAO: Rome, Italy, 2020. [CrossRef]
19. Blaylock, A.D. *Soil Salinity, Salt Tolerance and Growth Potential of Horticultural and Landscape Plants;* Co-Operative Extension Service, University of Wyoming, Department of Plant, Soil and Insect Sciences, College of Agriculture: Laramie, WY, USA, 1994.
20. Slama, I.; Ghnaya, T.; Messedi, D.; Hessini, K.; Labidi, N.; Savoure, A.; Abdelly, C. Effect of sodium chloride on the response of the halophyte species *Sesuvium portulacastrum* grown in mannitol-induced water stress. *J. Plant Res.* **2007**, *120*, 291–299. [CrossRef]
21. Nieves-Cordones, M.; Al Shiblawi, F.R.; Sentenac, H. Role and transport of sodium and potassium in plants. *Met. Ions Life Sci.* **2016**, *16*, 291–324. [CrossRef]
22. Taiz, L.; Zeiger, E. *Plant Physiology*, 5th ed.; Sinauer Associates Inc. Publishers: Massachusetts, UK, 2010.
23. Colmenero-Flores, J.M.; Franco-Navarro, J.D.; Cubero-Font, P.; Peinado-Torrubia, P.; Rosales, M. Chloride as a beneficial macronutrient in higher plants: New roles and regulation. *Int. J. Mol. Sci.* **2019**, *20*, 4686. [CrossRef]
24. Shaddad, M.A. The effect of proline application on the physiology of *Raphanus sativus* plant grown under salinity stress. *Biol. Plant.* **1990**, *32*, 104–112. [CrossRef]
25. Nuanjan, N.; Nghia, P.T.; Theerakulpisut, P. Exogenous proline and trehalose promote recovery of rice seedlings from salt-stress and differentially modulate antioxidant enzymes and expression of related genes. *J. Plant Physiol.* **2012**, *169*, 596–604. [CrossRef]
26. Wu, G.Q.; Jiao, Q.; Shui, Q.Z. Effect of salinity on seed germination, seedling growth and inorganic and organic solutes accumulation in sunflower (*Helianthus annuus* L.). *Plant Soil Environ.* **2015**, *61*, 220–226. [CrossRef]
27. Inayat, H.; Mehmood, H.; Danish, S.; Alharbi, S.A.; Ansari, M.J.; Datta, R. Impact of cobalt and proline foliar application for alleviation of salinity stress in radish. *BMC Plant Biol.* **2024**, *24*, 287. [CrossRef] [PubMed]
28. Blatt, M.R.; Chaumont, F.; Farquhar, G. Focus on Water. *Plant Physiol.* **2014**, *164*, 1553–1555. [CrossRef]

29. Ievinsh, G. Water content of plant tissues: So simple that almost forgotten? *Plants* **2023**, *12*, 1238. [CrossRef]
30. Kramer, P.J. *Water Relation of Plants*; New York Academic Press: New York, NY, USA, 1983. Available online: https://www.sciencedirect.com/book/9780124250406/water-relations-of-plants (accessed on 17 July 2024).
31. Gonzalez, L.; Gonzalez-Vilar, M. Chapter 14 Determination of relative water content. In *Handbook of Plant Ecophysiology Techniques*; Reigosa Roger, M.J., Ed.; Kluwer Academic Publishers: Dordrecht, The Netherlands; Springer: Berlin/Heidelberg, Germany, 2007; pp. 207–212. [CrossRef]
32. Suga, S.; Komatsu, S.; Maeshima, M. Aquaporin isoform responsive to salt and water stresses and phytohormones in radish seedlings. *Plant Cell Physiol.* **2002**, *43*, 1229–1237. [CrossRef]
33. Fricke, W.; Knipfer, T. Plant aquaporins and cell elongation. In *Plant Aquaporins: From Transport to Signaling*; Chaumont, F., Tyerman, S., Eds.; Springer: Berlin/Heidelberg, Germany, 2017; pp. 107–131.
34. Leng, H.; Jiang, C.; Song, X.; Lu, M.; Wan, X. Poplar aquaporin *PIP1;1* promotes *Arabidopsis* growth and development. *BMC Plant Biol.* **2021**, *21*, 253. [CrossRef]
35. Lopaz-Zaplana, A.; Martinez-Garcia, N.; Carvajal, M.; Bárzana, G. Relationships between aquaporins gene expression and nutrient concentrations in melon plants (*Cucumis melo* L.) during typical abiotic stresses. *Environ. Exp. Bot.* **2022**, *195*, 104759. [CrossRef]
36. Yaneff, A.; Sigaut, L.; Marquez, M.; Alleva, K.; Pietrasanta, L.I.; Amodeo, G. Heteromerization of PIP aquaporins affects their intrinsic permeability. *Proc. Natl. Acad. Sci. USA* **2013**, *111*, 231–236. [CrossRef]
37. Tripathi, D.K.; Singh, V.P.; Chauhan, D.K.; Prasad, S.M.; Dubey, N.K. Chapter 8 Role of macronutrients in plant growth and acclimation: Recent advances and future prospective. In *Improvement of Crops in the Era of Climatic Changes, Volume 2*; Ahmad, P., Wani, M.R., Azooz, M.M., Tran, L.P., Eds.; Springer: Berlin/Heidelberg, Germany, 2014; pp. 197–216. [CrossRef]
38. Mndi, O.; Sogoni, A.; Jimoh, M.O.; Wilmot, C.M.; Rautenbach, F.; Laubscher, C.P. Interactive Effects of salinity stress and irrigation intervals on plant growth, nutritional value, and phytochemical content in *Mesembryanthemum crystallinum* L. *Agriculture* **2023**, *13*, 1026. [CrossRef]
39. Wu, H.; Zhang, X.; Giraldo, J.P.; Shabala, S. It is not all about sodium: Revealing tissue specificity and signaling roles of potassium in plant responses to salt stress. *Plant Soil* **2018**, *431*, 1–17. [CrossRef]
40. Zarei, M.; Azizi, M. Evaluation of NaCl salinity tolerance of four fig genotypes based on vegetative growth and ion content in leaves, shoots, and roots. *J. Am. Soc. Hortic. Sci.* **2016**, *51*, 1427–1434. [CrossRef]
41. Efrida, E. Effect of NaCl and KCl to the growth and nutrient level of upland rice in an Ultisol. In Proceedings of the 7th International Conference on Multidisciplinary Research, Munich, Germany, 19–21 October 2018; pp. 143–148. [CrossRef]
42. Shi, Y.; Chen, J.; Hou, X. Similarities and differences of photosynthesis establishment related mRNAs and novel lncRNAs in early seedlings (coleoptile/cotyledon vs. true leaf) of rice and *Arabidopsis*. *Front. Genet.* **2020**, *11*, 565006. [CrossRef] [PubMed]
43. Santos, C.V. Regulation of chlorophyll biosynthesis and degradation by salt stress in sunflower leaves. *Sci. Hortic.* **2004**, *103*, 93–99. [CrossRef]
44. Turan, M.A.; Katkat, V.; Taban, S. Variation in proline, chlorophyll and mineral contents of wheat plant grown under salinity stress. *J. Agron.* **2007**, *6*, 137–141. [CrossRef]
45. M'rah, S.; Ouerghi, Z.; Berthomieu, C.; Havaux, M.; Jungas, C.; Hajji, M.; Grignin, C.; Lachaâl, M. Effects of NaCl on the growth, ion accumulation and photosynthetic parameters of *Thellungiella halophila*. *J. Plant Physiol.* **2006**, *163*, 1022–1031. [CrossRef]
46. Land Development Department. *Handbook for Analyzing Samples of Soil, Water, Fertilizer, Plants, and Soil Amendments: Ensuring Product Standards Through Rigorous Analysis*; Office of Science for Land Development, Land Development Department: Bangkok, Thailand, 2004. (In Thai)
47. Department of Agriculture. *A Handbook of Soil Analysis (Chemical and Physical Methods)*; Quick Print Offset: Bangkok, Thailand, 2010. (In Thai)
48. Smart, R.E.; Bingham, G.E. Rapid estimates of relative water content. *Plant Physiol.* **1974**, *53*, 258–260. [CrossRef]
49. Zeng, Y.; Li, L.; Yang, R.; Yi, X.; Zhang, B. Contribution and distribution of inorganic ions and organic compounds to the osmotic adjustment in *Halostachys caspica* response to salt stress. *Sci. Rep.* **2015**, *5*, 13639. [CrossRef]
50. Suga, S.; Imagawa, S.; Maeshima, M. Specificity of the accumulation of mRNA and proteins of the plasma membrane and tonoplast aquaporins in radish organs. *Planta* **2001**, *212*, 294–304. [CrossRef]
51. Tsuchihira, A.; Hanba, Y.T.; Kato, N.; Doi, T.; Kawazu, T.; Maeshima, M. Effect of overexpression of radish plasma membrane aquaporins water-use efficiency, photosunthesis and growth of *Eucayptus* trees. *Tree Physiol.* **2010**, *30*, 417–430. [CrossRef]
52. Higuchi, T.; Suga, S.; Tsuchiya, T.; Hisada, H.; Morishima, S.; Okada, Y.; Maeshima, M. Molecular Cloning, Water Channel Activity and Tissue Specific Expression of Two Isoforms of Radish Vacuolar Aquaporin. *Plant Cell Physiol.* **1998**, *39*, 905–913. [CrossRef]
53. Xu, Y.; Zhu, X.; Gong, Y.; Xu, L.; Wang, Y.; Liu, L. Evaluation of reference genes for gene expression studied in radish (*Raphanus sativus* L.) using quantitive real-time PCR. *Biochem. Biophys. Res. Commun.* **2012**, *424*, 398–403. [CrossRef] [PubMed]

54. Duan, M.; Wang, J.; Zhang, X.; Yang, H.; Wang, H.; Qui, Y.; Song, J.; Guo, Y.; Li, X. Identification of optimal reference genes for expression analysis in radish (*Raphanus sativus* L.) and its relatives based on expression stability. *Front. Plant Sci.* **2017**, *8*, 1605. [CrossRef] [PubMed]
55. Livak, K.J.; Schmittgen, T.D. Analysis of relative gene expression data using real-time quantitative PCR and the $2^{-\Delta\Delta CT}$ method. *Methods* **2001**, *25*, 402–408. [CrossRef] [PubMed]
56. Hoo, K.Y.; Shim, I.S.; Kobayashi, K.; Usui, K. Relationship between Na content or K/Na ratio in shoots and salt tolerance in several Gramineous plants. *J. Weed Sci. Technol.* **1999**, *44*, 293–299. [CrossRef]
57. Science Center of Land Development. *Manual for the Analysis of Soil Samples, Water, Fertilizers, Plants, Soil Improvement Materials and Analysis to Inspect and Certify Product Standards*; Land Development Department: Bangkok, Thailand, 2004. (In Thai)
58. Walkley, A.; Black, I.A. Acritical examination of a rapid method for determining organic carbon in soil-effect of variation in digestions and of inorganic soil constituents. *Soil Sci.* **1947**, *63*, 251–263. [CrossRef]
59. Dean, R.P.; Miskiewicz, E. Rates of electron transport in the thylakoid membranes of isolated, illuminated chloroplasts are enhanced in the presence of ammonium chloride. *Biochem. Mol. Biol. Educ.* **2003**, *31*, 410–417. [CrossRef]
60. Latkowska, E.; Lechowski, Z.; Bialczyk, J.; Pilarski, J. Photosynthesis and water relations in tomato plants cultivated long-term in media containing (+)-Usnic acid. *J. Chem. Ecol.* **2006**, *32*, 2053–2066. [CrossRef]
61. Kumari, S.; Phogat, D.; Sehrawat, K.D.; Choudhary, R.; Rajput, V.D.; Ahlawat, J.; Karunakaran, R.; Minkina, T.; Sehrawat, A.R. The effect of *Ascophyllum nodosum* extract on the nutraceutical antioxidant potential of *Vigna radiata* sprout under salt stress. *Plants* **2021**, *10*, 1216. [CrossRef]
62. Bates, L.S.; Waldren, R.P.; Teare, I.D. Rapid determination of free proline for water-stress studies. *Plant Soil* **1973**, *39*, 205–207. [CrossRef]
63. Ghoulam, C.; Foursy, A.; Fares, K. Effect of salt stress on growth, inorganic ions and proline accumulation in relation to osmotic adjustment in five sugar beet cultivars. *Environ. Exp. Bot.* **2002**, *47*, 39–50. [CrossRef]

plants

MDPI

Article

Mitigation of Oxidative Damage Caused by Salinity in the Halophyte *Crithmum maritimum* L. via Biostimulant-Enhanced Antioxidant Activity

Anastasia Evripidis Giannakoula [1,*], Michalis K. Stefanakis [2], Charikleia Papaioannou [3], Stavroula Tsimpliaraki [4], Sofia Kaftantzi [4], Alexandra Niniraki [1] and Athanasios Gertsis [4]

[1] Laboratory of Plant Physiology, Department of Agriculture, International Hellenic University, Sindos, 57400 Thessaloniki, Greece; alexandraninirak@gmail.com
[2] Department of Chemistry, University of Crete, Voutes, 71003 Heraklion, Greece; michstefanakis@yahoo.gr
[3] Laboratory of Applied Genetics & Breeding, Department of Biology, University of Patras, Rion, 26504 Patras, Greece; xpapaioannou@upatras.gr
[4] Department of Sustainable Agriculture and Management, Division of the American Farm School, Perrotis College, 57001 Thessaloniki, Greece; stavroulatsimpliaraki@gmail.com (S.T.); sofia.kaftantzi@gmail.com (S.K.); agerts@afs.edu.gr (A.G.)
* Correspondence: agianna@ihu.gr

Abstract: The xerophyte medicinal species *Crithmum maritimum* was investigated for its physiological and antioxidant responses under increasing salinity stress and foliar biostimulant application. At moderate salinity (10 dS/m sodium chloride NaCl), plant growth and photosynthetic activity were enhanced, whereas high salinity (20 dS/m) led to significant reductions in biomass, photosynthetic efficiency, and water use efficiency. Salinity-induced oxidative stress was confirmed by elevated levels of malondialdehyde (MDA) and hydrogen peroxide (H_2O_2), in upper plant tissues. Foliar application of two bioactive compounds—Aquamin and Cultisano—partially mitigated the detrimental effects of high salinity. Treated plants maintained higher photosynthetic parameters and exhibited improved antioxidant profile compared to untreated controls. Furthermore, GC–MS analysis of essential oils revealed that salinity altered the composition of key volatile components, notably increasing γ-terpinene and reducing *p*-cymene content. Biostimulant treatments counteracted these changes, enhancing terpene components such as *p*-cymene, and thymol-methyl-ether. Overall, the findings highlight the potential of biostimulants to enhance both salt tolerance as well as the phytochemical value of *C. maritimum*, suggesting promising applications in sustainable agriculture and high-value plant product development under saline conditions. It was concluded that the type of biostimulant significantly influenced the physiological and quality characteristics of sea fennel plants. Further study on this topic is proposed, aiming at the improvement of antioxidant activity, which is beneficial to human health.

Keywords: halophyte; salt stress; oxidative stress response; essential oils; GC-MS

1. Introduction

Soil salinization represents a major constraint in modern agriculture, posing a serious threat to crop productivity and global food security. With an estimated 20% of irrigated lands already affected by salinity—a figure expected to rise due to climate change—there is an increasing demand for sustainable crops capable of thriving under saline conditions [1–3]. *Crithmum maritimum* L. (Apiaceae family), commonly known as sea fennel, is a native of

Mediterranean coastal habitats and has recently emerged as a promising candidate for cultivation in saline environments due to its exceptional salt tolerance and high nutritional value [4–6]. Due to its natural salt tolerance, this species possesses significant economic and medicinal potential [7]. In addition, its adaptability to harsh coastal environments highlights its potential role in climate-resilient agricultural systems, where crop diversification is increasingly recognized as a key strategy for ensuring food security [8].

Halophytes are known to accumulate large amounts of bioactive metabolites under stressful conditions [9,10]. Moreover, the seeds of *C. maritimum* (rock samphire), a halophytic species native to coastal Mediterranean habitats and capable of completing its life cycle under a broad range of salinity regimes, contain a considerable amount of oil with a fatty acid composition similar to olive oil, making it a potential edible and industrial resource. Its leaves are rich in various compounds, including vitamin C, carotenoids, polyphenols, flavonoids, and other bioactive substances, with aromatic, medicinal, antimicrobial, and insecticidal applications [11,12]. The plant's secondary metabolites (PSMs) play a vital role in its defense against a wide range of biotic and abiotic stresses, including salinity [13,14]. Recent studies suggest that sea fennel exhibits a dynamic metabolic plasticity, modulating both primary and secondary metabolism in response to environmental stressors, which underscores its value as a model species for stress physiology research [15].

Nevertheless, high salinity still imposes considerable physiological and biochemical stress on this plant, resulting in reduced growth and pigment content at elevated NaCl concentrations. Under stress, plants often synthesize in defense secondary metabolites with pharmaceutical potential [1,13]. To mitigate oxidative stress through non-enzymatic defense mechanisms, plants tend to accumulate PSMs under adverse conditions. Consequently, while salinity can decrease plant growth and yield, it may also stimulate the production of a diverse range of bioactive compounds. This dual response—growth inhibition coupled with enhanced metabolite biosynthesis—renders halophytes particularly attractive for the production of high-value phytochemicals, especially in low-input agricultural systems.

Recent research emphasizes the role of biostimulant—substances or microorganisms that enhance plant growth, nutrient uptake, and tolerance to abiotic stresses, thus improving plant resilience [16]. To enhance plant tolerance under suboptimal environmental conditions, the application of biostimulants—alongside complementary strategies such as conventional breeding and genetic engineering—is regarded as a promising approach [17]. Biostimulants function as supplementary inputs to fertilizers, with the primary objective of improving nutrient-use efficiency and enabling reductions in overall nutrient application rates.

When applied to plants, seeds, or soil, biostimulants, which include a broad range of natural compounds and beneficial microbes, promote growth and development by stimulating the plant's intrinsic physiological processes, rather than acting as fertilizers [18,19]. Certain classes of biostimulants, such as seaweed extracts, humic substances, protein hydrolysates, and plant-growth-promoting rhizobacteria, have been shown to enhance antioxidant enzyme activity, osmotic adjustment, and ion homeostasis under saline conditions. Their eco-friendly nature and compatibility with sustainable agricultural practices make them promising alternatives to conventional chemical inputs [20].

This study investigates the effects of biostimulant treatments on *C. maritimum* under high salinity conditions. By evaluating physiological responses, antioxidant activity, and essential oil composition, we intend to provide evidence that biostimulants can mitigate the adverse effects of salt stress and enhance plant resilience. Furthermore, by analyzing the extracts and the chemical profile of its essential oils, we seek to identify the major bioactive constituents responsible for its distinctive aroma and/or therapeutic properties and to compare our data with those in previous studies. The results of this work may contribute to the development of sustainable cultivation strategies for sea fennel cultivation in marginal and saline soils.

2. Results

2.1. CO_2 Assimilation Rate

The CO_2 assimilation rate exhibited clear treatment-dependent responses across salinity levels (Figure 1a,b). At salinity 1 dS m^{-1}, control plants showed values around 4.8 µmol CO_2 m^{-2} s^{-1}, whereas Aquamin-treated plants showed higher assimilation. Cultisano-treated plants displayed values close to the control but with slightly higher variability. Increasing salinity resulted in a progressive reduction in assimilation in all treatments, most noticeable in the control group, which dropped to ca. 3.3 µmol CO_2 m^{-2} s^{-1} at salinity 10 dS m^{-1} and to 2.0–2.3 µmol CO_2 m^{-2} s^{-1} at salinity 20 dS m^{-1}. In contrast, Aquamin-treated plants maintained substantially higher values across salinity levels (3.68–3.98 at moderate salinity and 2.7 up to 4.19 at high salinity) while Cultisano treatments gave intermediate values. According to Tukey's post hoc test, Aquamin-treated plants consistently belonged to a separate statistical group either the control and Cultisano-treated, across all salinity levels, indicating a stronger effect on CO_2 assimilation. In contrast, Cultisano did not differ significantly from the control (Figure 1b). Mean values for all parameters are summarized in Supplementary Table S1.

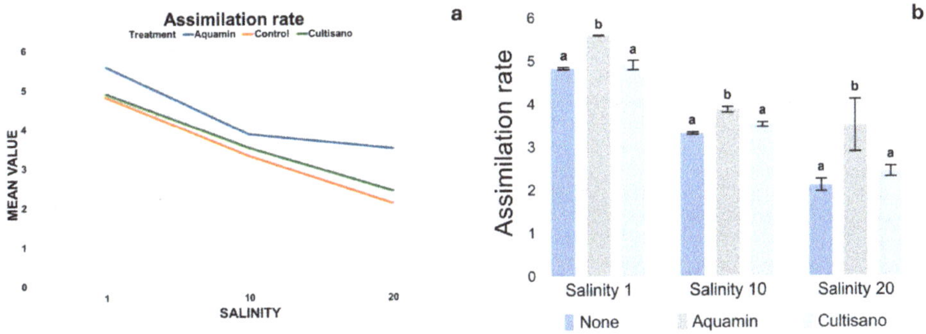

Figure 1. CO_2 assimilation rate under control, salinity and biostimulant treatments. (**a**) Linear plot showing the mean assimilation rate (±SE) across salinity levels (1, 10 and 20 dS m^{-1}). (**b**) Bar plots presenting treatment comparisons within each salinity level, with Tukey's HSD groupings indicated by different letters.

2.2. CO_2 Transpiration Rate

Transpiration followed an overall similar pattern to assimilation (Figure 2a,b). At salinity 1 dS m^{-1}, control plants exhibited transpiration values between 3.3 and 3.9 mmol H_2O m^{-2} s^{-1}, while Aquamin-treated plants displayed slightly higher rates (3.60–4.06). As salinity increased, the control group showed a uniform reduction (2.31 at moderate salinity; 1.74–1.92 at high salinity). Biostimulant treatments also showed a decline in transpiration values across salinity levels. Tukey's test revealed no treatment differences at control and moderate salinities. At high salinity, the control formed a distinct lower subset, whereas Cultisano and Aquamin grouped into higher and intermediate subsets, respectively.

A strong positive correlation between assimilation and transpiration (Figure 3) was supported by high Pearson R and R^2 values. Summary statistics for all physiological measurements are provided in Supplementary Table S1.

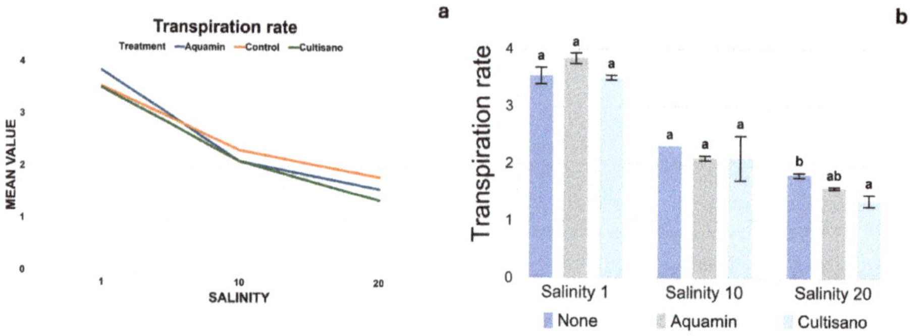

Figure 2. CO_2 transpiration rate across treatments and salinity levels. (**a**) Linear plot illustrating the mean transpiration rate across salinity levels. (**b**) Bar plots showing mean transpiration rate (±SE) and Tukey's HSD groupings indicated by different letters above the bars, show statistically significant differences between treatments within each salinity level.

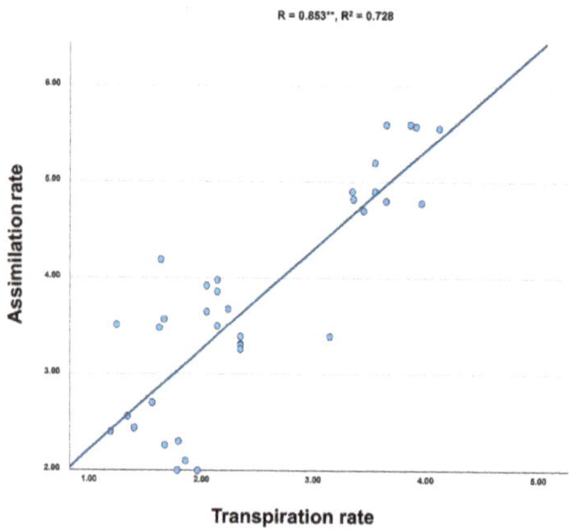

Figure 3. Scatter plot of CO_2 assimilation versus CO_2 transpiration across all treatments, showing the Pearson correlation coefficient (R) and coefficient of determination (R^2) displayed on the plot. Double asterisks (**) indicate statistically significant correlations ($p < 0.01$).

2.3. GC–MS Volatile Composition

GC–MS analysis revealed pronounced treatment-dependent shifts in volatile composition (Figure 4). For example, salinity increased several aliphatic constituents, with 1-hexadecanol reaching its highest levels in S20, while tetratriacontane declined markedly under salinity, reaching its lowest levels in S20. Several biostimulant treatments, particularly Cultisano, also exhibited reduced volatile levels relative to control. Monoterpene hydrocarbons presented distinct patterns: γ-terpinene and sabinene were elevated at both S10 and S20, whereas limonene reached its highest abundance in Aquamin-treated plants. Oxygenated monoterpenes, particularly terpinene-4-ol, thymol, and thymol methyl ether, varied distinctly, showing consistently higher values in the biostimulant treatments, especially in Cultisano. Sesquiterpene markers such as caryophyllene oxide, germacrene-B and E-caryophyllene also revealed pronounced treatment-dependent changes. The complete quantitative dataset of all identified volatile compounds, including retention times, chemical classes, retention indices, and relative abundances across treatments, is provided in Supplementary Table S2. These compound-level data underpin the heatmap clustering and subsequent multivariate analyses.

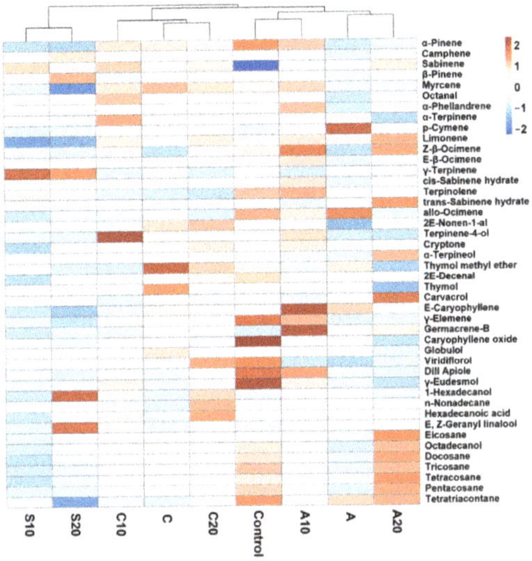

Figure 4. Heatmap of volatile compound abundances detected by GC–MS analysis across all treatments. Compounds and samples have been prioritized using hierarchical clustering based on Euclidean distance and Ward's linkage method. Nonanal, trans-Limonene oxide and Tridecanol values were constant and were removed from the analysis.

2.4. Oxidative Stress-Related VOC Analysis

The stacked bar plots of oxidative stress-related VOC classes (Figure 5) showed that salinity (S10, S20) led to a progressive reduction in oxygenated monoterpenes, oxygenated sesquiterpenes, and phenylpropanoid ether compared with the control. Biostimulant treatments displayed the opposite trend, with Aquamin maintaining class proportions close to control, while Cultisano produced the highest enrichment in oxygenated monoterpenes and moderate increases in oxygenated sesquiterpenes across salinity levels. Overall, salinity suppressed the synthesis of oxygenated VOC categories, while biostimulant application, especially Cultisano, resulted in higher overall proportions of oxygenated VOCs.

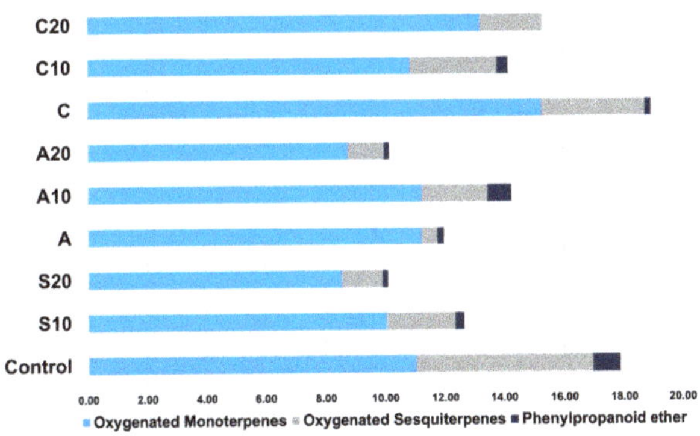

Figure 5. Stacked bar plots showing the relative proportions of oxidative stress-related volatile compound classes (oxygenated monoterpenes, oxygenated sesquiterpenes, and phenylpropanoid ether) across control, salinity (S10, S20), and biostimulant treatments (Aquamin: A, A10, A20; Cultisano: C, C10, C20). Values represent normalized relative abundances from GC–MS analysis.

2.5. PCA of Volatile Profiles

The PCA plot (Figure 6) showed clear differences among treatments. The salinity samples (S10, S20) grouped closely together in the lower left part of the plot, indicating similar VOC profiles. The control sample was positioned separately toward the right, showing an intermediate composition. Aquamin treatments were more spread out: A10 appeared higher on PC2, A was located near the center, and A20 shifted strongly to the right along PC1. Cultisano treatments formed a small cluster near the salinity samples, but were slightly higher on PC2. Overall, PC1 (23.24%) and PC2 (21.77%) captured most of the variation and separated the salinity, control, Aquamin and Cultisano groups.

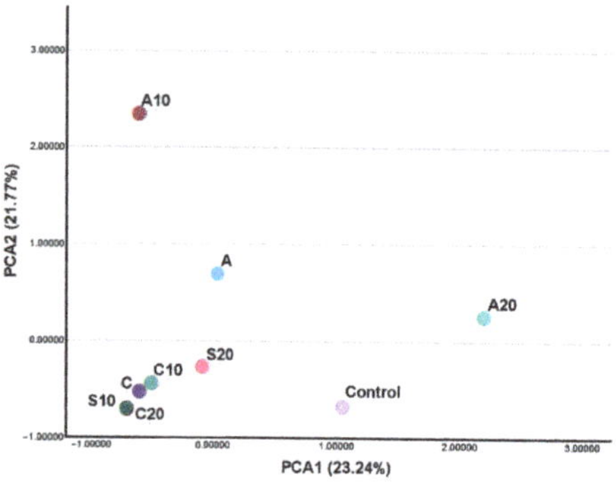

Figure 6. Principal Component Analysis (PCA) of volatile profiles across all treatments. The plot displays sample distribution along PC1 and PC2, which together explain 45.01% of total variance.

2.6. Malondialdehyde (MDA)

MDA concentrations showed a strong salinity-dependent increase, especially in the control plants (Figure 7a,b). At salinity 1 dS m^{-1}, control plants exhibited values of 5.0–5.6 nmol g^{-1} FW, whereas biostimulant-treated plants showed markedly lower MDA levels. At salinity 10 dS m^{-1}, MDA increased highly in the control (11.9–13.5), while both Aquamin and Cultisano maintained significantly lower values. At salinity 20 dS m^{-1}, the control reached the highest values in the dataset (15.8–17.6), while the biostimulant treatments retained substantially lower levels (6.4–8.7). Tukey's test showed that at salinity 1 and 10 dS m^{-1}, the control had higher MDA than both Aquamin and Cultisano values, which are no different from each other. At salinity 20 dS m^{-1}, all treatments were separated statistically, with Cultisano having the lowest MDA, Aquamin intermediate levels, and the control having the highest. Mean values are summarized inSupplementary Table S1.

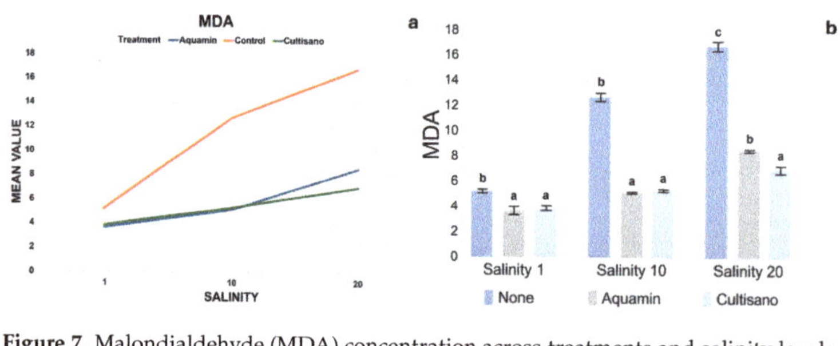

Figure 7. Malondialdehyde (MDA) concentration across treatments and salinity levels. (**a**) Linear plot showing MDA trends across salinity levels. (**b**) Bar plots showing mean MDA concentration (±SE) with Tukey's HSD groupings. Different letters above the bars indicate statistically significant differences between treatments within each salinity level ($p < 0.05$).

2.7. Total Phenolic Content (TPC)

TPC displayed an opposite pattern to MDA (Figure 8a,b). At 1 dS m^{-1}, TPC ranged between 9.6–10.8 mg GAE g^{-1} DW in control plants and was significantly higher under Aquamin (12.7–14.6) and Cultisano (11.1–12.5). Increasing salinity resulted in reductions in TPC in the control samples (6.6–8.8), whereas Aquamin maintained elevated values across salinity gradients. According to Tukey's HSD, all treatments differed significantly at control salinity, with Aquamin showing the highest TPC, Cultisano intermediate values, and the control the lowest. At moderate and high salinity, Aquamin again formed a separate higher statistical group, while the control and Cultisano did not differ from each other. A full summary of mean TPC values for all treatments appears in Supplementary Table S1.

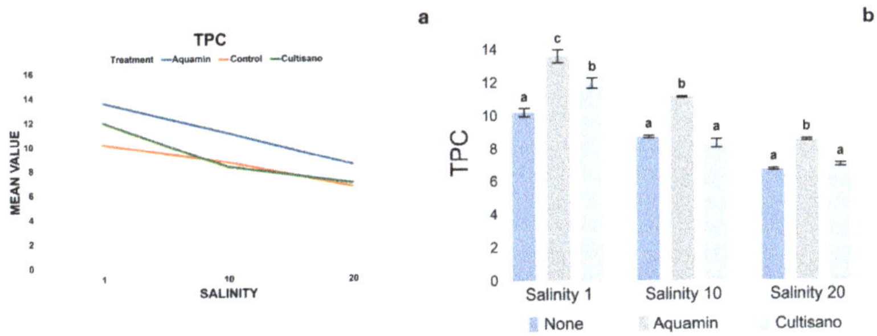

Figure 8. Total phenolic content (TPC) across treatments and salinity levels. (**a**) Linear plot showing TPC trends across salinity levels. (**b**) Bar plots showing mean TPC (±SE) with Tukey's HSD groupings. Different letters above the bars indicate statistically significant differences between treatments within each salinity level ($p < 0.05$).

2.8. Correlation Analysis

Scatter plot analyses established strong and consistent relationships among physiological and biochemical parameters (Figure 9a–e). CO_2 assimilation exhibited a strong positive correlation with both transpiration and TPC (Pearson R > 0.7), and a strong negative correlation with MDA. Transpiration followed similar patterns, showing positive associations with TPC and negative associations with MDA. TPC and MDA showed a clear inverse relationship across all treatments, supported by high R^2 values that indicate a strong and consistent correlation.

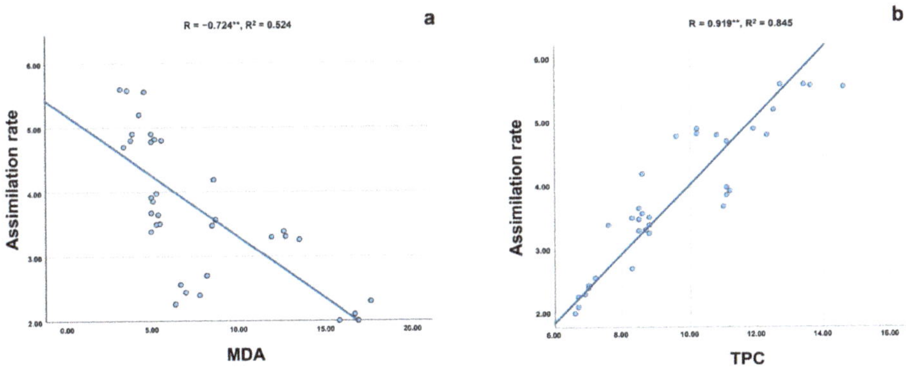

Figure 9. *Cont.*

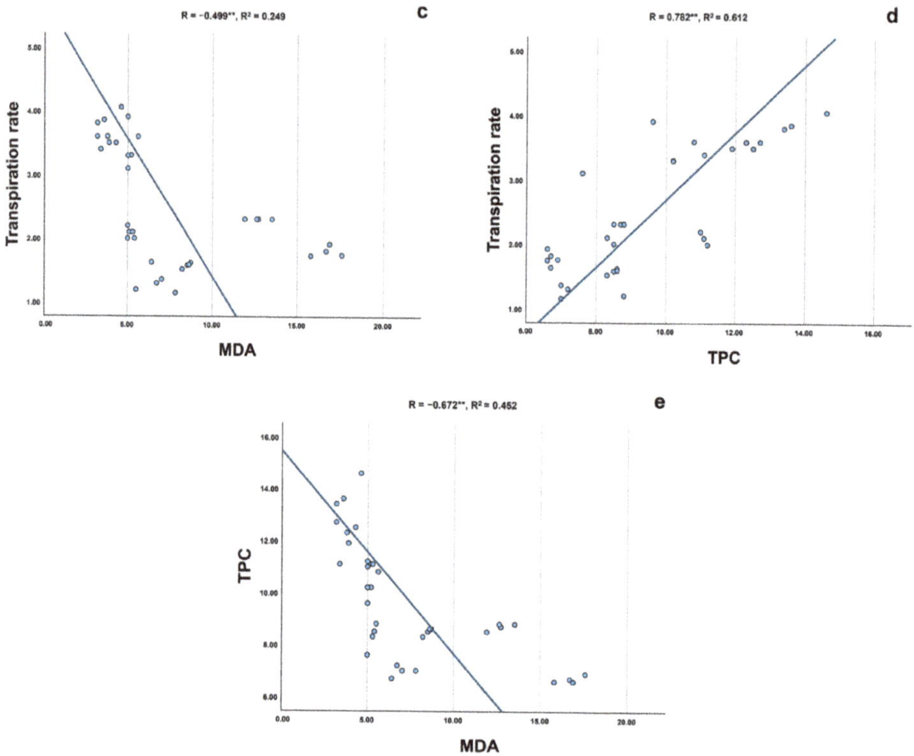

Figure 9. Correlation plots among physiological and biochemical parameters. (**a**) Scatter plot of CO_2 assimilation rate vs. MDA concentration, showing Pearson correlation coefficient (R) and coefficient of determination (R^2). (**b**) Scatter plot of CO_2 assimilation rate vs. total phenolic content (TPC), with R and R^2 displayed. (**c**) Scatter plot of CO_2 transpiration rate vs. MDA concentration, with R and R^2 displayed. (**d**) Scatter plot of CO_2 transpiration rate vs. total phenolic content (TPC), with R and R^2 displayed. (**e**) Scatter plot of MDA vs. total phenolic content (TPC), with R and R^2 displayed. Double asterisks (**) indicate statistically significant correlations ($p < 0.01$).

3. Discussion

Salinity stress exerts profound effects on both the physiological performance and secondary metabolism of *C. maritimum* L. In this study, moderate salinity stimulated the biosynthesis of several stress-related monoterpenes, reflecting an adaptive metabolic adjustment, whereas higher salinity levels markedly inhibited photosynthetic activity and secondary metabolite production. These trends align with previous studies [1,21], which report that salinity can initially activate defensive metabolic pathways before exceeding the plant's tolerance capacity and suppressing enzymatic and photosynthetic functions.

The application of biostimulants, particularly Aquamin and Cultisano, significantly enhanced the resilience of *C. maritimum* under saline conditions. Both products improved physiological performance by reinforcing antioxidant defenses, reducing lipid peroxidation, and maintaining pigment and metabolite stability. Aquamin, a marine-derived formulation rich in calcium and trace elements, likely stimulated enzymatic antioxidant pathways, contributing to the higher CO_2 assimilation and transpiration rates observed across salinity levels. In contrast, Cultisano, which contains organic acids and micronutrients, appeared to support chlorophyll retention and strongly promoted the biosynthesis of oxygenated monoterpenes and other essential oil constituents. Similar biostimulant-mediated improvements in stress tolerance have been reported by other researchers [16–18].

The physiological responses observed in this study clearly show that salinity levels exceeding 10 dS m^{-1} markedly impaired plant growth, photosynthetic performance, and biomass accumulation in *C. maritimum*. Both CO_2 assimilation and transpiration decreased significantly under increasing salinity, reflecting typical salt-induced restrictions in gas exchange. Such reductions are consistent with stomatal closure, declines in chlorophyll content, and potential inhibition of key photosynthetic enzymes. However, the application of Aquamin and Cultisano partially mitigated these effects, with Aquamin consistently sustaining the highest assimilation and transpiration rates across all salinity levels. These results suggest that the biostimulants enhance photosynthetic efficiency and carbon fixation under stress, likely through improved nutrient availability, strengthening of antioxidant defenses, and greater stabilization of chloroplast membranes as other researchers have previously reported [19].

These findings are consistent with previous reports indicating that elevated salinity disrupts the ionic and osmotic balance of *C. maritimum*, leading to impaired physiological performance [1]. In the present study, foliar application of Aquamin and Cultisano effectively mitigated these adverse effects, particularly at 10 dS m^{-1}, where both biostimulants enhanced antioxidant status, stabilized physiological traits, and reduced visible stress symptoms. Aquamin appeared to strengthen antioxidant capacity, potentially through stimulation of enzymatic defense pathways, whereas Cultisano contributed to greater pigment retention and overall metabolic stability. Even at 20 dS m^{-1}, biostimulant-treated plants maintained significantly higher chlorophyll content and biomass than untreated controls. These outcomes corroborate previous evidence that biostimulants enhance salinity tolerance by modulating antioxidant activity and improving nutrient uptake efficiency [22,23].

The recovery of *p*-cymene and terpinen-4-ol under biostimulant application suggests that these formulations not only alleviate salinity-induced oxidative stress but may also promote the biosynthesis of key bioactive metabolites [24]. Such compounds contribute to plant stress adaptation and enhance the medicinal and nutraceutical value of *C. maritimum*, highlighting the agronomic and economic relevance of incorporating biostimulants in halophyte cultivation.

Salinity stress and biostimulant treatments also induced pronounced modifications in the essential oil composition of *C. maritimum* leaves. Under control conditions, *p*-cymene (28.5%) and γ-terpinene (21.2%) were the dominant constituents. Exposure to salinity (10–20 dS m^{-1}) significantly altered this profile. At 10 dS m^{-1}, γ-terpinene increased to 37.1% while *p*-cymene declined to 16.9%, indicating a shift in monoterpene biosynthesis under moderate stress. At 20 dS m^{-1}, γ-terpinene remained elevated (33.5%), whereas β-pinene and terpinen-4-ol declined sharply, reflecting the suppressive effects of severe salinity on secondary metabolism, in agreement with previous studies [24,25].

Biostimulant application mitigated these alterations to varying extents. At 20 dS m^{-1}, Aquamin restored *p*-cymene levels to 36.4%, exceeding those of the control, while maintaining moderate concentrations of γ-terpinene (18.1%) and α-pinene (4.7%). Similarly, Cultisano sustained elevated *p*-cymene levels (32.5%) and markedly increased thymol-methyl-ether (11.2%), surpassing both control and salt-only treatments. Notably, Cultisano at 10 dS m^{-1} produced the highest terpinen-4-ol content (2.9%), a compound with recognized antimicrobial activity. These results indicate that biostimulants not only stabilize volatile biosynthesis under salt stress but may also selectively enhance the accumulation of valuable terpenoids, offering added potential for nutraceutical and essential oil applications. These findings align with previous reports showing that biostimulants can enhance antioxidant activity and stimulate secondary metabolism [26–28].

Compounds such as *β*-pinene and caryophyllene oxide, which declined markedly under salinity, were partially restored in plants treated with Aquamin or Cultisano, suggesting that these biostimulants exert a protective influence on terpene metabolism, potentially through modulation of enzymatic activity or gene expression. The detection of tetratriacontane (1.2%) in control plants—reduced under salinity and variably expressed following biostimulant application—further supports the notion that biostimulants can influence cuticular and defensive metabolic pathways under saline stress.

Total phenolic content (TPC) decreased progressively with increasing salinity, indicating that salt stress negatively affects either the biosynthesis or stability of phenolic compounds [25,29]. High salinity likely disrupts key metabolic pathways involved in secondary metabolite production, leading to reduced accumulation of antioxidant molecules. Similar declines in phenolics, terpenes and essential oil components under severe salinity have been reported in other aromatic species [30–33].

Both biostimulant treatments mitigated the reduction in TPC, with Aquamin showing the strongest positive effect across all salinity levels. The consistently higher TPC in Aquamin-treated plants suggests a stimulatory role in phenolic metabolism, potentially due to its high content of trace minerals and bioactive constituents that may enhance antioxidant enzyme activity and promote phenolic biosynthesis. Cultisano also improved phenolic retention relative to the control, albeit to a lesser extent, indicating a moderate protective effect.

The observed decline in phenolic content under salinity may reflect the diversion of metabolic energy toward essential survival processes and osmotic adjustment rather than secondary metabolite production. Nevertheless, the elevated phenolic levels maintained in both Aquamin- and Cultisano-treated plants indicate that these supplements may support antioxidant metabolism by enhancing redox balance or upregulating key enzymes in phenolic biosynthesis [34,35]. This interpretation is consistent with previous reports demonstrating that mineral-based products and biostimulants can increase stress tolerance through the stimulation of antioxidant pathways and the accumulation of phenolic compounds [19,20,34–39].

Overall, the results highlight the potential of targeted biostimulant applications as a sustainable agronomic strategy to alleviate salinity stress in halophytic crops such as *C. maritimum*. By partly restoring essential oil composition, enhancing phenolic metabolism, and improving physiological performance, Aquamin and Cultisano contribute to superior stress resilience. Nevertheless, further molecular and enzymatic studies are needed to elucidate the underlying mechanisms and to optimize application strategies for large-scale cultivation.

4. Materials and Methods

4.1. Plant Material and Experimental Design

The experiment was conducted at Perrotis College (Northern Greece; 40°34′ N, 22°58′ E) greenhouse facilities of the American Farm School in Thessaloniki, Greece and the biochemical analysis took place in the Laboratory of Plant Physiology of International Hellenic University in Sindos, Thessaloniki, Greece. One-year-old self-rooted *Crithmum maritimum* plants were grown in a greenhouse in 14-L pots filled with sandy-loam soil collected from the experimental site (65.2% sand, 29.6% silt, 15.2% clay; field capacity 2%). Inside the glasshouse, mean temperature ranged from 25 to 32 °C, relative humidity from 50 to 65%, and the photosynthetic photon flux density (PPFD) reached approximately 1200 μmol m^{-2} s^{-1} at midday on clear days.

A total of 45 black PVC pots (15 L each) were used, filled with a substrate (consisting of sand, silt, clay, and a small proportion of soil <2% v/v). The design followed a completely randomized block layout with two factors: salinity level and biostimulant treatment.

AQUAMIN, a product of Biosolids, is a water-soluble powder containing 62% plant-derived peptides. It enhances nutrient uptake and use efficiency while remaining a fully organic product.

CULTISANO contains chitosan (poly-D-glucosamine), a natural polymer derived from crustaceans, combined with amino acids obtained from enzymatic hydrolysis of plant proteins (Cultifort).

For foliar application, **AQUAMIN** was diluted in water according to the manufacturer's instructions (7 g L^{-1}). For each application, 750 mL per pot was manually sprayed using a 1-L plastic hand sprayer.

Similarly, **CULTISANO** was diluted in water following the recommended dose (2 mL L^{-1}) and applied foliarly in the same manner, with 750 mL sprayed per pot using a 1-L hand sprayer.

Three salinity levels were applied via irrigation water:

- 1 dS/m (control salinity);
- 10 dS/m (moderate salinity);
- 20 dS/m (high salinity).

At each salinity level, three biostimulant treatments were established:

1. Control (no biostimulant);
2. Foliar application of Aquamin;
3. Foliar application of Cultisano.

Each treatment was replicated five times, resulting in 15 pots per salinity level and a total of 45 pots. The pots were arranged in rows, with each row corresponding to a different salinity treatment. Within each row, the first five pots served as untreated controls, the next five received Aquamin, and the final five received Cultisano. The biostimulants were applied foliarly according to manufacturer instructions at regular intervals throughout the experimental period. Plants were grown under controlled greenhouse conditions. Data were collected on physiological and biochemical parameters, including biomass, pigment content, antioxidant activity, and stress indicators, following standard protocols.

4.2. Leaf Gas Exchange Measurements

A LI-COR 6410-XT portable infrared CO_2 analyzer (LI-COR Biosciences, Lincoln, NE, USA) was used to measure photosynthetic assimilation rate (A) and transpiration rate (E), on fully expanded leaves in both differential and open circuit modes. Sixty days after the experiment began, all measurements were made in sunny conditions at 22 °C. The CO_2 concentration inside the chamber was set at 400 μmol mol^{-1}, and the photon flux density was 1000 μmol m^{-2} s^{-1}. All photosynthetic parameters were calculated following Von Caemmerer and Farquhar [40].

4.3. Total Polyphenol Content (TPC)

The TPC was evaluated according to a previous methodology, Giannakoula et al. [39], with some modifications [41]. An Eppendorf tube was used to mix 100 μL of Folin–Ciocalteu reagent with 100 μL of a sample extract. After 2 min, 800 μL of 5% w/v Na_2CO_3 solution was added. After vortexing, the mixture was incubated for 30 min at room temperature. The absorbance at 760 nm was measured using a Shimadzu UV-1700 PharmaSpec Spectrophotometer from Kyoto, Japan. The results were expressed in gallic acid equivalents per gram of dry weight (GAE; mg/g DW) using a gallic acid standard curve.

4.4. GC–MS Analysis of Volatile Oils

GC–MS analysis of the isolated volatile oils was performed on a Shimadzu GC-17A gas chromatograph coupled with a Shimadzu GCMS-QP 5050 mass-selective Quadrupole Mass Spectrometer as a detector with the appropriate data system (Shimadzu Corp., Kyoto, Japan). The GC was equipped with a Grob-type split-splitless injector, the fused silica capillary column (Supelco, Bellefonte, PA, USA, SBP-5 with 0.25 μm film thickness, 30 m × 0.25 mm i.d.) was directly coupled to the ion source. Helium was used as a carrier gas with a back pressure of 0.8 Atm. The injector temperature was 250 °C, and the oven temperature program started at 50 °C for 5 min and then increased at a rate of 5 °C/min up to 150 °C, was retained at this temperature for 10 min, and increased again at a rate of 5 °C/min up to 280 °C, where it remained for 20 min. The scanning range was 30–700 *m*/*z*. A GS-MS detection electron ionization system was used with an ionization energy of 70 eV.

Identification and Quantification of Volatile Components

The quantification of the components was based on the total number of fragments (total ion count) of the metabolites, as detected by the mass spectrometer. The identification of the chemical components was carried out based on the retention time of each component (Rt) compared with those of commercially available compounds, by analysis of their mass spectra, by the use of the NIST21, NIST107 and PMW_TOX2 mass spectra libraries [42], as well as by comparison with literature data [43]. Calculation of retention indices was performed in accordance with the work of [44], in comparison to the retention times of standard hydrocarbons (C_8–C_{40}).

4.5. Determination of H_2O_2 Concentration

For the determination of H_2O_2 concentration in leaves of *C. maritinum* L. plants after 60 days of biostimulants treatment, leaf extraction was carried out according to Giannakoula et al. [45]. Hydrogen peroxide was measured spectrophotometrically after reaction with KI. The reaction mixture consisted of 0.5 mL leaf extract with 0.1% trichloroacetic acid (TCA), 0.5 mL of 100 mM K-phosphate buffer and 2 mL reagent (1 M KI in fresh double-distilled H_2O). The blank consisted of 0.1% TCA. The reaction was developed for 1 h in the dark. The absorbance of the solution was measured at 410 nm, and the H_2O_2 concentration was calculated using a standard curve ranging from 0.1 to 1 mm. H_2O_2 content and was expressed as nmol g^{-1} FW.

4.6. Malondialdehyde (MDA) Concentration

The level of lipid peroxidation of *C. maritinum* L. leaves at the end of the experiment was measured and the MDA content was determined by reaction with 2-thiobarbituric acid (TBA) according to [46]. Fresh leaf blade tissue, 0.1 g, was homogenised by adding 0.5 mL of 0.1% (*w*/*v*) trichloroacetic acid (TCA). The homogenate was centrifuged at 15,000× *g* and 4 °C for 10 min. From the supernatant, 0.5 mL was mixed with 1.5 mL of 0.5% TBA diluted in 20% TCA. Incubation follows at 95 °C for 25 min. The reaction stops by incubating in an ice bath. Afterward, the tubes were centrifuged at 10,000× *g* and 4 °C for 10 min and the absorbance of the supernatant was read at 532 and 600 nm. The value for the non-specific absorption at 600 nm was subtracted from the value at 532 nm. The concentration of MDA was calculated using Lambert–Beer's law using the MDA extinction coefficient of 155 mM^{-1} cm^{-1}. Results are presented as nmol MDA g^{-1} FW.

4.7. Statistical Analysis

Statistical analyses were performed using SPSS v29 (IBM, Armonk, NY, USA), GraphPad Prism v8 (GraphPad Software, Boston, MA, USA), and Microsoft Excel. Essential oil composition data derived from GC–MS analysis were visualized in ClustVis [47] using z-score normalization and hierarchical clustering based on Euclidean distances and Ward's method. Relative essential oil abundances, including key compounds associated with oxidative stress, were illustrated using stacked bar plots in Excel. Principal Component Analysis (PCA) was conducted in SPSS v29 to assess the separation among the nine treatment groups (Control, S10, S20; Aquamin: A, A10, A20; Cultisano: C, C10, C20) according to their volatile profiles.

Physiological and biochemical parameters (CO_2 assimilation rate, CO_2 transpiration rate, malondialdehyde—MDA concentration, and total phenolic content—TPC) were analyzed using two-way ANOVA, with Treatment categories (None, Aquamin, and Cultisano) and Salinity considered as fixed factors. When significant effects were detected, Tukey's HSD post hoc test was applied to determine differences among means. Data were visualized using bar plots with standard error bars, while linear plots were used to depict trends across salinity levels. Correlations among physiological and biochemical variables were evaluated using Pearson correlation analysis, and the corresponding correlation coefficients (R) and coefficients of determination (R^2) were reported [47].

5. Conclusions

In conclusion, salinity stress substantially affected the physiological performance, essential oil composition, and phenolic metabolism of *C. maritimum*. Increasing salinity levels, particularly above 10 dS m^{-1}, impaired growth, reduced photosynthetic activity, and suppressed the biosynthesis of several key volatile compounds. The foliar application of Aquamin and Cultisano effectively mitigated these detrimental effects, with Aquamin generally providing the greatest improvement across physiological parameters, phenolic content, and monoterpene stability.

Both biostimulants enhanced plant resilience by improving pigment retention, supporting antioxidant capacity, and partially restoring salinity-suppressed terpenoids such as *p*-cymene, *β*-pinene, and terpinen-4-ol. Notably, Aquamin at 20 dS m^{-1} restored *p*-cymene to levels higher than the control, while Cultisano selectively increased compounds such as thymol-methyl-ether and terpinen-4-ol, indicating a stimulatory effect on specific branches of monoterpene biosynthesis. These responses highlight the dual action of biostimulants in alleviating oxidative stress while enriching valuable bioactive metabolites.

Overall, the findings demonstrate that targeted biostimulant applications can significantly enhance the tolerance of halophytic crops to saline environments by stabilizing physiological function and promoting the production of antioxidant and antimicrobial secondary metabolites. Such strategies offer promising potential for the sustainable cultivation and valorization of *C. maritimum* in salt-affected agricultural systems. Further molecular analyses are warranted to elucidate the underlying regulatory mechanisms and optimize application regimes for large-scale production.

Supplementary Materials: The following supporting information can be downloaded at https://www.mdpi.com/article/10.3390/plants14243788/s1. **Supplementary Table S1:** Mean values (±SD) of CO_2 assimilation rate (μmol CO_2 m^{-2} s^{-1}), transpiration rate (mmol H_2O m^{-2} s^{-1}), malondialdehyde (MDA; nmol g^{-1} FW) and total phenolic content (TPC; mg GAE g^{-1} FW) measured in Crithmum maritimum plants across three salinity levels (1, 10 and 20 dS m^{-1}) and three treatments (Control, Aquamin, Cultisano). Each value represents the mean of n = 5 plants per treatment. **Supplementary Table S2:** Volatile organic compounds (VOCs) identified by GC–MS analysis in *Crithmum maritimum* across control, salinity (S10, S20) and biostimulant treatments (Aquamin: A, A10, A20; Cultisano: C, C10, C20); R.I.[a] (Retention Indices) from experimental using a SBP-5 column using a homologous series of n-alkanes (C8–C40); R.I.[b] (Retention Indices) from literature data, Adams (2007) and National Institute of Standards and Technology (2025).

Author Contributions: Conceptualization, A.E.G. and A.G.; methodology, A.E.G., M.K.S., C.P. and A.G.; software, A.E.G. and M.K.S.; validation, A.E.G. and S.T.; formal analysis, C.P.data analysis; A.E.G., S.T., A.N. and S.K.; investigation, A.E.G., S.T., S.K, and A.N.; resources, A.E.G., C.P. and M.K.S.; writing—original draft preparation, A.E.G. and C.P.; writing—review and editing, A.E.G., A.G. and M.K.S.; supervision and project administration, A.E.G. and A.G. All authors have read and agreed to the published version of the manuscript.

Funding: This research received no external funding.

Data Availability Statement: The data presented in this study are available on request from the corresponding author. The data are not publicly available due to privacy and ethical restrictions.

Conflicts of Interest: The authors declare no conflicts of interest.

References

1. Castillo, J.; Mancilla-Leyton, J.M.; Martins-Noguerol, R.; Moreira, X.; Moreno-Perez, A.J.; Munoz-Valles, S.; Pedroche, J.J.; Figueroa, M.E.; García-González, A.; Salas, J.J.; et al. Interactive effects between salinity and nutrient deficiency on biomass production and bio-active compounds accumulation in the halophyte *Crithmum maritimum*. *Sci. Hortic.* **2022**, *301*, 111136. [CrossRef]
2. Davazdahemami, S.; Allahdadi, M. Essential oil yield and composition of four annual plants (ajowan, dill, moldavian balm and black cumin) under saline irrigation. *Food Ther. Health Care* **2022**, *4*, 5. [CrossRef]
3. Nazari, J.; Tabatabaie, S.J. Influence of Nutrient Concentrations and NaCl Salinity on the Growth, Photosynthesis, and Essential Oil Content of Peppermint and Lemon Verbena. *Turk. J. Agric. For.* **2007**, *31*, 245–253. Available online: https://journals.tubitak.gov.tr/agriculture/vol31/iss4/4/ (accessed on 20 January 2022).
4. Atia, A.; Barhoumi, Z.; Mokded, R.; Abdelly, C.; Smaoui, A. Environmental eco-physiology and economical potential of the halophyte *Crithmum maritimum* L. (Apiaceae). *J. Med. Plants Res.* **2011**, *16*, 3564–3571.
5. Pereira, C.; Barreira, L.; Neng, N.; Nogueira, J.; Marques, C.; Santos, T.; Varela, J.; Custodio, L. Searching for new sources of innovative products for the food industry within halophyte aromatic plants: In vitro antioxidant activity and phenolic and mineral contents of infusions and decoctions of *Crithmum maritimum* L. *Food Chem. Toxicol.* **2017**, *107*, 581–589. [CrossRef]
6. Chatzimitakos, T.; Athanasiadis, V.; Makrygiannis, I.; Kalompatsios, D.; Bozinou, E.; Lalas, S.I. An Investigation into *Crithmum maritimum* L. Leaves as a Source of Antioxidant Polyphenols. *Compounds* **2023**, *3*, 532–551. [CrossRef]
7. Kraouia, M.; Nartea, A.; Maoloni, A.; Osimani, A.; Garofalo, C.; Fanesi, B.; Ismaiel, L.; Aquilanti, L.; Pacetti, D. Sea Fennel (*Crithmum maritimum* L.) as an Emerging Crop for the Manufacturing of Innovative Foods and Nutraceuticals. *Molecules* **2023**, *28*, 4741. [CrossRef]
8. Sahoo, S.; Singha, C.; Govind, A.; Moghimi, A. Review of climate-resilient agriculture for ensuring food security: Sustainability opportunities and challenges of India. *Environ. Sustain. Indic.* **2025**, *25*, 100544. [CrossRef]
9. Jallali, I.; Zaouali, Y.; Missaoui, I.; Smeoui, A.; Abdelly, C.; Ksouri, R. Variability of antioxidant and antibacterial effects of essential oils and acetonic extracts of two edible halophytes: *Crithmum maritimum* L. and *Inula crithmoïdes* L. *Food Chem.* **2014**, *145*, 1031–1038. [CrossRef]
10. Mekinic, I.G.; Šimat, V.; Ljubenkov, I.; Burčul, F.; Grga, M.; Mihajlovski, M.; Lončar, R.; Katalinic, V.; Skroza, D. Influence of the vegetation period on sea fennel, *Crithmum maritimum* L. (Apiaceae), phenolic composition, antioxidant and anticholinesterase activities. *Ind. Crops Prod.* **2018**, *124*, 947–953. [CrossRef]
11. Meot-Duros, L.; Magné, C. Antioxidant activity and phenol content of *Crithmum maritimum* L. leaves. *Plant Physiol. Biochem.* **2009**, *47*, 37–41. [CrossRef] [PubMed]
12. Correia, I.; Antunes, M.; Tecelão, C.; Neves, M.; Pires, C.L.; Cruz, P.F.; Rodrigues, M.; Peralta, C.C.; Pereira, C.D.; Reboredo, F.; et al. Nutritive Value and Bioactivities of a Halophyte Edible Plant: *Crithmum maritimum* L. (Sea Fennel). *Plants* **2024**, *13*, 427. [CrossRef] [PubMed]
13. Ben Hamed, K.; Castagna, A.; Salem, E.; Ranieri, A.; Abdelly, C. Sea fennel (*Crithmum maritimum* L.) under salinity conditions: A comparison of leaf and root antioxidant responses. *Plant Growth Regul.* **2007**, *53*, 185–194. [CrossRef]
14. Hamdani, F.; Derridj, A.; Roger, H.J. Diverse salinity responses in *Crithmum maritimum* tissues at different salinities over time. *J. Soil Sci. Plant Nutr.* **2017**, *17*, 716–734. [CrossRef]
15. Popović, M.; Radman, S.; Generalić Mekinić, I.; Ninčević RunjIć, T.; Urlić, B.; Veršić Bratinčević, M. A Year in the Life of Sea Fennel: Annual Phytochemical Variations of Major Bioactive Secondary Metabolites. *Appl. Sci.* **2024**, *14*, 3440. [CrossRef]
16. Ntanasi, T.; Karavidas, I.; Spyrou, G.P.; Giannothanasis, E.; Aliferis, K.A.; Saitanis, C.; Fotopoulos, V.; Sabatino, L.; Savvas, D.; Ntatsi, G. Plant Biostimulants Enhance Tomato Resilience to Salinity Stress: Insights from Two Greek Landraces. *Plants* **2024**, *13*, 1404. [CrossRef]

17. Kadoglidou, K.I.; Anthimidou, E.; Krommydas, K.; Papa, E.; Karapatzak, E.; Tsivelika, N.; Irakli, M.; Mellidou, I.; Xanthopoulou, A.; Kalivas, A. Effect of Biostimulants on Drought Tolerance of Greenhouse-Grown Tomato. *Horticulturae* **2025**, *11*, 601. [CrossRef]
18. Di Sario, L.; Boeri, P.; Matus, J.T.; Pizzio, G.A. Plant Biostimulants to Enhance Abiotic Stress Resilience in Crops. *Int. J. Mol. Sci.* **2025**, *26*, 1129. [CrossRef]
19. Chatzistathis, T.; Sarropoulou, V.; Papaioannou, E.; Giannakoula, A. Inorganic and Organic Fertilization Effects on the Growth, Nutrient Uptake, Chlorophyll Fluorescence and Fruit Quality in *Solanum melongena* L. Plants. *Agronomy* **2025**, *15*, 872. [CrossRef]
20. Win, P.P.; Park, H.-H.; Kuk, Y.-I. Integrated Approach of Using Biostimulants for Improving Growth, Physiological Traits, and Tolerance to Abiotic Stressors in Rice and Soybean. *Agronomy* **2025**, *15*, 2265. [CrossRef]
21. Hamed, K.; Debez, A.; Chibani, F.; Abdelly, C. Salt response of *Crithmum maritimum*, an oleagineous halophyte. *Trop. Ecol.* **2004**, *45*, 151–159.
22. Di Mola, I.; Ottaiano, L.; Cozzolino, E.; El-Nakhel, C.; Fiorentino, N.; Pelosi, M.E.; Rouphael, Y.; Mori, M. Impact of Salinity and Biostimulants on Cherry Tomato Yield and Quality. *Horticulturae* **2024**, *10*, 1239. [CrossRef]
23. Gedeon, S.; Ioannou, A.; Balestrini, R.; Fotopoulos, V.; Antoniou, C. Application of Biostimulants in Tomato Plants (*Solanum lycopersicum*) to Enhance Plant Growth and Salt Stress Tolerance. *Plants* **2022**, *11*, 3082. [CrossRef] [PubMed]
24. Pedreiro, S.; Figueirinha, A.; Cavaleiro, C.; Cardoso, O.; Donato, M.M.; Salgueiro, L.; Ramos, F. Exploiting the *Crithmum maritimum* L. Aqueous Extracts and Essential Oil as Potential Preservatives in Food, Feed, Pharmaceutical and Cosmetic Industries. *Antioxidants* **2023**, *12*, 252. [CrossRef]
25. Siracusa, L.; Kulisic-Bilusic, T.; Politeo, O.; Krause, I.; Dejanovic, B.; Ruberto, G. Phenolic composition and antioxidant activity of aqueous infusions from *Capparis spinosa* L. and *Crithmum maritimum* L. before and after submission to a two-step in vitro digestion model. *J. Agric. Food Chem.* **2011**, *59*, 12453–12459. [CrossRef]
26. Zafeiropoulou, V.; Tomou, E.-M.; Ioannidou, O.; Karioti, A.; Skaltsa, H. Sea fennel: Phytochemical analysis of Greek wild and cultivated *Crithmum maritimum* L. populations, based on HPLC-PDA-MS and NMR methods. *J. Pharmacogn. Phytochem.* **2020**, *9*, 998–1004.
27. Dzhoglova, V.; Ivanova, S.; Shishmanova-Doseva, M.; Saracheva, K. *Crithmum maritimum* L.: Phytochemical Profile, Bio logical Activities, and Therapeutic Potential. *Molecules* **2025**, *30*, 2832. [CrossRef]
28. Hancioglu, N.E.; Kurunc, A.; Tontul, I.; Topuz, A. Irrigation water salinity effects on oregano (*Origanum onites* L.) water use, yield and quality parameters. *Sci. Hortic.* **2019**, *247*, 327–334. [CrossRef]
29. Porrello, A.; Vaglica, A.; Savoca, D.; Bruno, M.; Sottile, F. Variability in *Crithmum maritimum* L. Essential Oils' Chemical Composition: PCA Analysis, Food Safety, and Sustainability. *Sustainability* **2024**, *16*, 2541. [CrossRef]
30. Alves-Silva, J.M.; Guerra, I.; Gonçalves, M.J.; Cavaleiro, C.; Cruz, M.T.; Figueirinha, A.; Salgueiro, L. Chemical composition of *Crithmum maritimum* L. essential oil and hydrodistillation residual water by GC-MS and HPLC-DAD-MS/MS, and their biological activities. *Ind. Crop. Prod.* **2020**, *149*, 112329. [CrossRef]
31. Baser, K.H.C.; Özek, T.; Demirci, B.; Saritas, Y. Essential Oil of *Crithmum maritimum* L. from Turkey. *J. Essent. Oil Res.* **2000**, *12*, 424–426. [CrossRef]
32. Marongiu, B.; Maxia, A.; Piras, A.; Porcedda, S.; Tuveri, E.; Gonçalves, M.J.; Cavaleiro, C.; Salgueiro, L. Isolation of *Crithmum maritimum* L. Volatile Oil by Supercritical Carbon Dioxide Extraction and Biological Assays. *Nat. Prod. Res.* **2007**, *21*, 1145–1150. [CrossRef] [PubMed]
33. Mustapha, M.B.; Zardi-Bergaoui, A.; Chaieb, I.; Flamini, G.; Ascrizzi, R.; Jannet, H.B. Chemical Composition and Insecticidal Activity of *Crithmum maritimum* L. Essential Oil against Stored-Product Beetle i. *Chem. Biodivers.* **2020**, *17*, e1900552. [CrossRef] [PubMed]
34. Cunsolo, F.; Ruberto, G.; Amico, V.; Piattelli, M. Bioactive metabolites from Sicilian marine fennel, i. *J. Nat. Prod.* **1993**, *56*, 1598–1600. [CrossRef] [PubMed]
35. Lakhdar, A.; Trigui, M.; Montemurro, F. An Overview of Biostimulants' Effects in Saline Soils. *Agronomy* **2023**, *13*, 2092. [CrossRef]
36. Franzoni, G.; Cocetta, G.; Prinsi, B.; Ferrante, A.; Espen, L. Biostimulants on Crops: Their Impact under Abiotic Stress Conditions. *Horticulturae* **2022**, *8*, 189. [CrossRef]
37. Khan, W.; Rayirath, U.P.; Subramanian, S.; Jithesh, M.N.; Rayorath, P.; Hodges, D.M.; Critchley, A.T.; Craigie, J.S.; Norrie, J.; Prithiviraj, B. Seaweed Extracts as Biostimulants of Plant Growth and Development. *J. Plant Growth Regul.* **2009**, *28*, 386–399. [CrossRef]
38. Van Oosten, M.J.; Pepe, O.; De Pascale, S.; Silletti, S.; Maggio, A. The Role of Biostimulants and Bioeffectors as Alleviators of Abiotic Stress in Crop Plants. *Chem. Biol. Technol. Agric.* **2017**, *4*, 5. [CrossRef]
39. Giannakoula, A.; Ouzounidou, G.; Stefanou, S.; Daskas, G.; Dichala, O. Effects of Biostimulants on the Eco-Physiological, traits and Fruit Quality of Black Chokeberry (*Aronia melanocarpa* L.). *Plants* **2024**, *13*, 3014. [CrossRef]
40. Von Caemmerer, S.; Farquhar, G.D. Some relationships between the biochemistry of photosynthesis and the gas exchange of leaves. *Planta* **1981**, *153*, 376–387. [CrossRef]

41. Karastergiou, A.; Gancel, A.-L.; Jourdes, M.; Teissedre, P.-L. Transforming winemaking waste: Grape pomace as a sustainable source of bioactive compounds. *OENO One* **2025**, *59*. [CrossRef]
42. National Institute of Standards and Technology (NIST). 2017. Available online: http://webbook.nist.gov/chemistry/name-ser.html (accessed on 21 December 2021).
43. Adams, R.P. *Identification of Essential Oil Components by Gas Chromatography/Mass Spectrometry*, 4th ed.; Allured Publishing Corp.: Carol Stream, IL, USA, 2007.
44. Van den Dool, H.; Kratz, P.D. A generalization of the retention index system including linear temperature programmed as-liquid partition chromatography. *J. Chromatogr. A* **1963**, *11*, 463–471. [CrossRef]
45. Giannakoula, A.; Therios, I.; Chatzissavvidis, C. Effect of Lead and Copper on Photosynthetic Apparatus in Citrus (*Citrus aurantium* L.) Plants. The Role of Antioxidants in Oxidative Damage as a Response to Heavy Metal Stress. *Plants* **2021**, *10*, 155. [CrossRef]
46. Heath, R.L.; Packer, L. Photoperoxidation in isolated chloroplasts. I. Kinetics and stoichiometry of fatty acid peroxidation. *Arch. Biochem. Biophys.* **1968**, *125*, 189–198. [CrossRef]
47. Metsalu, T.; Vilo, J. ClustVis: A web tool for visualizing clustering of multivariate data using Principal Component Analysis and heatmap. *Nucleic Acids Res.* **2015**, *43*, W566–W570. [CrossRef]

plants

MDPI

Article

Physiological and Yield Responses of Pepper (*Capsicum annuum* L.) Genotypes to Drought Stress

Theodora Ntanasi [1,†], Ioannis Karavidas [1,†], Dimitrios Savvas [1], George P. Spyrou [1], Evangelos Giannothanasis [1], Beppe Benedetto Consentino [2], Vasileios Papasotiropoulos [3], Leo Sabatino [2] and Georgia Ntatsi [1,*]

1 Laboratory of Vegetable Production, Department of Crop Science, Agricultural University of Athens, Iera Odos 75, 11855 Athens, Greece; ntanasi@aua.gr (T.N.); karavidas@aua.gr (I.K.); dsavvas@aua.gr (D.S.); gspyrou@aua.gr (G.P.S.); giannothanasis@aua.gr (E.G.)

2 Department of Agricultural, Food and Forest Sciences, University of Palermo, 90128 Palermo, Italy; beppebenedetto.consentino@unipa.it (B.B.C.); leo.sabatino@unipa.it (L.S.)

3 Laboratory of Plant Breeding and Biometry, Department of Crop Science, Agricultural University of Athens, Iera Odos 75, 11855 Athens, Greece; vpapasot@aua.gr

* Correspondence: ntatsi@aua.gr

† These authors contributed equally to this work.

Abstract: Drought stress is a critical abiotic constraint on agricultural productivity, particularly affecting crops like pepper (*Capsicum annuum* L.), which are highly susceptible to water deficits due to their physiological characteristics. The present study investigated the impact of a 40% reduction in irrigation on yield, macronutrient concentrations, and fruit quality across several pepper genotypes. The cultivars evaluated included two landraces, namely 'JO109' and 'JO204' (*Capsicum annuum* var. *grossum*), as well as the California cultivar 'Yolo Wonder' and the commercial F1 hybrid 'Sammy RZ', which served as controls. The experiment was conducted at the greenhouse facilities of the Laboratory of Vegetable Production, Agricultural University of Athens. Under reduced irrigation, most of the cultivars studied exhibited a decline in yield, which was attributed to a decrease in fruit number in 'Yolo Wonder' and a reduction in fruit weight in both 'JO204' and 'Sammy'. In contrast, the landrace 'JO109' exhibited consistent yields under both growing conditions, a response likely attributed to elevated K concentration in the leaves and lower Na accumulation in the fruit, indicating enhanced tolerance to water deficit. A decline in leaf K concentration was observed in response to drought stress, while concomitantly increased concentrations of Na, Ca and Mg were recorded. Among fruit macronutrients, only Ca showed a significant decrease under reduced irrigation. Furthermore, fruit firmness (FF), titratable acidity (TA) and total soluble solids content (TSSC) exhibited higher levels under drought stress, particularly in 'JO109', while TA remained unaltered. These findings highlight the potential of landraces such as 'JO109' to be utilized in breeding programs aimed at enhancing resilience, while maintaining pepper fruit quality under limited water availability.

Keywords: *Capsicum annuum* L.; abiotic stress; productivity; quality; nutrient concentration; soilless culture; osmotic stress

1. Introduction

The long-term effects of global climate change are expected to exacerbate abiotic stresses, leading to more frequent and prolonged droughts, elevated temperatures, and unpredictable rainfall patterns across many regions of the world [1]. A global assessment conducted by the Food and Agriculture Organization [2] projected that feeding a growing global population would require approximately 70% more food production by 2050. In the

Mediterranean region, water availability is of particular concern, as both human activity and ongoing climate change have led to an intensification of drought conditions [3]. It is well established that evident that drought stress leads to substantial reduction in plant growth, primarily due to the disruption of vital processes such as cell division and cell expansion. These processes are governed by complex interactions among anatomical, physiological, and genetic factors. These interactions are adversely affected by limited water availability [4]. Drought is recognized as one of the most critical abiotic stresses affecting agricultural productivity, with the potential to reduce average crop yields by over 50% [5,6].

Drought stress poses a significant challenge to the growth, development, and yield of vegetable crops, as water accounts for more than 90% of the fresh weight of most vegetables [7,8]. It is well documented that vegetables typically exhibit sensitivity to drought when water content reaches approximately 20% of their biomass [9]. Water deficiency directly hinders photosynthesis by limiting CO_2 diffusion into the drought stressed leaves due to reduced conductance, primarily in the stomatal and mesophyll tissues [10]. Furthermore, in conditions of drought, the uptake of nutrients by roots is reduced due to decreased soil water content, which slows the diffusion of nutrients from the soil to the root surface [11]. In addition to its adverse effects on plant growth, drought stress can influence the quality of the harvested fruits. Research on hydroponically grown vegetable crops under water limited conditions has shown that such stress can increase the concentration of sugars and acids in fruits, potentially improving their nutritional value, flavor, and taste [12].

Pepper (*Capsicum annuum* L.) is a key crop cultivated in Mediterranean regions, where limited water availability poses a significant challenge to its production [13]. The primary appeal of pepper is attributable to its flavour, nutritional content, and contribution to a healthy diet [14]. Bell pepper is highly sensitive to drought stress, primarily due to its extensive leaf surface area and elevated stomatal conductance [15]. It has been demonstrated that the occurrence of drought stress during the early developmental and reproductive phases leads to a decline in both the quantity and size of buds and fruits [16]. Furthermore, drought stress significantly impacts the yield of harvested pepper fruits [17]. The increased need for drought tolerant pepper varieties that can withstand reduced water availability, while maintaining acceptable yields and fruit quality, is evident by numerous studies [18]. Therefore, in agroecological and organic farming systems, maintaining a diverse range of crop varieties—particularly traditional and locally adapted ones that can be effectively used for breeding and as rootstocks—is predicted to enhance the resilience of vegetable production in the face of increasingly unpredictable future weather patterns [19,20].

Landraces frequently manifest as components of a more extensive metapopulation dispersed across landscapes exhibiting diverse environmental conditions, including variations in water availability [21]. However, it should be noted that these varieties may also be subject to certain limitations, including extended maturation times and relatively low yields. [22]. Notwithstanding the aforementioned disadvantages, landraces are considered to be of significant value as genetic resources for the purpose of crop improvement, particularly with regard to traits associated with quality [23]. It is evident that local landraces have adapted to specific environments, thus demonstrating the potential to exhibit superior performance in comparison to non-local populations under local conditions. This phenomenon can be attributed to the evolutionary selection of particular traits in response to their respective environments [24].

To address this, the present study aimed to assess the performance of various pepper cultivars—either native to or commonly grown in the Mediterranean region—under conditions of limited irrigation. Our evaluation focused on yield, fruit quality, and the nutritional status of both leaves and fruits. A thorough analysis of the findings will help identify traits associated with drought tolerance in pepper plants. This information can support breeding initiatives aimed at developing (a) new varieties or hybrids capable of thriving in drought—prone environments without compromising yield, or (b) improved rootstocks for grafted plant production. The novelty of this study lies in the use of a fully controlled open hydroponic system to apply precise water stress treatments. This allows the differentiation of drought responses and the identification of agronomic and quality characteristics unique to each Mediterranean landrace.

2. Results

The yield of different pepper genotypes grown under either normal or reduced irrigation is illustrated in Figure 1. The commercial variety 'Yolo Wonder' and the hybrid 'Sammy' were the most productive, exhibiting a high yield per plant. However, under water stress conditions, their yield declined by 28% and 15%, respectively. A more pronounced decline, reaching 33%, was observed in the landrace 'JO204'. In contrast, the traditional Jordanian landrace 'JO109' did not show a statistically significant yield reduction under limited water availability.

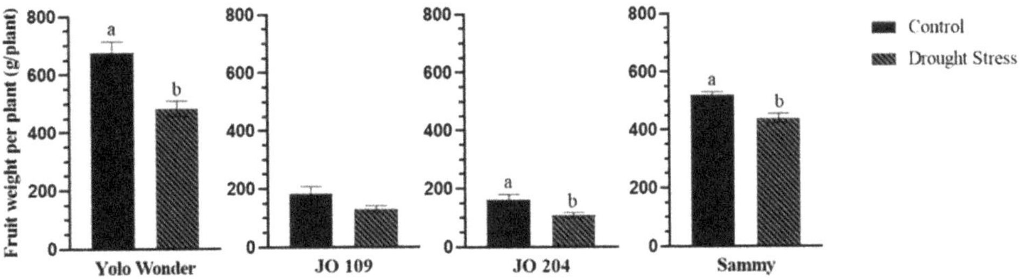

Figure 1. Effect of drought stress on fruit weight per plant (g/plant) across different pepper genotypes. The label 'Control' represents well-irrigated conditions, while 'Drought stress' indicates exposure to limited water availability (40% reduction). Different letters above each bar signify statistically significant differences based on Duncan's multiple range test ($p < 0.05$). Vertical bars represent the standard error of the mean (n = 4).

As shown in Table 1, the decline in yield observed for the varieties 'JO204' and 'Sammy' under drought stress, was attributed to a reduction in fruit weight rather than fruit number. Specifically, average fruit weight decreased by 28% in 'JO204' and by 9% in 'Sammy'. In contrast, the yield reduction in the variety 'Yolo Wonder' under water shortage conditions for was associated with a 23% decrease in fruit number.

Table 2 shows the concentrations of the macronutrients potassium (K), sodium (Na), calcium (Ca) and magnesium (Mg) in the leaves of different pepper cultivars. Water stress significantly affected all four nutrients. Specifically, K increased by 12% under drought conditions, while the concentrations of Na, K/Na, Ca, and Mg decreased by 16%, 24%, 17%, and 14%, respectively. There was significant variation among cultivars in the concentrations of K, Na, and Ca, whereas no significant differences were observed for Mg. The cultivar 'JO109' demonstrated the highest concentrations of K, K/Na and Na in its leaves, and together with 'JO204', the lowest Ca content. Moreover, a significant interaction between drought stress and cultivar was found only for K. It is noteworthy that K concentration in 'JO204' leaves remained unaffected under drought stress.

Table 1. Effect of drought stress on the average fruit number per plant (FN) and the mean fruit weight (MFW) across the tested genotypes.

	FN (No Plant^{-1})				MFW (g)			
Water Availability	**Yolo Wonder**	**JO109**	**JO204**	**Sammy**	**Yolo Wonder**	**JO109**	**JO204**	**Sammy**
Control	4.75	14.17	13.88	12.08	142.22	12.81	11.61	43.00
Drought stress	3.67	10.92	13.00	11.17	132.29	11.96	8.31	39.33
Statistical significance	**	NS	NS	NS	NS	NS	**	*

Mean values (n = 4) represent statistically significant differences according to Duncan's multiple range test ($p < 0.05$). Asterisks indicate significance levels: ** ($p < 0.01$), and * ($p < 0.05$), while NS signifies a non-significant difference.

Table 2. Effect of drought stress and cultivar on leaf macronutrient concentration in the tested genotypes.

		Leaves				
Water Availability	**Variety**	**K (mg/g)**	**Na (mg/g)**	**K/Na**	**Ca (mg/g)**	**Mg (mg/g)**
		Main effects (Drought stress)				
Control		43.75	0.25	176.06	30.04	9.20
Drought stress		38.38	0.29	134.47	35.15	10.48
		Main effects (Variety)				
	Yolo Wonder	38.75 b	0.23 c	171.32 a	35.34 a	10.73
	JO 109	52.00 a	0.33 a	166.86 a	27.10 b	9.40
	JO 204	39.50 b	0.29 ab	140.03 b	30.03 b	9.58
	Sammy	34.00 c	0.25 bc	142.84 b	37.91 a	9.64
		Interaction				
Control	Yolo Wonder	41.50 c	0.23	180.68 ab	33.27	10.52
	JO 109	55.00 a	0.29	197.30 a	24.23	8.48
	JO 204	39.00 cd	0.26	148.97 bc	27.13	8.36
	Sammy	39.50 cd	0.23	177.29 ab	35.54	9.42
Drought stress	Yolo Wonder	36.00 d	0.23	161.96 bc	37.40	10.94
	JO 109	49.00 b	0.38	136.43 cd	29.96	10.33
	JO 204	40.00 cd	0.31	131.10 cd	32.93	10.79
	Sammy	28.50 e	0.27	108.39 d	40.29	9.86
		Statistical significance				
Drought Stress		***	**	***	***	***
Variety		***	***	**	***	NS
Drought Stress × Variety		**	NS	*	NS	NS

Mean values (n = 4) followed by different letters within the same column represent statistically significant differences according to Duncan's multiple range test ($p < 0.05$). Asterisks indicate significance levels: *** ($p < 0.001$), ** ($p < 0.01$) and * ($p < 0.05$), while NS signifies a non-significant difference.

Table 3 presents the concentrations of the macronutrients K, Na, Ca and Mg in pepper fruits. Drought stress showed a significant decrease of 14% on Ca levels, compared to the control conditions. In contrast, significant differences among pepper cultivars were observed for all four macronutrients. Specifically, the cultivar 'JO204' exhibited the highest concentrations of K, Na, K/Na and Mg, and the lowest of Ca. 'JO109' exhibited the lowest Na concentration and the highest K/Na ratio and Mg concentration in its fruits. The cultivars 'Yolo Wonder' and 'Sammy' exhibited the highest Na concentrations in fruit tissue. A significant interaction was identified between the effects of drought stress and the variety for both Na and Ca, with 'JO204' displaying notable changes, including a 55%

increase in fruit Na concentration and a 40% decrease in Ca concentration under drought conditions.

Table 3. Effect of drought stress and cultivar on fruit macronutrient concentration in the tested genotypes.

		Fruit				
Drought Stress	**Variety**	**K (mg/g)**	**Na (mg/g)**	**K/Na**	**Ca (mg/g)**	**Mg (mg/g)**
		Main effects (Drought stress)				
Control		47.88	0.18	279.56	11.28	5.59
Drought stress		49.75	0.19	269.59	9.66	5.58
		Main effects (Variety)				
	Yolo Wonder	45.25 c	0.21 a	226.47 b	10.91 b	5.15 b
	JO 109	49.50 b	0.15 b	341.04 a	9.33 c	5.96 a
	JO 204	59.25 a	0.20 a	306.26 a	9.09 c	6.23 a
	Sammy	41.25 d	0.19 a	224.54 b	12.56 a	4.99 b
		Interaction				
Control	Yolo Wonder	43.00	0.22 ab	204.65 b	11.35 abc	5.27
	JO 109	49.00	0.15 cd	322.41 a	9.29 c	6.07
	JO 204	57.50	0.16 cd	358.69 a	11.37 abc	5.88
	Sammy	42.00	0.19 bcd	232.50 b	13.11 a	5.13
Drought stress	Yolo Wonder	47.50	0.20 abc	248.30 b	10.46 bc	5.02
	JO 109	50.00	0.14 d	359.67 a	9.37 c	5.85
	JO 204	61.00	0.25 a	253.83 b	6.82 d	6.58
	Sammy	40.50	0.19 bcd	216.58 b	12.01 ab	4.85
		Statistical significance				
Drought Stress		NS	NS	NS	**	NS
Variety		***	**	***	***	***
Drought Stress × Variety		NS	**	**	**	NS

Mean values (n = 4) followed by different letters within the same column represent statistically significant differences according to Duncan's multiple range test ($p < 0.05$). Asterisks indicate significance levels: *** ($p < 0.001$) and ** ($p < 0.01$), while NS signifies a non-significant difference.

Fruit firmness was significantly increased (by 20%) under drought stress (Figure 2). Statistically significant differences were also found among the different pepper cultivars. The varieties 'Yolo Wonder', 'JO109', and 'JO204' exhibited similar firmness values, while the 'Sammy' hybrid showed the lowest fruit firmness compared to the other cultivars. However, no statistical significance was observed in the interaction of the two factors (drought × variety).

Regarding total soluble solids content (TSSC), both drought stress and cultivar had a significant effect, with TSSC increasing by 6% under drought conditions (Table 4). Among the cultivars, 'JO109' exhibited the highest value for TSSC, while the commercial variety 'Yolo Wonder' recorded the lowest. A significant interaction between drought stress and variety was also observed (Figure 3). Notably 'JO109', which had the highest TSSC among all varieties, exhibited a 17% increase in TSSC under drought conditions, the greatest increase recorded among the tested cultivars.

Finally, statistically significant differences were observed for titratable acidity (TA) only among the cultivated varieties (Figure 4). Specifically, the landrace 'JO109' exhibited the highest TA value, while the commercial variety 'Yolo Wonder' exhibited the lowest. Neither drought stress nor the interaction drought × variety showed any statistically significant effect on TA.

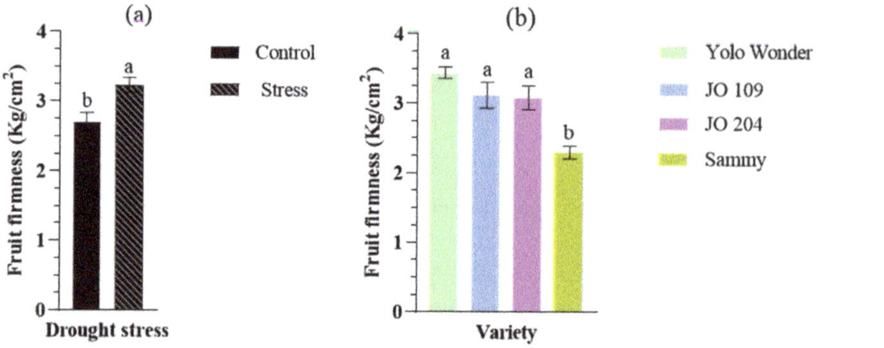

Figure 2. Effect of drought stress (**a**) and genotype (**b**) on fruit firmness in different cultivated pepper genotypes. (**a**) 'Control' refers to well-irrigated conditions, while 'Stress' indicates exposure to limited water availability. Genotype names are provided in the legend (**b**). Different letters above bars indicate statistically significant differences according to Duncan's multiple range test ($p < 0.05$). Error bars represent the standard error of the mean (n = 10).

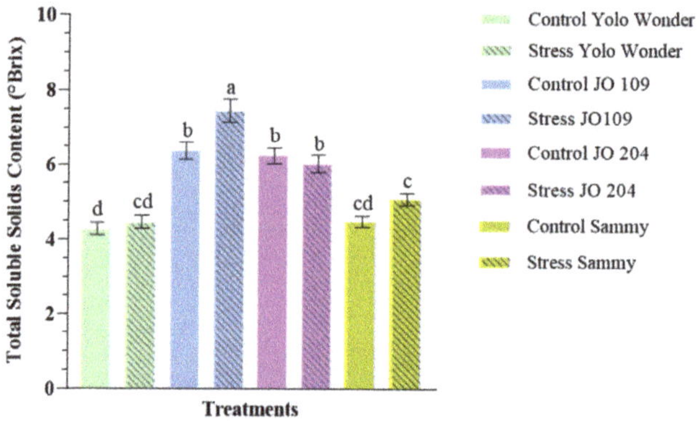

Figure 3. Impact of drought stress and genotype on the total soluble solid content in pepper fruits. Different letters above the bars indicate statistically significant differences among treatments based on Duncan's multiple range test ($p < 0.05$). Error bars represent the standard error of the mean (n = 10).

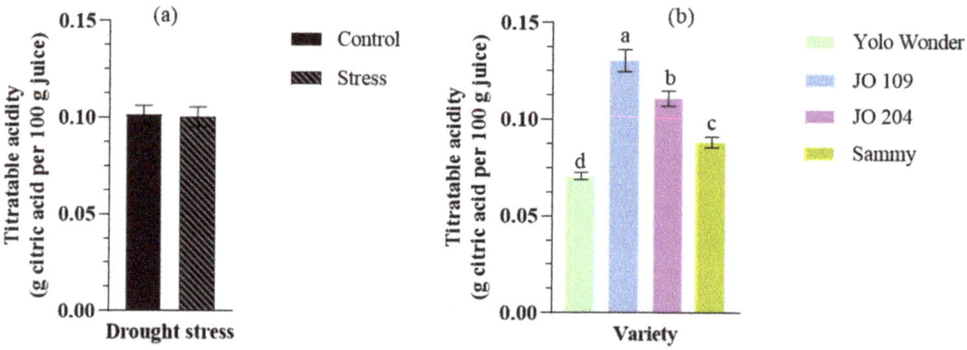

Figure 4. Effect of drought stress (**a**) and genotype (**b**) on fruit titratable acidity in different pepper cultivars. (**a**) 'Control' refers to well-watered conditions, while 'Stress' indicates limited water availability. (**b**) Genotype names are provided in the legend. Different letters above the lines indicate statistically significant differences according to Duncan's multiple range test ($p < 0.05$). Error bars represent the standard error of the mean (n = 10).

Table 4. Impact of drought stress and cultivar on the total soluble solids content (TSSC) in the fruits of the evaluated pepper genotypes.

Water Stress	Variety	TSSC (°Brix)
	Main effects (Drought stress)	
Control		5.41
Drought stress		5.74
	Main effects (Variety)	
	Yolo Wonder	4.37 d
	JO 109	6.81 a
	JO 204	6.12 b
	Sammy	4.81 c
	Statistical significance	
Drought Stress		**
Variety		***
Drought Stress × Variety		*

Mean values (n = 10) followed by different letters within the same column represent statistically significant differences according to Duncan's multiple range test ($p < 0.05$). Asterisks indicate significance levels: *** ($p < 0.001$), ** ($p < 0.01$) and * ($p < 0.05$).

3. Discussion

Drought stress is a well-studied abiotic factor that has been demonstrated to substantially limit crop growth and yield [25]. According to Earl et al. [26], the reduction in yield under water stress conditions is primarily associated to decreased light interception, reduced photosynthetic efficiency, lower water use efficiency (WUE), and a decline in the harvest index. Pepper plants exhibit increased sensitivity to water shortage, primarily attributed to their pronounced leaf surface area and elevated stomatal conductance, which amplify water loss during periods of drought [15]. In the present study, a 40% reduction in irrigation compared to well-watered plants led to a noticeable decrease in yield, with some varieties exhibiting reductions of up to 20–25%. Comparable findings were reported by Dorji et al. [27], who observed a 35% reduction in hot pepper yield when irrigation was decreased to 50% of normal levels. Furthermore, as demonstrated in numerous studies on drought stress in different crops [6,28,29], yield losses are frequently attributed to the suppression of plant growth resulting from decreased photosynthetic activity, reduced chlorophyll content in leaves, and limited nutrient and water uptake by the roots due to inadequate irrigation.

In the present study, the yield reduction under drought conditions was attributed to a decrease in fruit number in the bell pepper cultivar 'Yolo Wonder', whereas in cultivars 'JO 204' and 'Sammy', it was primarily due to a reduction in the mean fruit weight. Similar findings were documented by Ficiciyan et al. [17], who noted a 10% decline in the yield of sweet pepper fruits subjected to drought stress conditions. In a study conducted by Kurunc et al. [30], it was observed that the mean fresh fruit weight of long peppers was reduced under water deficit, as was also the case in the varieties examined in the present study. However, the 'JO109' genotype exhibited resilience to drought stress, as evidenced by its stable yield performance under water-limited conditions—a trait presumably associated with its landrace origin. A similar observation was reported by Fullana-Pericàs et al. [31] in their study on tomato cultivars, where several local varieties demonstrated comparable or even superior performance to modern ones in mitigating yield reduction under drought stress. Concurrently, Techawongstien et al. [32] investigated the long-term effects of short-term water stress on key physiological traits at the pre-anthesis stage in four chilli pepper cultivars. They reported that genotypes exhibiting higher leaf water potential demonstrated greater drought tolerance, as evidenced by their ability to swiftly regain photosynthetic capacity upon rehydration, thereby minimizing the adverse effect on final fruit yield.

Drought stress significantly impairs the uptake of nutrients by roots and their transport to the shoots, primarily by restricting potassium (K) diffusion in the soil toward the root zone, which in turn limits K absorption [33]. As for the concentration of macronutrients (K, Na, Ca, Mg) in the leaves, it was observed that drought conditions resulted in a reduction in potassium levels in pepper plants. This finding is consistent with previous studies, such as Kirnak et al. [34], which reported reduced K concentration in leaves under water deficit conditions. In a similar study, Sadak et al. [35], observed a decline in K concentration of approximately 18% in the leaves of drought stressed pepper seedlings in comparison to the control plants. Among the different varieties tested, the 'JO109' landrace exhibited the highest K concentration in its leaves under drought stress conditions, despite the decline in leaf K levels across all varieties. The absence of yield reduction in the 'JO109' may be explained by its higher K levels, as K is known to alleviate various stresses experienced by plants, including drought, freezing, and high light intensity [36]. Conversely, K deficiency has been shown to disrupt CO_2 assimilation, transport, and utilization processes during photosynthesis. [36]. The consistent yield of the landrace 'JO109' could be attributed to a high K/Na ratio in the leaves. As Hu and Schmidhalter [37] has demonstrated, the high K/Na ratio is also subject to regulation by the roots, indicating a selective uptake of K over Na and a preferential transport of K into the xylem. Concerning the remaining macronutrients, Na, Ca, and Mg, an increase in their concentrations was observed in pepper leaves under reduced irrigation conditions. Consistent findings were reported by Delfine et al. [38], who observed elevated levels of Ca and Na in drought stressed pepper plants in comparison to those under regular irrigation. Moreover, higher Na concentrations were observed in the leaves of landraces 'JO109' and 'JO204', whereas the commercial varieties, such as 'Yolo Wonder' and 'Sammy', exhibited the highest Ca concentrations. Similar findings were reported for these cultivars in the study by Ntanasi et al. [39].

In fruits, significant variations in the concentrations of K, Na, Ca and Mg macronutrients were observed among the different pepper varieties. However, reduced irrigation significantly affected only the Ca concentration, which exhibited a substantial decrease in pepper fruits. Similar findings were also reported by Martins et al. [40], who observed a 13.2% reduction in fruit Ca concentration under drought conditions in green peppers. No significant differences were observed for the other macronutrients (K, Na), a finding consistent with the study by De Pascale et al. [41]. Furthermore, the studies conducted by Birgin et al. [42] and De Pascale et al. [41] revealed no significant differences in Mg fruit concentration between drought-stressed and control plants. However, a notable observation from the present study is that, despite varietal differences in all measured macronutrients, the only variety that maintained stable yield under reduced irrigation conditions (JO109) was also the one that maintained a stable fruit K/Na under drought stress. In addition, this landrace possessed the lowest Na concentration in its fruits, a condition not observed in the leaves. Conversely, the remaining varieties demonstrated comparable Na levels in their fruit.

A number of studies have indicated that water deficit conditions may positively influence fruit quality by activating primary and secondary metabolic pathways [43]. Regarding fruit quality traits, reduced irrigation levels in pepper plants led to a substantial increase of fruit firmness compared to those cultivated under standard irrigation. This improvement may be attributed to notable differences in Ca concentration between the two treatments, reinforcing the established correlation between Ca levels and fruit firmness [44]. However, unlike the findings of Abdelkhalik et al. [45], who investigated the effects of varying deficit irrigation levels on sweet pepper, thid study did not observe any statistically significant differences in fruit firmness. Furthermore, a reduction in irrigation had a significant impact on total soluble solids content (TSSC), which increased under drought stress. Notably, varietal disparities in TSSC were recorded, with the 'JO109' landrace exhibiting the highest value. This landrace exhibited a substantial increase in TSSC under reduced irrigation compared to well-watered conditions, and was the only one that did not experience a decline in yield. These findings are consistent with those reported in previous research. For instance, Abdelkhalik et al. [45] documented a 26% increase in °Brix—which is closely related to TSSC- in sweet pepper when irrigation was reduced by 50% of the crop's water requirements. Similarly, Maduranga Bandara Rathnayaka et al. [46] observed elevated °Brix levels in chili peppers under drought stress, with irrigation reduced by approximately 60%. Demir et al. [47] also reported significant increases in °Brix of pepper fruit, with values rising by approximately 10% under 75% irrigation and 16% under 50% of water requirements. The observed increase in TSSC under limited irrigation may be attributed to a reduced water accumulation in the fruit [48]. One possible explanation for this phenomenon is that when fruit development is hindered, the sugars produced through photosynthesis are distributed among fewer sinks, thereby increasing their concentration in individual fruits [49]. Additionally, Renquist et al. [50] proposed that the elevated soluble solids content observed under water-deficit conditions is a result of reduced water uptake, which leads to a concentration effect within the tomato fruit. However, no significant effect on TA was observed in pepper fruits, which is consistent with the findings of Abdelkhalik et al. [45], who reported similar results in sweet pepper under 50% water shortage of the plant's requirements. In contrast, Demir et al. [47] reported a 5% increase in TA under a 50% reduction in irrigation compared to full irrigation, while no notable difference was observed when irrigation was reduced to 75% of normal levels.

4. Materials and Methods

4.1. Experimental Design

The experiment was carried out in the glasshouse facilities of the Laboratory of Vegetable Production at the Agricultural University of Athens (AUA), situated at 37°59′2″ N and 23°42′19″ E. The study comprised two Jordanian landraces, named 'JO 109' and 'JO 204', along with the 'Yolo Wonder' commercial variety and the 'Sammy RZ' hybrid, which were used as reference. Table 5 provides a comprehensive overview of the names and origins of the pepper seeds utilized in the experiment. The cultivation took place in an open soilless system with two irrigation levels, and the experimental design was based on a randomized complete block design (RCBD).

Table 5. Details on the origins and names of the pepper seeds cultivated.

Variety	Type	Provider
Yolo Wonder	Reference	INRA [1]
JO109 (*Capsicum anuum* var. *grossum*)	Landrace	NARC [2]
JO204 (*Capsicum anuum* var. *grossum*)	Landrace	NARC [2]
Sammy RZ (F1-Hybrid)	Reference	Rijk Zwaan [3]

[1] Institut National de la Recherche Agronomique [2] The National Agricultural Research Center [3] Vegetable breeding company.

4.2. Growing the Seedlings

The seedlings were produced by the Laboratory of Vegetable Production at the Agricultural University of Athens. To ensure the viability of the initial material, all seeds underwent an initial disinfection process on 20 September. This process involved soaking the seeds in a 15% v/w solution of Na_3PO_4 for 20 min. Following disinfection, they were placed in Petri dishes and incubated in a temperature-controlled chamber at 25 °C for 5 days. On 27 September, the germinated seeds were transferred to sowing trays, using turf as a substrate, enabling seedlings growth.

4.3. Cultivation Practice

On November 13, the seedlings that had reached the stage of four or five true leaves were transplanted into 33-L perlite bags, thus commencing cultivation in an open hydroponic system in perlite bags that had been previously soaked to saturation with a starter solution (Table 2) for 24 h. At this stage, the application of a reduced nutrient solution flow rate was initiated. Specifically, the first treatment was applied to the control, i.e., plants receiving the recommended amount of irrigation, and the second treatment was applied to plants that were irrigated with reduced amounts, with irrigation requirements reduced by up to 40% compared to well-irrigated plants. This percentage was chosen because it represents the moderate level of stress that plants [51,52] commonly encounter in Mediterranean environments. This level of stress enables the plants to grow and produce fruit, which allows us to evaluate the effects of water stress. After dipping the perlite bags in the starter solution for 24 h, the bottom of each bag was perforated to allow proper drainage of excess nutrient solution. Each treatment was repeated four times, with four bags of perlite utilized for each treatment. Three plants of the same variety were placed in each bag of perlite. Throughout the growing season, the average temperatures maintained were 21 °C during the day and 17 °C during the night.

4.4. Nutrient Solution Formula

The formulation of the nutrient solution required for the cultivation of pepper crop was conducted using a specially designed computer tool, known as NUTRISENSE, accessible at https://nutrisense.online/ (accessed on 23 June 2025). This program utilizes advanced algorithms to accurately determine the required quantities of macro- and micronutrients, taking into consideration the specific growth stage of the plant, the current season, and the type of cultivation system employed. Prior to the transplantation of the pepper crop, a starter solution was applied to the growing substrate. During cultivation, two distinct nutrient solutions were prepared, customized to align with the vegetative and reproductive phases of plant development, respectively. The nutrient solutions dispensed to the plants were initially prepared by formulating concentrated stock solutions, which were then diluted to a ratio of 1:100. In the experiment, all plants were provided with the same standard nutrient solution, while the amount of irrigation varied. Half of the plants received the recommended amount of water, thus serving as the control group. The remaining half were subjected to deficit irrigation, receiving up to 40% less NS than the well-irrigated control plants. The pH of the nutrient solution was meticulously adjusted daily to 5.6, employing an appropriate volume of 1 N HNO_3 solution. The specific concentrations of macro- and micronutrients for each treatment and growth stage are presented in Table 6.

Table 6. The nutrient concentrations in the starter solution and within the nutrient solution supplied to the pepper plants throughout the vegetative and reproductive growth stages.

Nutrient	Starter Solution (13 November 2021)	Vegetative Phase (14 November 2021)	Reproductive Phase (7 February 2022)	Unit
NO_3^-	16.05	15.79	16.64	mM
K^+	5.70	5.86	6.94	mM
Ca^{2+}	6.15	5.60	5.55	mM
Mg^{2+}	2.50	1.63	1.66	mM
SO_4^{2-}	3.27	2.10	2.08	mM

Table 6. *Cont.*

Nutrient	Starter Solution (13 November 2021)	Vegetative Phase (14 November 2021)	Reproductive Phase (7 February 2022)	Unit
$H_2PO_4^-$	1.25	1.35	1.35	mM
NH_4^+	1.05	1.22	1.00	mM
Fe	20.00	20.00	16.20	μM
Mn^{2+}	12.00	12.00	10.80	μM
Zn^{2+}	7.00	6.00	5.40	μM
B	50.00	32.00	32.40	μM
Cu^{2+}	0.80	0.80	0.86	μM
Mo	0.60	0.60	0.54	μM
Cl^-	0.40	0.40	0.40	μM

4.5. *Sampling of Leaves and Fruits for Macro- and Micronutrient Evaluation*

At the final stage of the experiment, samples were collected from pepper plants from all treatments. The third, fourth and fifth leaves were collected from the uppermost portion of each plant. Furthermore, fruit samples were collected for analysis. Four replicates per treatment were used for nutrient analyses. Subsequently, both leaf and fruit samples were dried at 65 °C until a constant weight was achieved, ensuring consistency.

4.6. *Assessment of Crop Yield*

The first fruit harvest was conducted on 11 March, while the final harvest took place on 7 April. Harvesting was carried out approximately once or twice weekly, and the fruits harvested were selected based on the criterion of achieving commercial maturity. The process involved manual collection or the use of scissors, ensuring that the fruits were removed from the plant along with their calyx and a portion of the pedicel. Following each harvest, a meticulous enumeration and weighing of the fruits was conducted. The data collection process encompassed various parameters, including the total number of fruits per plant, the total fresh weight of fruit per plant (g plant^{-1}), and the average fresh weight of individual fruit (g).

4.7. *Nutrient Content in Leaves and Fruit*

After the drying process, all samples were ground using an MF 10 Basic Micro Fine Grinder (IKA Werke, Staufen, Germany). The extraction process was performed through the dry ashing method, and the potassium (K) and sodium (Na) levels were assessed using a flame photometer (Sherwood Model 410, Cambridge, UK). Concentrations of calcium (Ca) and magnesium (Mg) were measured using an atomic absorption spectrophotometer (AA-7000, Shimadzu Co., Tokyo, Japan).

4.8. *Organoleptic Characteristics*

Quality attributes, including titratable acidity (TA), total soluble solids content (TSSC) and fruit firmness (FF), were evaluated using ten ripe fruits per treatment. Fruit acidity (FA) was analyzed by potentiometric titration, where 10 mL of juice was titrated with 0.02 M NaOH until reaching a pH of 8.1. Firmness was assessed with a Mechanical Force Gauge (Chatillon penetrometer—DPP5KG).

4.9. *Statistical Analysis*

In this study, one-way analysis of variance (ANOVA) was conducted to assess the primary effects of water stress on yield. Furthermore, a two-way ANOVA was conducted to assess the organoleptic characteristics and nutrient composition in plant tissues across different pepper cultivars. The statistical analysis was conducted using the STATISTICA software package (version 12.0 for Windows). In instances where drought stress exerted a significant influence on a measured parameter in the cultivars, mean comparisons were conducted employing the Duncan's Multiple Range Test at a significance level of $p \leq 0.05$.

5. Conclusions

The present study confirms the significant impact of drought stress on pepper (*Capsicum annuum* L.) yield, nutrient uptake, and fruit quality, highlighting the varying degrees of sensitivity among different cultivars. A 40% reduction in irrigation resulted in substantial yield losses across most cultivars, primarily attributable to a decline in fruit number or size. However, the landrace 'JO109' exhibited notable resilience, maintaining stable yields under water limited conditions. This tolerance appears to be linked to the higher leaf K concentration and lower Na concentration in fruit, reinforcing the crucial role of K in mitigating drought induced physiological stress. In relation to fruit quality, water deficit conditions were found to enhance TSSC and FF, particularly in 'JO109', without exerting a negative effect on TA. These enhancements in quality are presumably associated with physiological adjustments that concentrate solutes due to reduced water content and altered metabolic activity under stress. The findings of this study highlight the potential of landraces, such as 'JO109', for incorporation into breeding programs that are focused on enhancing drought resilience and fruit quality in pepper.

Author Contributions: Conceptualization, G.N.; methodology, T.N., I.K., D.S., V.P., L.S. and G.N.; software, T.N., I.K., B.B.C. and L.S.; validation, T.N., I.K., D.S., G.P.S., E.G. and G.N.; formal analysis, T.N., I.K., G.P.S. and E.G.; investigation, T.N., I.K., V.P., L.S. and G.N.; resources, G.N. and D.S.; data curation, T.N., I.K., L.S. and G.N.; writing—original draft preparation, T.N., I.K. and G.N.; writing—review and editing, T.N., I.K., D.S., G.P.S., E.G., B.B.C., V.P., L.S. and G.N.; visualization, T.N., I.K., G.P.S., E.G., B.B.C., V.P. and L.S.; supervision, G.N.; project administration, G.N.; funding acquisition, G.N. and D.S. All authors have read and agreed to the published version of the manuscript.

Funding: This research is supported by PRIMA 2018-11 within the project 'VEGADAPT: AdaptingMediterranean vegetable crops to climate change-induced multiple stress' Available online: https://www.veg-adapt.unito.it/ (accessed on 23 June 2025), a Research and Innovation Action funded by the GreekGeneral Secretariat for Research and Innovation (GSRI) and supported by the European Union.

Data Availability Statement: The data are included in the manuscript.

Acknowledgments: We would like to thank our project partners Maxime Guillaume from Gautier Semences and Naem Mazahrirh from National Agricultural Research Center, for providing us with the seeds ('Yolo Wonder', JO109', 'JO204').

Conflicts of Interest: The authors declare no conflicts of interest.

References

1. Redden, R. New approaches for crop genetic adaptation to the abiotic stresses predicted with climate change. *Agronomy* **2013**, *3*, 419–432. [CrossRef]
2. FAO. *The State of Food and Agriculture 2022, Leveraging Automation in Agriculture for Transforming Agrifood Systems*; FAO: Rome, Italy, 2022; ISBN 9789251360439.
3. Hoerling, M.; Eischeid, J.; Perlwitz, J.; Quan, X.; Zhang, T.; Pegion, P. On the increased frequency of mediterranean drought. *J. Clim.* **2012**, *25*, 2146–2161. [CrossRef]
4. Wach, D.; Skowron, P. An overview of plant responses to the drought stress at morphological, physiological and biochemical levels. *Pol. J. Agron.* **2022**, *50*, 25–34. [CrossRef]
5. Anjum, S.A.; Ashraf, U.; Tanveer, M.; Khan, I.; Hussain, S.; Shahzad, B.; Zohaib, A.; Abbas, F.; Saleem, M.F.; Ali, I.; et al. Drought induced changes in growth, osmolyte accumulation and antioxidant metabolism of three maize hybrids. *Front. Plant Sci.* **2017**, *8*, 69. [CrossRef]

6. Hussain, S.; Hussain, S.; Qadir, T.; Khaliq, A.; Ashraf, U.; Parveen, A.; Saqib, M.; Rafiq, M. Drought stress in plants: An overview on implications, tolerance mechanisms and agronomic mitigation strategies. *Plant Sci. Today* **2019**, *6*, 389–402. [CrossRef]
7. Li, J.; Abbas, K.; Wang, W.; Gong, B.; Wang, L.; Hou, S.; Xia, H.; Wu, X.; Chen, L.; Gao, H. Drought Tolerance Evaluation and Verification of Fifty Pakchoi (*Brassica rapa* ssp. chinensis) Varieties under Water Deficit Condition. *Agronomy* **2023**, *13*, 2087. [CrossRef]
8. Dhiman, S.; Singh, D.B.; Singh, S.P.; Sohi, A. Abiotic Stress in Vegetable Crops: Challenges and Strategies. *Agri Artic.* **2024**, *4*, 582–586.
9. Razi, K.; Muneer, S. Drought stress-induced physiological mechanisms, signaling pathways and molecular response of chloroplasts in common vegetable crops. *Crit. Rev. Biotechnol.* **2021**, *41*, 669–691. [CrossRef]
10. Flexas, J.; Bota, J.; Loreto, F.; Cornic, G.; Sharkey, T.D. Diffusive and metabolic limitations to photosynthesis under drought and salinity in C3 plants. *Plant Biol.* **2004**, *6*, 269–279. [CrossRef]
11. Farooq, M.; Basra, S.M.A.; Wahid, A.; Ahmad, N.; Saleem, B.A. Improving the drought tolerance in rice (*Oryza sativa* L.) by exogenous application of salicylic acid. *J. Agron. Crop Sci.* **2009**, *195*, 237–246. [CrossRef]
12. Auerswald, H.; Schwarz, D.; Kornelson, C.; Krumbein, A.; Brückner, B. Sensory analysis, sugar and acid content of tomato at different EC values of the nutrient solution. *Sci. Hortic.* **1999**, *82*, 227–242. [CrossRef]
13. Penella, C.; Nebauer, S.G.; Bautista, A.S.; López-Galarza, S.; Calatayud, Á. Rootstock alleviates PEG-induced water stress in grafted pepper seedlings: Physiological responses. *J. Plant Physiol.* **2014**, *171*, 842–851. [CrossRef] [PubMed]
14. Azlan, A.; Sultana, S.; Huei, C.S.; Razman, M.R. Antioxidant, Anti-Obesity, Nutritional and Other Beneficial Effects of Different Chili Pepper: A Review. *Molecules* **2022**, *27*, 898. [CrossRef] [PubMed]
15. Alvino, A.; Centritto, M.; De Lorenzi, F. Photosynthesis response of sunlit and shade pepper (*Capsicum annuum*) leaves at different positions in the canopy under two water regimes. *Aust. J. Plant Physiol.* **1994**, *21*, 377–391. [CrossRef]
16. Campos, H.; Trejo, C.; Peña-Valdivia, C.B.; García-Nava, R.; Conde-Martínez, F.V.; Cruz-Ortega, M.R. Stomatal and non-stomatal limitations of bell pepper (*Capsicum annuum* L.) plants under water stress and re-watering: Delayed restoration of photosynthesis during recovery. *Environ. Exp. Bot.* **2014**, *98*, 56–64. [CrossRef]
17. Ficiciyan, A.M.; Loos, J.; Tscharntke, T. Similar Yield Benefits of Hybrid, Conventional, and Organic Tomato and Sweet Pepper Varieties Under Well-Watered and Drought-Stressed Conditions. *Front. Sustain. Food Syst.* **2021**, *5*, 628537. [CrossRef]
18. Gisbert-Mullor, R.; Martín-García, R.; Bažon Zidarić, I.; Pascual-Seva, N.; Pascual, B.; Padilla, Y.G.; Calatayud, Á.; López-Galarza, S. A Water Stress-Tolerant Pepper Rootstock Improves the Behavior of Pepper Plants under Deficit Irrigation through Root Biomass Distribution and Physiological Adaptation. *Horticulturae* **2023**, *9*, 362. [CrossRef]
19. Conti, V.; Mareri, L.; Faleri, C.; Nepi, M.; Romi, M.; Cai, G.; Cantini, C. Drought stress affects the response of italian local tomato (*Solanum lycopersicum* L.) varieties in a genotype-dependent manner. *Plants* **2019**, *8*, 336. [CrossRef]
20. Kumar, S.; Kumar, R.; Kumar, D.; Gautam, N.; Dogra, R.K.; Mehta, D.K.; Sharma, H.D.; Kansal, S. Parthenocarpic Gynoecious Parental Lines of Cucumber Introduced from Netherlands for Developing High-Yielding, Quality Hybrids. *J. Crop Improv.* **2016**, *30*, 352–369. [CrossRef]
21. Kessler, A. Inducible plant defences and the environmental context. *Funct. Ecol.* **2016**, *30*, 1738–1739. [CrossRef]
22. Quartey, E.K.; Nunekpeku, W.; Owusu-Ansah, M.; Appiah, A.S.; Ofori, E.S.K.; Amoatey, H.M. Agronomic Evaluation of Eight Genotypes of Hot Pepper (*Capsicum* spp. L.) in a Coastal Savanna Zone of Ghana. *J. Biol. Agric. Healthc.* **2014**, *4*, 16–28.
23. Ilić, Z.S.; Kevrešan, Ž.; Mastilović, J.; Zorić, L.; Tomšik, A.; Belović, M.; Pestorić, M.; Karanović, D.; Luković, J. Evaluation of Mineral Profile, Texture, Sensory and Structural Characteristics of Old Pepper Landraces. *J. Food Process. Preserv.* **2017**, *41*, 1–11. [CrossRef]
24. Kawecki, T.J.; Ebert, D. Conceptual issues in local adaptation. *Ecol. Lett.* **2004**, *7*, 1225–1241. [CrossRef]
25. Prasad, P.V.V.; Staggenborg, S.A. Impacts of Drought and/or Heat Stress on Physiological, Developmental, Growth, and Yield Processes of Crop Plants. In *Response of Crops to Limited Water: Understanding and Modeling Water Stress Effects on Plant Growth Processes*; Advances in Agricultural System Modeling; American Society of Agronomy: Madison, WI, USA, 2008; pp. 301–355.
26. Earl, H.J.; Davis, R.F. Effect of drought stress on leaf and whole canopy radiation use efficiency and yield of maize. *Agron. J.* **2003**, *95*, 688–696. [CrossRef]
27. Dorji, K.; Behboudian, M.H.; Zegbe-Domínguez, J.A. Water relations, growth, yield, and fruit quality of hot pepper under deficit irrigation and partial rootzone drying. *Sci. Hortic.* **2005**, *104*, 137–149. [CrossRef]
28. Moustafa, K.A.; Saleh, M.; Al-Doss, A.A.; Elshafei, A.A.; Salem, A.K.; Al-Qurainy, F.H.; Barakat, M.N. Identification of TRAP and SRAP markers linked with yield components under drought stress in wheat (*Triticum aestivum* L.). *Plant Omics* **2014**, *7*, 253–259.
29. Namaki, A.; Ghahremani, Z.; Aelaei, M.; Barzegar, T.; Ranjbar, M.E. The First Report of Drought Tolerance Assessment of Iranian Asparagus. *Gesunde Pflanz.* **2022**, *74*, 141–149. [CrossRef]
30. Kurunc, A.; Unlukara, A.; Cemek, B. Salinity and drought affect yield response of bell pepper similarly. *Acta Agric. Scand. Sect. B Soil Plant Sci.* **2011**, *61*, 514–522. [CrossRef]

31. Fullana-Pericàs, M.; Conesa, M.; Douthe, C.; El Aou-ouad, H.; Ribas-Carbó, M.; Galmés, J. Tomato landraces as a source to minimize yield losses and improve fruit quality under water deficit conditions. *Agric. Water Manag.* **2019**, *223*, 105722. [CrossRef]
32. Techawongstien, S.; Nawata, E.; Shigenaga, S. After-effect of short-term water stress at the pre-anthesis stage on physiological characteristics in four chilli pepper cultivars. *Jpn. J. Trop. Agric.* **1992**, *36*, 88–93.
33. Coskun, D.; Britto, D.T.; Kronzucker, H.J. The physiology of channel-mediated K^+ acquisition in roots of higher plants. *Physiol. Plant.* **2014**, *151*, 305–312. [CrossRef] [PubMed]
34. Kirnak, H.; Kaya, C.; Higgs, D.; Tas, I. Responses of drip irrigated bell pepper to water stress and different nitrogen levels with or without mulch cover. *J. Plant Nutr.* **2011**, *26*, 263–277. [CrossRef]
35. Sadak, A.; Akkopru, A.; Sensoy, S. Effects of Endophytic Bacteria on Some Physiological Traits and Nutrient Contents in Pepper Seedlings under Drought Stress. *Yuz. Yil Univ. J. Agric. Sci.* **2021**, *31*, 237–245. [CrossRef]
36. Waraich, E.A.; Ahmad, R.; Halim, A.; Aziz, T. Alleviation of temperature stress by nutrient management in crop plants: A review. *J. Soil Sci. Plant Nutr.* **2012**, *12*, 221–244. [CrossRef]
37. Hu, Y.; Schmidhalter, U. Drought and salinity: A comparison of their effects on mineral nutrition of plants. *J. Plant Nutr. Soil Sci.* **2005**, *168*, 541–549. [CrossRef]
38. Delfine, S.; Tognetti, R.; Loreto, F.; Alvino, A. Physiological and growth responses to water stress in field-grown bell pepper (*Capsicum annuum* L.). *J. Hortic. Sci. Biotechnol.* **2002**, *77*, 697–704. [CrossRef]
39. Ntanasi, T.; Savvas, D.; Karavidas, I.; Papadopoulou, E.A.; Mazahrirh, N.; Fotopoulos, V.; Aliferis, K.A.; Sabatino, L.; Ntatsi, G. Assessing Salinity Tolerance and Fruit Quality of Pepper Landraces. *Agronomy* **2024**, *14*, 309. [CrossRef]
40. Martins, B.L.R.; Ferreira, K.N.; Rocha, J.L.A.; Araujo, R.H.C.R.; Lopes, G.; Santos, L.C.d.; Bezerra Neto, F.; Sá, F.V.d.S.; Silva, T.I.d.; da Silva, W.I.; et al. Nano ZnO and Bioinoculants Mitigate Effects of Deficit Irrigation on Nutritional Quality of Green Peppers. *Horticulturae* **2024**, *10*, 969. [CrossRef]
41. De Pascale, S.; Ruggiero, C.; Barbieri, G.; Maggio, A. Physiological responses of pepper to salinity and drought. *J. Am. Soc. Hortic. Sci.* **2003**, *128*, 48–54. [CrossRef]
42. Birgin, Ö.; Akhoundnejad, Y.; Dasgan, H.Y. The effect of foliar calcium application in tomato (*Solanum lycopersicum* L.) under drought stress in greenhouse conditions. *Appl. Ecol. Environ. Res.* **2021**, *19*, 2971–2982. [CrossRef]
43. Ripoll, J.; Urban, L.; Staudt, M.; Lopez-Lauri, F.; Bidel, L.P.R.; Bertin, N. Water shortage and quality of fleshy fruits-making the most of the unavoidable. *J. Exp. Bot.* **2014**, *65*, 4097–4117. [CrossRef] [PubMed]
44. Belakbir, A.; Ruiz, J.M.; Romero, L. Yield and fruit quality of pepper (*Capsicum annuum* L.) in response to bioregulators. *HortScience* **1998**, *33*, 85–87. [CrossRef]
45. Abdelkhalik, A.; Pascual, B.; Nájera, I.; Domene, M.A.; Baixauli, C.; Pascual-Seva, N. Effects of deficit irrigation on the yield and irrigation water use efficiency of drip-irrigated sweet pepper (*Capsicum annuum* L.) under Mediterranean conditions. *Irrig. Sci.* **2020**, *38*, 89–104. [CrossRef]
46. Maduranga Bandara Rathnayaka, R.M.S.; Minami, M.; Nemoto, K.; Prabandaka, S.S.; Matsushima, K. Relationship between water supply and sugar and capsaicinoids contents in fruit of Chili pepper (*Capsicum Annuum* L.). *Hortic. J.* **2021**, *90*, 58–67. [CrossRef]
47. Demir, Z.; Özbahçe, A. Evaluating the effects of different irrigation and nitrogen applications on soil water content and yield quality parameters of pepper using surface and subsurface drip irrigation. *Irrig. Drain.* **2021**, *70*, 1039–1055. [CrossRef]
48. Ehret, D.L.; Ho, L.C. The effects of salinity on dry matter partitioning and fruit growth in tomatoes grown in nutrient film culture. *J. Hortic. Sci.* **1986**, *61*, 361–367. [CrossRef]
49. Bakr, J.; Daood, H.G.; Pék, Z.; Helyes, L.; Posta, K. Yield and quality of mycorrhized processing tomato under water scarcity. *Appl. Ecol. Environ. Res.* **2017**, *15*, 401–413. [CrossRef]
50. Renquist, A.R.; Reid, J.B. Processing tomato fruit quality: Influence of soil water deficits at flowering and ripening. *Aust. J. Agric. Res.* **2001**, *52*, 793–799. [CrossRef]
51. Mohammed, S.; Hussen, A. Influence of deficit irrigation levels on agronomic performance of pepper (*Capsicum annuum* L.) under drip at alage, central rift valley of Ethiopia. *PLoS ONE* **2023**, *18*, e0280639. [CrossRef]
52. Mardani, S.; Tabatabaei, S.H.; Pessarakli, M.; Zareabyaneh, H. Physiological responses of pepper plant (*Capsicum annuum* L.) to drought stress. *J. Plant Nutr.* **2017**, *40*, 1453–1464. [CrossRef]

plants

MDPI

Article

Root-Zone Heating Boosts the Production of Mini Romaine Lettuce Grown in Nutrient Film Technique and Aeroponics Systems

Filippos Bantis [1,*], Nikolaos Tostsidis [2], George Zervoudakis [2], Athanasios Koukounaras [3] and Athanasios Koulopoulos [2]

1 Department of Agriculture, University of Western Macedonia, 53100 Florina, Greece
2 Department of Agriculture, University of Patras, 30200 Messolonghi, Greece; up1082677@ac.upatras.gr (N.T.); gzerv@upatras.gr (G.Z.); tkoulop@upatras.gr (A.K.)
3 School of Agriculture, Aristotle University of Thessaloniki, 54124 Thessaloniki, Greece; thankou@agro.auth.gr
* Correspondence: fbantis@uowm.gr

Abstract

Root-zone temperature is a critical environmental parameter affecting the development, physiology, and nutritional status of leafy vegetables in soilless systems such as the nutrient film technique (NFT) and aeroponics. In the present article, we report on responses of mini Romaine lettuce (*Lactuca sativa* L.) upon cultivation using heated nutrient solution targeting minimum temperatures of 14, 18, and 22 °C versus ambient (control; 11–12 °C), both in the NFT and in the aeroponics system. In both systems, the higher temperatures (i.e., 18 and 22 °C) led to considerably higher leaf mass per system area (127–232% in NFT; 54–75% in aeroponics) and leaf length (more than 21% in all cases). Root dry weight and total soluble solids were positively affected by increasing temperatures only in the NFT. Performance indices of the photosynthetic mechanism (PI_{abs} and PI_{tot}) were increased in the lower temperatures in the NFT. Antioxidant activity and total phenolics were not affected in either soilless system. Total chlorophylls and carotenoids were enhanced by 18 and 22 °C in the NFT and aeroponics, respectively, while anthocyanins were also variably affected. Finally, nitrate content was significantly reduced (−42%) in 18 °C in the NFT. Sub-optimal root-zone temperatures constrained root development and biomass accumulation, indicating that growth limitation was mainly driven by sink-related processes rather than carbon assimilation. Overall, heating the nutrient solution to a minimum of 22 °C in low- and mid-tech greenhouses during cool months can increase the production efficiency of mini Romaine lettuce in the NFT and aeroponics.

Keywords: soilless systems; hydroponics; NFT; root-zone temperature; nutrient solution

1. Introduction

Mini leafy vegetables show a continuously increasing demand due to their high nutritional content, excellent taste, tender texture, and convenient nature when it comes to ready-to-eat vegetables [1]. Mini or miniature is an intermediate stage where heads are harvested at a size between the baby leaf and the mature type. This is a common practice in hydroponic systems and soil-based experiments due to its compact size, short cultivation cycle, and suitability for controlled-environment agriculture [2,3]. In particular, lettuce (*Lactuca sativa* L.) is among the most important leafy vegetables cultivated worldwide until mini stage, constituting a specialty type of lettuce. The crop shows elevated economic and nutritional value due to its high demand for immediate consumption as fresh-cut salads. Lettuce holds a special place in the Mediterranean diet since it is a key ingredient in salads and is related to the consumption of fresh, light, and nutritious meals. In many countries, lettuce covers large areas and contributes significantly to farmers' income and food security. Increased emphasis on matters related to food quality, high nutritional content, and sustainable production processes makes lettuce a highly essential food as well as a strategically important vegetable product. Lettuce cultivation shows high adaptability to various production models, from outdoor fields to greenhouses and vertical farms, and from soil to hydroponic systems, thus constituting it as a product with high market availability. The Romaine (or Cos; var. Longifolia) type in particular is characterized by elongated leaves, crispy texture, and overall high nutritional content; thus, it is commercially very popular [4]. On a worldwide scale, lettuce yield surpasses 28 million tons, with China being the leading producer [5]. In Europe, the leading producer countries are Spain, Italy, and France, which export large quantities mainly within Europe [5].

Soilless (hydroponic) cultivation of lettuce nowadays is gaining increasing attention. The cultivation of leafy vegetables in hydroponic systems is considered one of the most efficient strategies of intensive food production employed in confined environments (i.e., greenhouses and vertical farms) under controlled conditions. Due to the increasing demand for fresh, safe, high-quality food, soilless production systems are crucial for the future of agriculture, especially in regions with limited water and land availability and high urbanization [6]. Within hydroponic production, the potential of precise root-environment control offers substantial benefits over traditional soil agriculture, which shows significant nutrient losses and an inadequate ecological footprint. Among the various soilless systems, the nutrient film technique (NFT) and aeroponics allow for precise control of the root zone environment, high water- and nutrient-use efficiency, and an enhanced overall capacity for cultivation under relatively controlled conditions [6]. Briefly, in the NFT, a thin layer of nutrient solution flows through a horizontal canal where the root system directly receives water and nutrients [7]. Aeroponics can be described as a soilless system where roots are exposed in a tube (usually vertically placed) and periodically moisturized with nutrient solution [8]. Both systems vary considerably with respect to the environment where the root system is exposed. The discrepancies between soilless systems emphasizes the need for targeted management strategies depending on the cultural characteristics of each system. Moreover, soilless cultivation is considered a tool critical to achieving the Sustainable Development Goals established by the United Nations, involving the cultivation of traditional or underutilized leafy vegetables [9], as well as the implementation of organically produced compounds such as biostimulants and liquid digestates, among others [10,11]. The success of soilless cultivation largely relies on the stability of critical parameters such as root-zone temperature and air temperature [12].

Nowadays, root-zone temperature control is a hot topic if we take into account the adaptation of agricultural production to the climate crisis. The extreme temperature fluctuations projected in the following decades will require the development of strategies to control and stabilize the root-zone microenvironment in a similar manner as the air environment. The use of novel technologies such as optimized soilless systems, real-time sensors, and precision-heating systems will be essential parts of sustainable and efficient crop production. Root-zone temperature not only alters the root system growth and architecture affecting root cell elongation and division [13,14] but also affects several physiological processes, including nutrient absorption and distribution, enzyme activity, transpiration, photosynthesis [15], reactive oxygen species metabolism, plasma membrane structure, and osmotic balance [16]. A study with lettuce and rocket baby leaves revealed enhanced productivity when root-zone heating was applied during winter production [17].

https://doi.org/10.3390/plants15030422

Moreover, root-zone temperature is often overlooked since most attention is given to aerial parameters (i.e., temperature, relative humidity, light, CO_2 etc.). Several articles focused on the optimum air temperature for lettuce, which was defined in the range of 17–28 °C during the day and 3–12 °C during the night [18]. In a low- or mid-tech greenhouse with an absence of heating systems, cold conditions (i.e., winter months) can be very challenging, decreasing the nutrient solution temperature to suboptimal levels and leading to decelerated or even halted development, biomass decrease, and alterations in plant nutritional content [19]. For example, red leaf lettuce showed reduced leaf area, stem diameter, and shoot and root fresh weight when the root system was exposed to 10 °C [20]. Root-zone temperatures of approximately 18–22 °C are generally considered optimal for lettuce growth in soilless systems, supporting efficient root metabolism, nutrient uptake, and balanced shoot development. The abovementioned constraints may also negatively affect the taste and marketability of produced crops [21]. On a practical level, root-zone temperature control can be an essential tool for optimized production efficiency, especially during the colder months, in low- to mid-tech greenhouses where environmental conditions are characterized by sharp fluctuations. The difference between maximum and minimum, also known as the diurnal temperature range (ΔT), is a critical index for explaining various developmental and physiological observations. Smaller fluctuations can be correlated with better balance between biomass production and other qualitative characteristics.

Chlorophyll fluorescence parameters, including PI_{abs} (performance index on absorption basis) and PI_{tot} (total performance index), reflect the overall functionality of photosystem II (PSII) and photosynthetic performance under different thermal conditions [22]. Specifically, PI_{abs} shows the overall functionality of PSII and the efficiency with which absorbed light energy is converted into photochemical processes, while PI_{tot} extends this assessment to the combined performance of both PSII and PSI [23]. These indices respond rapidly to environmental constraints, including sub-optimal root-zone temperature, and therefore provide an effective means of detecting early impairments in the photosynthetic apparatus. Nitrate concentration was assessed as a key indicator of nitrogen metabolism due to its sensitivity to root-zone temperature. Moreover, nitrate content is an indicator of product quality linked with consumer safety under current EU regulations [17]. In addition, photosynthetic pigments (i.e., chlorophylls and carotenoids) and secondary metabolites (e.g., phenolics, such as anthocyanins) are excellent indicators of photosynthetic capacity and photoprotection, usually associated with stress responses such as low temperature [24].

To that end, our objective was to test the effect of heated root-zone temperature on mini Romaine lettuce growth, physiological attributes, and nutritional content, upon cultivation in an NFT and an aeroponics system. We hypothesized that (i) root-zone temperatures above 18 °C would significantly increase leaf mass per system area (LMSA) and morphological attributes compared to the non-heated control, and (ii) sub-optimal root-zone temperatures below 14 °C would limit growth parameters in both soilless systems. The literature usually deals with comparing normal with extremely high (e.g., 35 °C) or low temperatures (e.g., 10 °C) [19,25]. In our case, temperature regimes starting from a minimum of 14, 18, and 22 °C were compared to a respective non-treated control treatment to identify the minimum required nutrient solution temperature within a normal non-stressful range for optimum plant growth, physiological status, and nutritional quality during cultivation in a mid-tech greenhouse in winter months. According to the above, the NFT treatments are hereby labeled as NFT-C, NFT-14, NFT-18, and NFT-22, while the aeroponics treatments are hereby labeled as Aero-C, Aero-14, Aero-18, and Aero-22. It should be noted that the two soilless systems were not directly compared to each other but were selected to examine the responses of root-zone heating and showcase possible discrepancies within each system.

https://doi.org/10.3390/plants15030422

2. Results

2.1. Growth and Biomass Parameters

Plants grew normally during the experimental period. No physiological or phytopathological disorders were observed (Figure 1). LMSA was significantly affected by root zone temperature. In the NFT, LMSA was greater at NFT-22 compared to NFT-C and NFT-14, while NFT-18 was greater than NFT-C ($p < 0.001$). In aeroponics, LMSA was significantly greater in Aero-18 compared to Aero-14 ($p = 0.041$) (Figure 2A). Leaf length in the NFT was significantly greater in NFT-22 and NFT-18 compared to NFT-C ($p = 0.004$). Quite similarly, in aeroponics, leaf length was greater in Aero-22 and Aero-18 compared to Aero-C and Aero-14 ($p = 0.010$) (Figure 2B). In the NFT, root dry weight was significantly greater in NFT-22 compared to NFT-C ($p = 0.027$). On the contrary, root dry weight was not affected by temperature in aeroponics (Figure 2C).

Figure 1. Representative images at harvest of mini Romaine lettuce cultivated in NFT (nutrient film technique) or Aero (aeroponics) soilless systems and treated with different root-zone temperatures in the nutrient solutions. C: control (mean minimum temperature of 12.2 and 10.9 °C for NFT and Aero, respectively).

2.2. Physiological Parameters

Regarding chlorophyll fluorescence parameters, PI_{abs} and PI_{tot} in the NFT were significantly greater in NFT-C compared to NFT-18 and NFT-22 ($p = 0.045$ and $p = 0.001$, respectively). No significant effects were recorded in aeroponics (Figure 3A,B).

2.3. Nutritional Safety and Quality

In the NFT, greater nitrate content was found in NFT-14 than NFT-18 ($p = 0.047$), while different temperatures did not impose specific effects in aeroponics (Figure 3C). Nitrate concentrations reached 768 mg/kg of fresh weight across all treatments and remained below the maximum limit established for lettuce by the European Commission (5000 mg/kg of fresh weight).

https://doi.org/10.3390/plants15030422

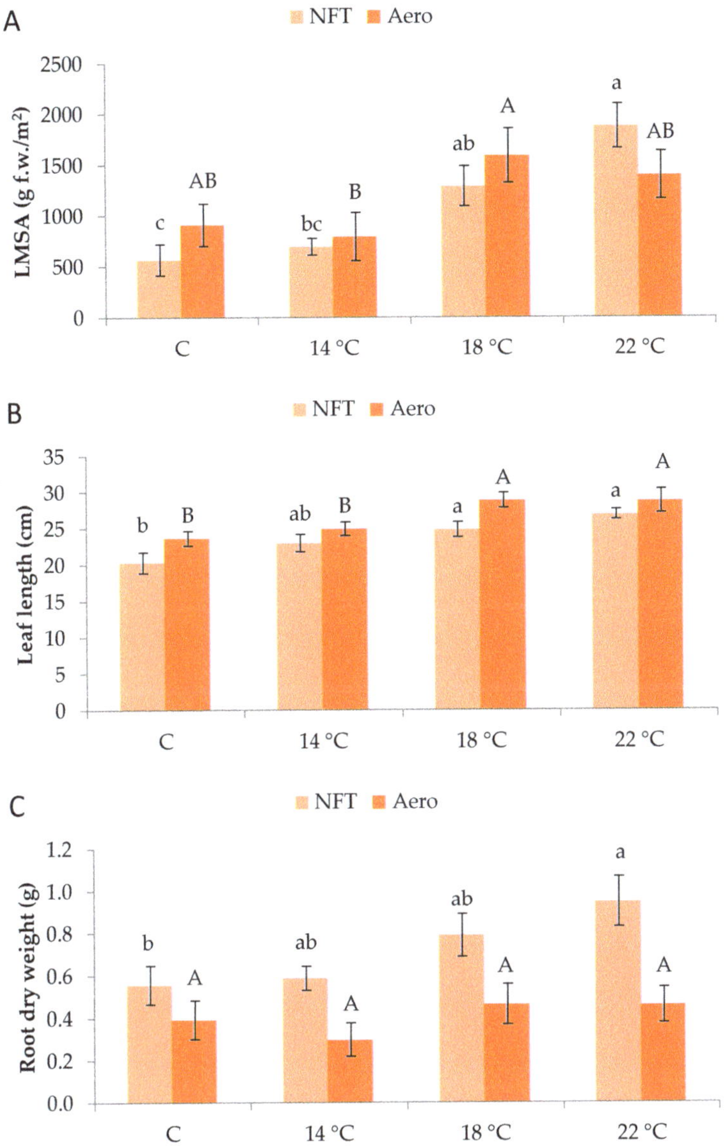

Figure 2. (**A**) Leaf mass per system area (LMSA), (**B**) leaf length, and (**C**) root dry weight of mini Romaine lettuce cultivated in NFT (nutrient film technique) or Aero (aeroponics) soilless systems and treated with different root-zone temperatures in the nutrient solutions. LMSA represents leaf fresh mass per unit growing area, adjusted for the plant density of each soilless system (i.e., 15.2 plants/m^2 in the NFT and 22.2 plants/m^2 in Aero). Within each soilless system, bars (N = 7 for NFT; N = 8 for Aero; ± SE) followed by different lowercase (NFT) or capital (Aero) letters are significantly different ($p \leq 0.05$). C: control (mean minimum temperature of 12.2 and 10.9 °C for NFT and Aero, respectively). No direct statistical comparison between the soilless systems was conducted.

https://doi.org/10.3390/plants15030422

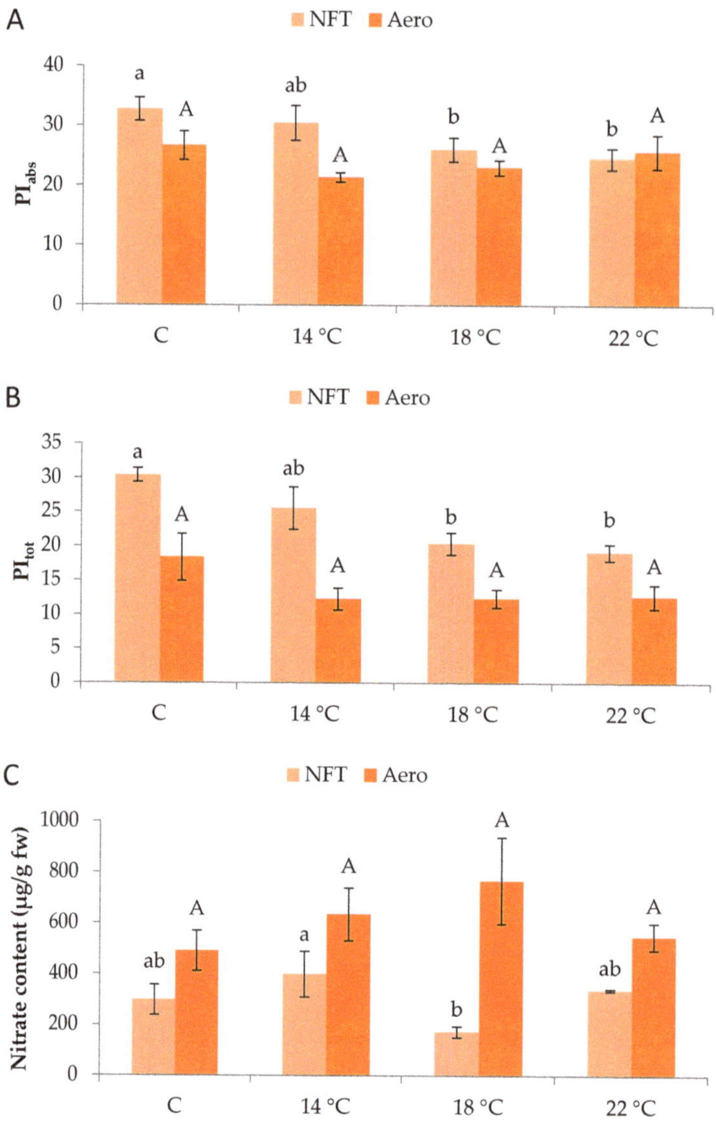

Figure 3. (**A**) PI_{abs}, (**B**) PI_{tot}, and (**C**) nitrate content of mini Romaine lettuce cultivated in NFT (nutrient film technique) or Aero (aeroponics) soilless systems and treated with different root-zone temperatures in the nutrient solutions. PI_{abs}: performance index on absorption basis (PSII performance); PI_{tot}: total performance index integrating PSII and PSI functionality. Within each soilless system, bars (N = 7 for NFT; N = 8 for Aero; ± SE, or N = 3 for nitrate content) followed by different lowercase (NFT) or capital (Aero) letters are significantly different ($p \leq 0.05$). C: control (mean minimum temperature of 12.2 and 10.9 °C for NFT and Aero, respectively). No direct statistical comparison between the soilless systems was conducted.

Regarding pigment quantification, total chlorophylls were greater in NFT-18 and NFT-C compared to NFT-22 ($p = 0.012$). On the contrary, in aeroponics, total chlorophylls were greater in Aero-22 than Aero-18 ($p = 0.042$) (Figure 4A). Similar results were obtained for carotenoid content. In the NFT, greater values were recorded in NFT-18 compared to NFT-22 ($p = 0.013$), while in aeroponics, greater values were found in Aero-22 than

 https://doi.org/10.3390/plants15030422

Aero-18 and Aero-14 ($p = 0.039$) (Figure 4B). In the NFT, relative anthocyanin content was significantly greater in NFT-C, NFT-14, and NFT-18 compared to NFT-22, which showed the lowest values ($p < 0.001$). Regarding aeroponics, relative anthocyanin content was greater in Aero-C compared to Aero-14 ($p = 0.019$) (Figure 4C).

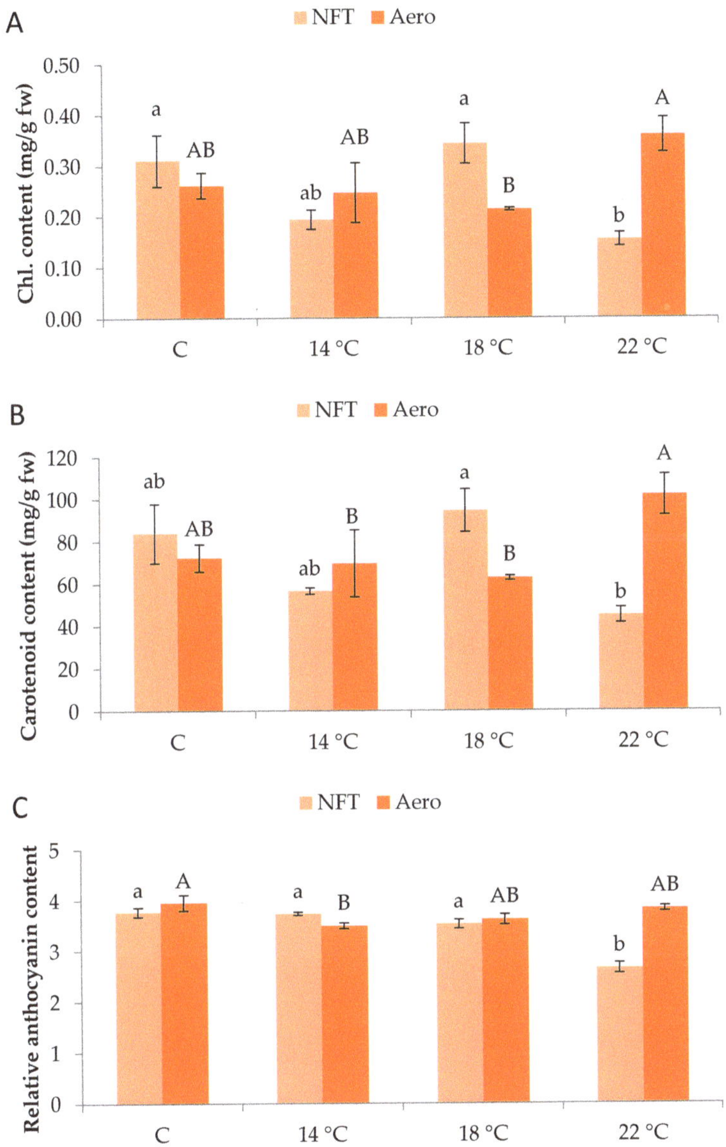

Figure 4. (**A**) Total chlorophyll content, (**B**) total carotenoid content, and (**C**) relative anthocyanin content of mini Romaine lettuce cultivated in NFT (nutrient film technique) or Aero (aeroponics) soilless systems and treated with different root-zone temperatures in the nutrient solutions. Within each soilless system, bars (N = 3 ± SE, or N = 6 for anthocyanins) followed by different lowercase (NFT) or capital (Aero) letters are significantly different ($p \leq 0.05$). C: control (mean minimum temperature of 12.2 and 10.9 °C for NFT and Aero, respectively). Anthocyanin content is expressed as relative index values. No direct statistical comparison between the soilless systems was conducted.

https://doi.org/10.3390/plants15030422

Regarding total soluble solids (Brix°), values were significantly greater in NFT-22 compared to NFT-18 and NFT-14 ($p = 0.046$), while Aeroponics did not show significant differences between temperature regimes (Figure 5A). Moreover, total phenolics content and antioxidant activity (displayed by FRAP) were not significantly affected by temperature, in either the NFT or aeroponics (Figure 5B,C).

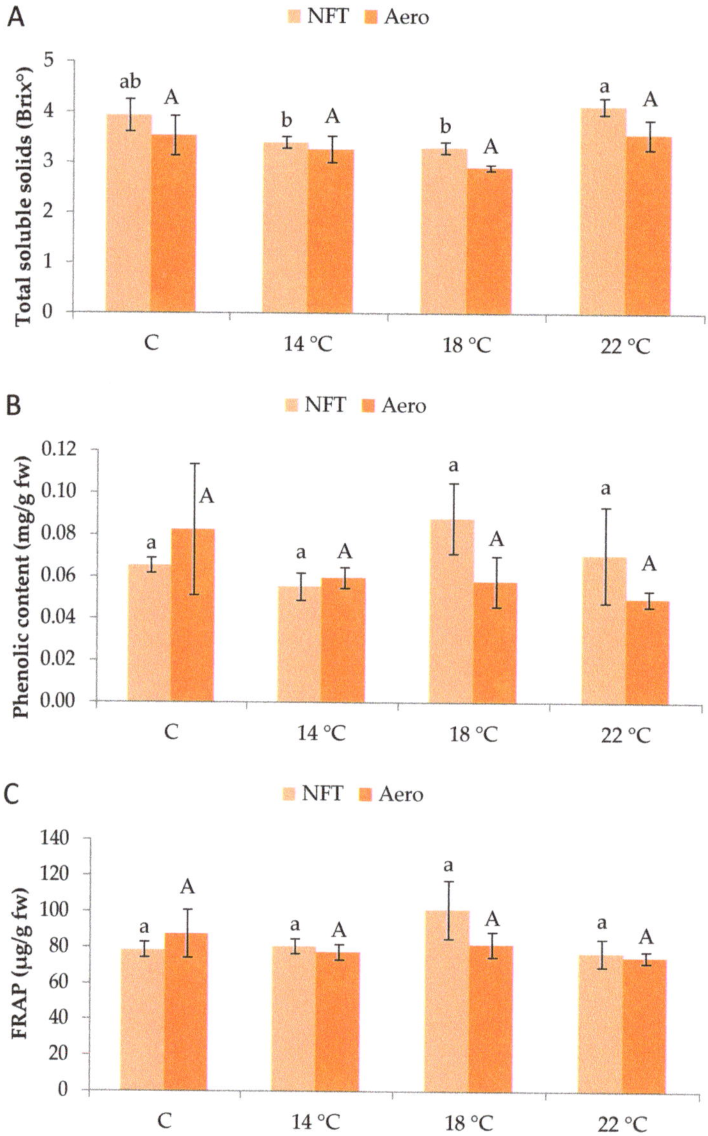

Figure 5. (**A**) Total soluble solids, (**B**) total phenolic content, and (**C**) antioxidant activity (displayed by ferric-reducing antioxidant power-FRAP) of mini Romaine lettuce cultivated in NFT (nutrient film technique) or Aero (aeroponics) soilless systems and treated with different root-zone temperatures in the nutrient solutions. Within each soilless system, bars (N = 3 ± SE) followed by different lowercase (NFT) or capital (Aero) letters are significantly different ($p \leq 0.05$). C: control (mean minimum temperature of 12.2 and 10.9 °C for NFT and Aero, respectively). No direct statistical comparison between the soilless systems was conducted.

https://doi.org/10.3390/plants15030422

3. Discussion

Our results help clarify the sensitivity range of the lettuce root system to thermal regimes under soilless cultivation, particularly during winter production. The decline in growth and photosynthetic indices observed under minimum 14 °C indicates that Romaine lettuce exhibits a narrow tolerance window, with root metabolism and hydraulic conductance significantly constrained below approximately 18 °C [15]. These findings contribute to refining species-specific thermal thresholds relevant to controlled-environment production. Moreover, it should be noted that although treatments were defined by minimum root-zone temperatures, plants were exposed to mean root-zone temperatures approaching values (c.f. Table 1) previously reported as optimal for lettuce growth in hydroponic systems, e.g., Ref. [26].

Table 1. Daily average, maximum, and minimum temperatures monitored inside the nutrient solution reservoir of each treatment, as well as in the surrounding environment (air). NFT: nutrient film technique. Aero: aeroponics. C: control.

Treatment	Average (°C)	Mean Maximum (°C)	Mean Minimum (°C)
NFT-C	16.7	22.2	12.2
NFT-14	17.3	22.1	14.4
NFT-18	20.7	23.2	18.6
NFT-22	23.8	25.4	22.2
Aero-C	16.1	22.6	10.9
Aero-14	17.8	22.8	14.0
Aero-18	21.2	24.8	18.2
Aero-22	24.9	26.5	22.5
Air	15.8	26.2	8.4

In particular, elevated nutrient solution temperature led to a considerable increase in aboveground characteristics related to leaf biomass, such as LMSA and leaf length in both soilless systems, as well as enhanced root system development in the NFT. Along with minimum temperature, the diurnal temperature range also affects the root activity. In our case, treatments with higher ΔT (i.e., control and 14 °C treatments) inhibited plant growth compared to treatments with lower ΔT values. In the present study, ΔT was not a controlled factor; thus, the observed association between increased ΔT and lower plant growth should be interpreted as correlative. Our findings reaffirm previous results showing the positive effect of about 20 °C root-zone temperature for leafy vegetables' growth [20,27]. For example, several lettuce cultivars exhibited greater fresh weight under a root-zone temperature of 21.1 °C than ambient temperatures of 18.3 °C and higher in an NFT system [28]. *Eruca sativa* (rocket) plants grown in an aeroponics system under 20 °C root-zone temperature showed considerably greater productivity compared to higher ambient temperatures in the root zone [27]. In another experiment with a deep water culture hydroponic system, lettuce and arugula showed higher yield when grown under elevated root-zone temperature (about 15 °C daily minimum and 20 °C daily average) powered by a photovoltaic system compared to ambient temperature (2–6 °C lower root-zone temperature) during winter [17]. Low root-zone temperatures are known to decelerate enzyme and transport protein activities, limiting water and nutrient absorption, and ultimately inhibiting plant growth [19]. The production of new root tissue includes cell division, enlargement, and differentiation [29]. Root growth depends on the meristem tissue activity, generating new cells that subsequently expand and elongate. Root cell development is influenced by temperature since both the duration of mitosis and the cell-doubling time are shortened when the temperature is increased from 3 to 25 °C [13]. For decades, there have been reports that growth of the length of roots is optimum around

https://doi.org/10.3390/plants15030422

20 °C, decreasing towards 10 °C and 30 °C. Moreover, it is considered that root cell elongation takes place in two steps. The first step should involve a loosening of the cell wall and a cell stretching, while during the second one, a new cell wall formation is conducted by new microfibrils [30]. In the expansion zone of the roots, cells undergo several fold size enlargements that cannot be achieved with the initial primary wall. Secondary wall formation is accompanied by cell enlargement and differentiation, a resource-demanding process. Research data from plants that have been evolutionarily adapted to very low-temperature environments—for example, alpine species—imply that cell differentiation (including lignification) is a likely cause of root growth cessation at very low temperatures that otherwise still enable photosynthesis [29].

In our aeroponics system, root development was similar under all treatments due to specific conditions in the root environment. It should be mentioned that the actual temperature in the root system may have differed slightly from the reservoir temperature, particularly in the aeroponics system due to its lower thermal inertia. However, the controlled nutrient solution temperature remains the primary determinant of root thermal conditions and allows meaningful comparison among treatments. This approach (temperature control in the nutrient solution reservoir) is common in such experiments, e.g., Refs. [15,26]. In particular, in low-pressure aeroponics systems, the roots are exposed to an environment with high humidity but relatively low heat capacity, possibly meaning that root-zone temperature control is difficult to manage with respect to root growth. This suggests that root-zone temperature effects in aeroponics may not necessarily be expressed as changes in root biomass, but may instead be reflected in aboveground growth and physiological responses. On the other hand, root-zone temperature in the NFT can efficiently be adjusted and impose its effects related to the root system development. Quite similarly, in a study involving 17 lettuce cultivars grown in an NFT system, root dry biomass was more pronounced under a 21.1 °C root-zone temperature compared to 18.3 °C and ambient temperature [28]. In another study, red leaf lettuce had greater root dry weight when treated with 25 °C compared to 15 and 35 °C in an NFT [26]. It seems that elevated root-zone temperature facilitates the distribution of photosynthetic assimilates towards the leaves, providing an explanation for the LMSA and leaf length results. In addition, beneficial root-zone temperature may promote the expression of genes related to cell differentiation in roots, which was observed in the NFT but not in aeroponics. For example, previous studies showed that variations in root-zone temperature can alter gene expression profiles in roots, which may lead to changes in root architecture [31,32].

Noticeably, the parameters related to the performance of the photosynthetic apparatus, PI_{abs} and PI_{tot}, were higher in the lowest temperature regimes of the NFT. These values remained within the normal range for healthy plants, suggesting that the photosynthetic apparatus maintained functionality under sub-optimal root-zone temperatures. Low or high root-zone temperature can impose an adjusted increase in the photosystem II performance, leading to a temporary enhancement of the photochemical efficiency triggered by environmental stresses [25]. On the other hand, extreme stress in terms of intensity and duration may lead to temporary impairment of the photosynthetic apparatus, as displayed for rocket and lettuce, which were treated with 42 °C root-zone temperature compared to 25 °C [25]. The absence of significant differences in aeroponics indicates the lower sensitivity of this system when it comes to variable temperature regimes. The increase in PI_{abs} and PI_{tot} when plants were exposed to lower root-zone temperatures suggests that plants maintain photosynthetic performance under suboptimal conditions, which may be reflected in the measured aboveground parameters. This possibly explains why NFT plants presented normal levels in both tested indices even in treatments with low-temperature regimes. Earlier studies have indicated that tissue formation becomes

https://doi.org/10.3390/plants15030422

very slow below 5 °C and was never observed below 0 °C, a temperature that still permits CO_2 uptake at ca. 30% of photosynthetic capacity. Hence, it seems that at low temperatures plant growth is not carbon limited. Lettuce plants seem to be able to maintain relatively sufficient rates of photosynthetic carbon assimilation even at low temperatures, while the process of tissue formation (i.e., cell division, elongation, and differentiation) is rather more sensitive. As previously mentioned, plant growth in cold climates is not limited by carbon assimilation (source activity) but rather by reduced carbon investment into new tissues (sink limitation) [29].

Nitrate content was variably affected within the two soilless systems, with no significant differences observed in aeroponics. Quite similarly, Karnoutsos et al. [17] reported no effect of two root-zone temperatures (20 °C versus ambient) in lettuce and arugula grown in deep water culture. However, in the NFT of our experiment, a low temperature of 14 °C led to greater nitrate amounts, possibly due to the limited activity of the nitrate-reducing enzyme nitrate reductase, which is reportedly heat-sensitive [33]. On the other hand, the enhanced nitrate content at low temperature could be the result of increased absorption since it has been mentioned that roots facing short-duration low-temperature stress respond by increasing root water uptake, thus also increasing nutrient absorption [14]. Nitrate accumulation in the leaves of leafy vegetables is a crucial matter for consumers' safety as well as plant physiology. It should be noted that nitrate content in all soilless systems and temperature regimes was well within the legal amounts defined by the European Commission Regulation No. 1258/2011 for lettuce (up to 5000 mg/kg of fresh weight). The compliance of our samples with the EU regulation under all treatments is very encouraging, as it proves that even in sub-optimum root-zone temperatures, soilless systems can lead to safe products with low nitrate levels. This is particularly important for markets such as in Central and Northern Europe, where temperatures are lower and leafy vegetables tend to produce more nitrates.

The different photosynthetic pigment accumulation observed under each soilless system possibly mirrors the variable microclimatic conditions surrounding the root systems. A study involving *Amaranthus tricolor* revealed higher photosynthetic pigment and bioactive compound content under a root-zone temperature of 20 °C for 7 days [34]. Meanwhile, the reduction in anthocyanins in the higher NFT temperature in our experiment was possibly due to favorable conditions that limit the oxidative stress and the accumulation of secondary metabolites, leading more assimilates towards leaf growth. Such pigments are typically produced as a stress response due to the presence of reactive oxygen species (ROS), while they are associated with enhanced antioxidant enzyme activity [35]. A similar observation but with elevated root-zone temperature was reported for red leaf lettuce, which produced more pigments (i.e., chlorophylls, carotenoids, and anthocyanins) under the higher extreme temperature (35 °C), a typical response to heat stress [26].

Total soluble solids were increased with elevated temperature in the NFT, showcasing the enhanced photosynthetic activity as well as the greater rate of sugar flow. This could be due to more intense photosynthetic activity leading to a greater production of carbohydrates such as sugars, a characteristic linked to higher organoleptic value and greater acceptance by consumers. On the other hand, when plants are exposed to low temperatures, water loss occurs in cells. Hence, the increased soluble solid content of the control treatment compared with the 14 and 18 °C ones could be a result of osmotic adjustment, accumulating various organic substances (or even inorganic ones such as nitrate) to lower the osmotic potential of cells, thus maintaining a certain osmotic concentration in cells [16]. In aeroponics, the effect of temperature was insignificant for total soluble solid accumulation. The same was observed for total phenolics and antioxidant activity (i.e., FRAP) in both soilless systems, indicating that the tested temperature regimes did not impose

https://doi.org/10.3390/plants15030422

high oxidative stress to trigger the production and accumulation of protective secondary metabolites. A similar observation was made for lettuce and arugula grown in spring, summer, and winter experiments, when total phenolics and antioxidant activity were not affected by the controlled (about 20 °C) compared to the ambient temperature [17]. On the other hand, a study involving arugula and lettuce (canasta) in aeroponics reported an increase in total phenolic compounds and ascorbic acid at harvest when plants were treated with increased root-zone temperature up to 42 °C [25]. That response was attributed to ROS such as hydrogen peroxide produced due to heat stress, which triggers the accumulation of antioxidant compounds (such as phenolics) to alleviate the adverse effects. It seems that in our case, a significant stress response was not imposed since the tested root-zone temperatures lie within the normal range for optimum growth of lettuce in soilless systems. This is reaffirmed by the PI_{abs} and PI_{tot} values in both soilless systems, indicating healthy plants.

4. Materials and Methods

4.1. Growth Conditions and Treatments

The experiment took place in a plastic, non-heated greenhouse in Messolonghi, Greece (N 38.366; E 21.476), during the winter of 2023–2024. Romaine lettuce seedlings (cv. Parris Island) were obtained from a certified nursery. Seedlings were produced in peat plugs (3 × 3 cm). Prior to transplantation, peat was gently removed by rinsing the root system with running water to minimize mechanical damage of the soilless systems. All seedlings were visually inspected to ensure uniformity in size, leaf number (four leaves), and root integrity, and placed in plastic net pots. The seedlings were transplanted either to an NFT hydroponic system or an aeroponics system, described below.

The experiment was conducted using eight independent closed-loop hydroponic subsystems operated either as the nutrient film technique (four subsystems) or as aeroponics (four subsystems). All subsystems shared the same electromechanical infrastructure (nutrient solution reservoirs, pumps, temperature-control and monitoring systems), with structural modifications applied to create the distinct growing environments for the NFT and aeroponics. Specifically, the NFT units (Figure 6A) consisted of PVC pipes (100 mm diameter, 1400 mm length) placed horizontally. A nutrient solution reservoir supplied each NFT pipe via a pump (6 W, 60 L h^{-1}), delivering a continuous thin film of nutrient solution along the bottom of the pipe before returning to the reservoir. The aeroponics units (Figure 6B) used the same pipe–reservoir architecture, but the pipes were mounted vertically to create an open root chamber. The nutrient solution was delivered from the top of the vertical pipe, allowing the solution to trickle downward towards the roots. Excess solution was collected at the bottom of the pipe and returned to the reservoir. In both systems, each pipe included seven planting holes (55 mm diameter) spaced at 200 mm intervals.

In each system, three of the four subsystems were equipped with 350 W heating elements installed in the reservoirs, allowing for temperature increase of the nutrient solution. Temperature regulation relied on DS18B20 sensors (Analog Devices Inc./Maxim Integrated, Wilmington, MA, USA) connected to an Arduino Uno microcontroller, which activated heating elements via relays to maintain temperature above the target values. A separate data-logging system recorded the nutrient solution temperatures inside each hydroponic pipe. Ambient temperature was also monitored (MX1104 HOBO Data Logger, Cape Cod, MA, USA) and is presented in Table 1. Air relative humidity was 55 ± 10%.

https://doi.org/10.3390/plants15030422

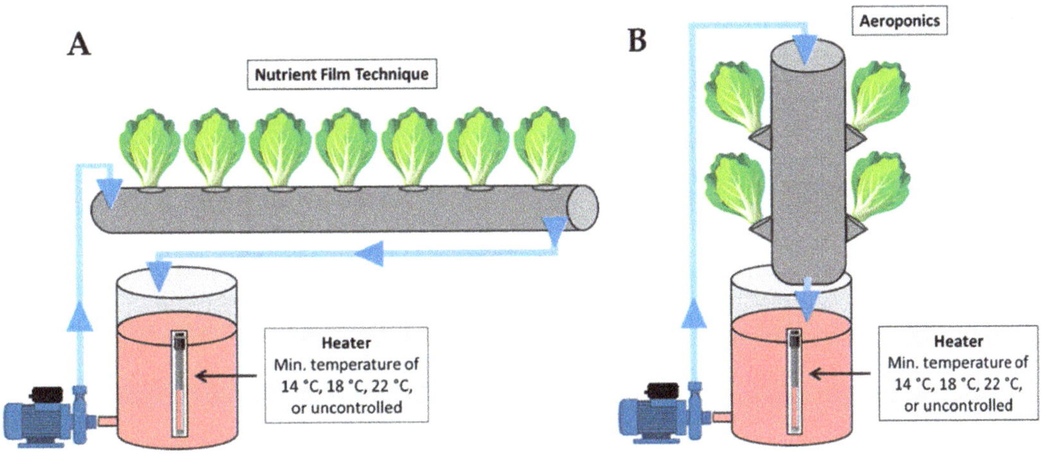

Figure 6. Schematic representation of the experimental setup used for the cultivation of mini Romaine lettuce in the two soilless systems: (**A**) nutrient film technique and (**B**) aeroponics.

Root-zone temperature was controlled indirectly through precise regulation of the minimum nutrient solution temperature in the reservoirs supplying each sub-system. The reported temperatures refer to the nutrient solution at the point of delivery to the growing units [15,26]. In both soilless systems, we tested the control treatment with no temperature adjustments, as well as minimum root-zone temperatures of 14, 18, and 22 °C. According to the above, the NFT treatments are hereby labeled as NFT-C, NFT-14, NFT-18, and NFT-22, while the aeroponics treatments are hereby labeled as Aero-C, Aero-14, Aero-18, and Aero-22. Due to natural diurnal fluctuations, the actual root-zone temperatures were higher and are shown in Table 1. Within a soilless system, each root-zone temperature treatment included one subsystem containing seven plants in the NFT or eight plants in aeroponics. In addition, each temperature treatment was applied to a single independent cultivation unit, where individual plants were considered replicates since all plants were exposed to identical conditions. Statistical analyses were conducted within each soilless system, while avoiding direct comparisons between the soilless systems.

The two soilless systems were selected because of the differences they provide in the root environment related to the temperature control. The NFT is nowadays the most employed hydroponic system throughout the world, while aeroponics is promoted as a novel approach, with increasing research interest due to the constant root exposure in the air. In addition, these temperature regimes (14, 18, and 22 °C) were selected to test a range of realistic winter conditions taking place in the Mediterranean region.

The nutrient solution composition was obtained by Nutrisense Decision Support System (https://nutrisense.online/; accessed on 5 December 2023), specialized software used to calculate nutrient concentrations in various crops [36]. Briefly, electrical conductivity (EC) was 2.0 dS/m, pH was 5.6 (adjusted with nitric acid 68%), NO_3^- was 11.6 mM, NH_4^+ was 1.6 mM, K^+ was 6.4 mM, Ca^{2+} was 4.7 mM, Mg^{2+} was 1.7 mM, $H_2PO_4^-$ was 1.1 mM, and Fe was 16.0 µM. The solution also contained trace elements (i.e., Mn^{2+}, Zn^{2+}, Cu^{2+}, B, and Mo). The nutrient solution was prepared using fertigation-grade fertilizers (water-soluble salts suitable for hydroponic application), dissolved directly into tap water. The water used for solution preparation had an electrical conductivity of 0.33 dS/m and a pH of 7.8. The nutrient solution was replenished weekly due to plant consumption, as well as to keep EC and pH at the desired levels. Mixing of the nutrient solution was

https://doi.org/10.3390/plants15030422

ensured through continuous pumping and recirculation in the reservoirs. All nozzles were inspected regularly to ensure that no clogging occurred.

4.2. Measurements and Analyses

Plants were harvested after 29 days of cultivation. Measurements included a range of biomass-related, morphological, physiological, nutritional, and safety-related characteristics to properly describe the overall plant development under the effect of the different treatments. Specifically, leaf fresh weight was measured to calculate the LMSA depending on the planting density of each soilless system (i.e., 15.2 plants/m^2 in the NFT and 22.2 plants/m^2 in aeroponics). Root dry weight (after 3 days in a drying oven at 72 °C) and maximum leaf length were also measured. Relative anthocyanin content was recorded with an ACM-200+ (Opti-Sciences, Hudson, NH, USA), which provides an anthocyanin index based on leaf optical properties. Chlorophyll fluorescence parameters (PI_{abs} and PI_{tot}) were recorded with Pocket PEA (Hansatech Instruments, Amesbury, MA, USA).

Later, leaves were homogenized and placed in a freezer (−20 °C) for analysis of the nutritional content (3 replications per treatment). Total soluble solids (Brix°) were recorded with a PAL-α refractometer (Atago, Tokyo, Japan). Total phenolic content was quantified with the Folin–Ciocalteu assay [37] upon extraction with 80% methanol. Absorbance of the colored reagent was measured with a spectrophotometer at 760 nm. Antioxidant activity was quantified with the ferric-reducing antioxidant power (FRAP) assay [38] upon extraction with 80% methanol. Absorbance of the colored reagent (sample with CH_3COONa, TPTZ, and $FeCl_3$) was measured with a spectrophotometer at 593 nm. Photosynthetic pigments (chlorophylls and carotenoids) were quantified upon extraction with 80% aqueous acetone [39]. Absorbance of the colored acetone was measured with a spectrophotometer at 470, 647, and 663 nm. Finally, nitrate content was quantified after extraction in 25 mL water [40]. Absorbance of the colored reagent (sample, H_2SO_4, and NaOH) was measured with a spectrophotometer at 410 nm.

4.3. Statistical Analysis

Analysis of variance (ANOVA) was conducted with SPSS 25.0 (IBM Corp., Armonk, NY, USA) software. Post-hoc comparisons between temperatures and within soilless systems was performed with Tukey's HSD method at significance level $\alpha = 0.05$.

5. Conclusions

Our findings highlight that controlling root-zone temperature is a valuable tool for the optimization of the growth and development of mini Romaine lettuce, especially in NFT systems. A minimum root-zone temperature of 22 °C led to increased yield and a greater amount of total soluble solids, but a lower pigment content and photosynthetic capacity (though plants were healthy) compared to the lowest temperature regimes. This can be attributed to the lowest ΔT recorded among all treatments. In our aeroponics system, produced yield was positively affected by increased temperature regimes, but differences were less profound regarding the physiological and nutritional profile. Romaine lettuce exhibited limited capacity to sustain root growth at suboptimal root-zone temperatures, despite preserved photosynthetic activity, showing that growth limitation is primarily due to sink-related characteristics rather than carbon assimilation. Compared with the optimal root-zone temperature, lower temperatures resulted in reduced growth efficiency, highlighting the narrow thermal tolerance window of romaine lettuce during soilless cultivation. Overall, the production efficiency of soilless systems can be enhanced by heating the nutrient solution, especially in low- and mid-tech greenhouses during

https://doi.org/10.3390/plants15030422

cool months with a view to reduced ΔT. The utilization of such knowledge can contribute significantly to the sustainable production of fresh leafy vegetables in such systems and climates, and even in future climate scenarios where extreme temperature fluctuations are predicted. The variable responses manifested by the two soilless systems emphasize the necessity for further research to determine the optimum temperature regimes for each crop species and even genotype.

Author Contributions: Conceptualization, methodology, and data analysis: F.B., G.Z., A.K. (Athanasios Koukounaras), and A.K. (Athanasios Koulopoulos); experimental measurements: F.B., N.T., and G.Z.; writing—original draft preparation: F.B.; editing: F.B., N.T., G.Z., A.K. (Athanasios Koukounaras), and A.K. (Athanasios Koulopoulos). All authors have read and agreed to the published version of the manuscript.

Funding: This research received no external funding.

Data Availability Statement: The original contributions presented in this study are included in the article. Further inquiries can be directed to the corresponding author.

Acknowledgments: The authors would like to thank Dimitrios Savvas and Georgia Ntatsi for providing us with access to the Nutrisense Decision Support System (https://nutrisense.online/), as well as Georgios Salachas for critically reviewing the manuscript.

Conflicts of Interest: The authors declare no conflicts of interest.

References

1. O'Connell, S. Miniature head lettuce yield and anthocyanin concentration under high tunnels and the field in Georgia. *HortTechnology* **2021**, *31*, 53–63. [CrossRef]
2. Lee, J.H.; Kwon, Y.B.; Choi, I.-L.; Yoon, H.S.; Kim, J.; Kim, Y.; Kang, H.-M. Changes in spectral reflectance, photosynthetic performance, chlorophyll fluorescence, and growth of mini green romaine lettuce according to various light qualities in indoor cultivation. *Horticulturae* **2024**, *10*, 860. [CrossRef]
3. Clemente, A.A.; Maciel, G.M.; Siquieroli, A.C.S.; Gallis, R.B.d.A.; Luz, J.M.Q.; Sala, F.C.; Pereira, L.M.; Yada, R.Y. Nutritional characterization based on vegetation indices to detect anthocyanins, carotenoids, and chlorophylls in mini-lettuce. *Agronomy* **2023**, *13*, 1403. [CrossRef]
4. Kim, M.J.; Moon, Y.; Tou, J.C.; Mou, B.; Waterland, N.L. Nutritional value, bioactive compounds and health benefits of lettuce (*Lactuca sativa* L.). *J. Food Compos. Anal.* **2016**, *49*, 19–34. [CrossRef]
5. FAOSTAT. Lettuce Production (Combined with Chicory) in 2023, Crops/Regions/World List/Production Quantity/Year (Pick Lists). UN Food and Agriculture Organization, Corporate Statistical Database (FAOSTAT). 2025. Available online: https://nutrisense.online/ (accessed on 5 December 2023).
6. Savvas, D.; Gruda, N.S. Application of soilless culture technologies in the modern greenhouse industry—A review. *Eur. J. Hortic. Sci.* **2018**, *83*, 280–293. [CrossRef]
7. Hendrickson, T.; Dunn, B.L.; Goad, C.; Hu, B.; Singh, H. Effects of elevated water temperature on growth of basil using nutrient film technique. *HortScience* **2022**, *58*, 925–932. [CrossRef]
8. Salachas, G.; Savvas, D.; Argyropoulou, K.; Tarantillis, P.A.; Kapotis, G. Yield and nutritional quality of aeroponically cultivated basil as affected by the available root-zone volume. *Emir. J. Food Agric.* **2015**, *27*, 911–918. [CrossRef]
9. Platis, D.P.; Papoui, E.; Bantis, F.; Katsiotis, A.; Koukounaras, A.; Mamolos, A.P.; Mattas, K. Underutilized vegetable crops in the Mediterranean region: A literature review of their requirements and the ecosystem services provided. *Sustainability* **2023**, *15*, 4921. [CrossRef]
10. İkiz, B.; Dasgan, H.Y.; Balik, S.; Kusvuran, S.; Gruda, N.S. The use of biostimulants as a key to sustainable hydroponic lettuce farming under saline water stress. *BMC Plant Biol.* **2024**, *24*, 808. [CrossRef] [PubMed]
11. Ntinas, G.K.; Bantis, F.; Koukounaras, A.; Kougias, P.G. Exploitation of liquid digestate as the sole nutrient source for floating hydroponic cultivation of baby lettuce (*Lactuca sativa*) in greenhouses. *Energies* **2021**, *14*, 7199. [CrossRef]
12. Yamori, N.; Levine, C.P.; Mattson, N.S.; Yamori, W. Optimum root zone temperature of photosynthesis and plant growth depends on air temperature in lettuce plants. *Plant Mol. Biol.* **2022**, *110*, 385–395. [CrossRef]
13. Creber, H.M.C.; Davies, M.S.; Francis, D. Effects of temperature on cell division in root meristems of natural populations of *Dactylis glomerata* of contrasting latitudinal origin. *Environ. Exp. Bot.* **1993**, *33*, 433–442. [CrossRef]

 https://doi.org/10.3390/plants15030422

14. Tiwari, M.; Kumar, R.; Subramanian, S.; Doherty, C.J.; Krishna Jagadish, S.V. Auxin–cytokinin interplay shapes root functionality under low-temperature stress. *Trends Plant Sci.* **2023**, *28*, 447–459. [CrossRef]
15. Hayashi, S.; Levine, C.P.; Yu, W.; Usui, M.; Yukawa, A.; Ohmori, Y.; Kusano, M.; Kobayashi, M.; Nishizawa, T.; Kurimoto, I.; et al. Raising root zone temperature improves plant productivity and metabolites in hydroponic lettuce production. *Front. Plant Sci.* **2024**, *15*, 1352331. [CrossRef] [PubMed]
16. Zhou, L.; Ullah, F.; Zou, J.; Zeng, X. Molecular and physiological responses of plants that enhance cold tolerance. *Int. J. Mol. Sci.* **2025**, *26*, 1157. [CrossRef] [PubMed]
17. Karnoutsos, P.; Karagiovanidis, M.; Bantis, F.; Chatzistathis, T.; Koukounaras, A.; Ntinas, G.K. Controlled root-zone temperature effect on baby leaf vegetables yield and quality in a floating system under mild and extreme weather conditions. *J. Sci. Food Agric.* **2020**, *101*, 3933–3941. [CrossRef]
18. Holmes, S.C.; Wells, D.E.; Pickens, J.M.; Kemble, J.M. Selection of heat tolerant lettuce (*Lactuca sativa* L.) cultivars grown in deep water culture and their marketability. *Horticulturae* **2019**, *5*, 50. [CrossRef]
19. Zhang, H.; Sun, X.; Song, W. Physiological and growth characteristics of tomato seedlings in response to low root-zone temperature. *HortScience* **2023**, *58*, 442–448. [CrossRef]
20. Sakamoto, M.; Suzuki, T. Effect of root-zone temperature on growth and quality of hydroponically grown red leaf lettuce (*Lactuca sativa* L. cv. Red Wave). *Am. J. Plant Sci.* **2015**, *6*, 2350–2360. [CrossRef]
21. Adebooye, O.C.; Schmitz-Eiberger, M.; Lankes, C.; Noga, G.J. Inhibitory effects of sub-optimal root zone temperature on leaf bioactive components, photosystem II (PS II) and minerals uptake in *Trichosanthes cucumerina* L. Cucurbitaceae. *Acta Physiol. Plant.* **2010**, *32*, 67–73. [CrossRef]
22. Bantis, F.; Radoglou, K.; Brüggemann, W. Differential ecophysiological responses to seasonal drought of three co-existing oak species in northern Greece. *Plant Biosyst.* **2019**, *153*, 378–384. [CrossRef]
23. Guimarães, Z.T.M.; dos Santos, V.A.H.F.; Ferreira, M.J. Chlorophyll a fluorescence parameters are related to the leaf economics spectrum of tropical tree species in a mixed plantation. *Trees* **2022**, *36*, 763–775. [CrossRef]
24. Oh, M.-M.; Carey, E.E.; Rajashekar, C.B. Environmental stresses induce health-promoting phytochemicals in lettuce. *Plant Physiol. Biochem.* **2009**, *47*, 578–583. [CrossRef]
25. He, J.; Tan, C.; Qin, L. Root-zone heat priming effects on maximum quantum efficiency of PSII, productivity, root morphology and nutritional quality of two aeroponically grown leafy greens in a tropical greenhouse. *Plants* **2022**, *11*, 1684. [CrossRef]
26. Levine, C.P.; Hayashi, S.; Ohmori, Y.; Kusano, M.; Kobayashi, M.; Nishizawa, T.; Kurimoto, I.; Kawabata, S.; Yamori, W. Controlling root zone temperature improves plant growth and pigments in hydroponic lettuce. *Ann. Bot.* **2023**, *132*, 455–470. [CrossRef] [PubMed]
27. He, J.; See, X.E.; Qin, L.; Choong, T.W. Effects of root-zone temperature on photosynthesis, productivity and nutritional quality of aeroponically grown salad rocket (*Eruca sativa*) vegetable. *Am. J. Plant Sci.* **2016**, *7*, 1993–2005. [CrossRef]
28. Thakulla, D.; Dunn, B.; Hu, B.; Goad, C.; Maness, N. Nutrient solution temperature affects growth and °Brix parameters of seventeen lettuce cultivars grown in an NFT hydroponic system. *Horticulturae* **2021**, *7*, 321. [CrossRef]
29. Nagelmüller, S.; Hiltbrunner, E.; Körner, C. Low temperature limits for root growth in alpine species are set by cell differentiation. *AoB PLANTS* **2017**, *9*, plx054. [CrossRef]
30. Burström, H. Temperature and root cell elongation. *Physiol. Plant.* **1956**, *9*, 682–692. [CrossRef]
31. Boter, M.; Pozas, J.; Jarillo, J.A.; Piñeiro, M.; Pernas, M. *Brassica napus* roots use different strategies to respond to warm temperatures. *Int. J. Mol. Sci.* **2023**, *24*, 1143. [CrossRef] [PubMed]
32. Fei, Q.; Liang, L.; Li, F.; Zhang, L.; Li, Y.; Guo, Y.; Wu, L.; Meng, X.; Gao, H.; Li, X. Transcriptome profiling and phytohormone responses of Arabidopsis roots to different ambient temperatures. *J. Plant Interact.* **2019**, *14*, 314–323. [CrossRef]
33. Feng, L.; Shi, X.; Chen, Y.; Tang, H.; Wang, L. Effects of temperature on the nitrate reductase activity and growth of *Ulva prolifera*. *J. Phycol.* **2021**, *57*, 955–966. [CrossRef]
34. Wittayathanarattana, T.; Wanichananan, P.; Supaibulwatana, K.; Goto, E. A short-term cooling of root-zone temperature increases bioactive compounds in baby leaf *Amaranthus tricolor* L. *Front. Plant Sci.* **2022**, *13*, 944716.
35. Ding, X.; Jiang, Y.; He, L.; Zhou, Q.; Yu, J.; Hui, D.; Huang, D. Exogenous glutathione improves high root-zone temperature tolerance by modulating photosynthesis, antioxidant and osmolytes systems in cucumber seedlings. *Sci. Rep.* **2016**, *6*, 35424. [CrossRef]
36. Savvas, D.; Giannothanasis, E.; Ntanasi, T.; Karavidas, I.; Drakatos, S.; Panagiotakis, I.; Neocleous, D.; Ntatsi, G. Improvement and validation of a decision support system to maintain optimal nutrient levels in crops grown in closed-loop soilless systems. *Agric. Water Manag.* **2023**, *285*, 108373. [CrossRef]
37. Singleton, V.L.; Rossi, J.A. Colorimetry of total phenolics with phosphomolybdic-phosphotungstic acid reagents. *Am. J. Enol. Vitic.* **1965**, *16*, 144–158. [CrossRef]
38. Benzie, I.F.F.; Strain, J.J. The ferric reducing ability of plasma (FRAP) as a measure of 'antioxidant power': The FRAP assay. *Anal. Biochem.* **1996**, *239*, 70–76. [CrossRef] [PubMed]

 https://doi.org/10.3390/plants15030422

39. Sumanta, N.; Haque, C.I.; Nishika, J.; Suprakash, R. Spectrophotometric analysis of chlorophylls and carotenoids from commonly grown fern species by using various extracting solvents. *Res. J. Chem. Sci.* **2014**, *4*, 63–69.
40. Cataldo, D.A.; Maroon, M.; Schrader, L.E.; Youngs, V.L. Rapid colorimetric determination of nitrate in plant tissue by nitration of salicylic acid. *Commun. Soil Sci. Plant Anal.* **1975**, *6*, 71–80. [CrossRef]

https://doi.org/10.3390/plants15030422

plants

MDPI

Article

Transient Overexpression of *VvMYBPA1* in Grape Berries Enhances Susceptibility to *Botrytis cinerea* Through ROS Homeostasis Modulation

Lihong Hao *, Yuxin Zhang, Zeying Ge, Xinru Meng, Yu Sun and Huilan Yi *

School of Life Science, Shanxi University, Taiyuan 030006, China; zhangyuxin39@sxu.edu.cn (Y.Z.); gezeying@sxu.edu.cn (Z.G.); mengxinru@sxu.edu.cn (X.M.); sunyu3@sxu.edu.cn (Y.S.)

* Correspondence: haolihong@sxu.edu.cn (L.H.); yihl@sxu.edu.cn (H.Y.)

Abstract: Gray mold disease, caused by *Botrytis cinerea*, severely impacts grape production worldwide. Although proanthocyanidins (PAs) contribute to fungal pathogen resistance, their role in grape defense against *B. cinerea* remains unclear. Here, we demonstrate that VvMYBPA1, a key transcriptional regulator of PA biosynthesis, negatively modulates *B. cinerea* resistance in grape berries. While infection suppressed endogenous VvMYBPA1, its agroinfiltration-mediated transient overexpression in berries elevated susceptibility, paralleling reduced β-1,3-glucanase (BGL) and polyphenol oxidase (PPO) activities. Additionally, VvMYBPA1 overexpression elevated *VvRBOHs'* expression and reduced peroxidase (POD) activity, resulting in excessive hydrogen peroxide (H_2O_2) accumulation and more cell death. Our results reveal that VvMYBPA1 negatively regulates *B. cinerea* resistance by disrupting antioxidant enzyme activity and ROS homeostasis, providing new insights into the interplay between PA biosynthesis and fungal defense mechanisms.

Keywords: *Botrytis cinerea* stress; grape; MYBPA1; ROS; proanthocyanidin

1. Introduction

Grape (*Vitis vinifera* L.) is a globally vital fruit crop with high economic impact but its production faces severe threats from multiple diseases, notably *Botrytis cinerea*-induced gray mold [1]. *B. cinerea* is a broad-host-range necrotrophic fungal pathogen, causing significant pre- and post-harvest losses in grape berries [2,3]. The grapefruit-*B. cinerea* pathosystem exhibits a dynamic molecular interplay characterized by host pattern-triggered immunity and fungal effector-mediated virulence [4]. Elucidating the molecular basis of this interaction is essential for developing grape resistance strategies and mitigating economic impacts.

B. cinerea triggers lipid peroxidation and ROS bursts in host tissues [5]. ROS exhibit dual functionality in plant–pathogen systems [6]. ROS exhibit a biphasic role: at low levels, they mediate defense signaling (hypersensitive response, callose deposition); at high levels, they provoke oxidative injury, aiding necrotrophy proliferation [5,7]. The homeostatic regulation between ROS generation and detoxification is tightly regulated by a network of enzymes, including NADPH oxidases (RBOHs), peroxidase (POD), superoxide dismutase (SOD), and catalase (CAT). Emerging evidence underscores the pivotal regulatory role of ROS homeostasis in plant defense against *B. cinerea*. Molecular studies have demonstrated that Sly-miR398b-mediated disruption of ROS equilibrium compromises tomato resistance to this pathogen [8]. Similarly, genetic evidence confirms SlWRKY46 as a pivotal transcriptional controller of *B. cinerea* resistance through ROS metabolism reprogramming [9].

Supporting these findings, a comparative transcriptomic study between resistant and susceptible grape cultivars uncovered that perturbations in ROS metabolism and redox regulation significantly influence host susceptibility [10]. In grape berries, the regulation of ROS homeostasis during *B. cinerea* infection remains poorly understood, particularly in the context of transcriptional regulators such as MYB transcription factors.

MYB transcription factors play critical roles in plant secondary metabolism, stress responses, and defense mechanisms [11]. Among them, VvMYBPA1 has been functionally characterized as a transcriptional regulator of proanthocyanidin (PA) biosynthesis in grapevine through direct activation of anthocyanidin reductase (ANR) and leucoanthocyanidin reductase (LAR) promoters [12]. PAs, or condensed tannins, significantly influence fruit taste and wine flavor, underscoring the economic importance of their biosynthesis regulation. Research shows PAs also play key roles in plant–pathogen interactions. For example, cucumber CsMYB60 enhances resistance to *Fusarium solani* by promoting PA synthesis [13], while poplar MYB115 directly activates PA-specific genes *ANR1* and *LAR3*, boosting resistance to *Dothiorella gregaria* [14]. In grapevines, the VvPUB26-VvWRKY24-VvDFR/VvLAR and VqWRKY56-VqbZIPC22-VvCHS3/VvLAR1/VvANR modules regulate PA biosynthesis, enhancing leaf resistance to *Erysiphe necator* [15,16]. Additionally, *V. amurensis* VvbHLH137 increases fruit resistance to *Colletotrichum gloeosporioides* by promoting PA accumulation [17]. Emerging evidence demonstrates that MYB transcription factors can influence ROS production and scavenging, but the underlying mechanisms remain elusive. For instance, besides the role in the biosynthesis of anthocyanidin, SlMYB75 improved the scavenging of excess H_2O_2 to resist *B. cinerea*, suggesting a potential trade-off between secondary metabolite biosynthesis and ROS-mediated defense responses [18]. Despite its known functions, VvMYBPA1's precise role in orchestrating ROS metabolism and immune responses against *B. cinerea* infection has not been systematically elucidated.

In this study, we systematically examined VvMYBPA1's mechanistic involvement in orchestrating ROS metabolism and defense activation during *B. cinerea* pathogenesis, using both constitutively overexpressing *Arabidopsis thaliana* and agroinfiltration-mediated transient overexpression in grape berries. Our findings not only advance the knowledge of MYB transcription factors in plant–pathogen interactions but also offer potential strategies for enhancing grape resistance to *B. cinerea* through targeted manipulation of ROS dynamics and secondary metabolism.

2. Results

2.1. Overexpression of VvMYBPA1 Enhances the Susceptibility of Arabidopsis Leaves to B. cinerea

To investigate the role of PA biosynthesis in resistance to *B. cinerea*, we first analyzed VvMYBPA1 expression in mature 'Kyoho' grape berries at harvesting stage (E-L 38) during infection (Figure 1a). At 6 hpi, VvMYBPA1 expression decreased by nearly half, and by 48 hpi, it dropped to less than two-thirds of the initial level. To further explore cultivar-specific responses, we analyzed VvMYBPA1 expression across different varieties using the Botrytis Stress Atlas Explorer. While 'Pinot', 'Semillon', and 'Garganega' (G) berries showed negligible VvMYBPA1 induction upon infection, 'Furmint' exhibited progressive upregulation correlating with disease severity, suggesting a negative association between VvMYBPA1 expression and resistance (Figure S1) [19,20]. To elucidate VvMYBPA1's involvement in defense mechanisms, we established overexpression Arabidopsis lines, from which three transgenic T3 progenies (OE#1, OE#10, and OE#23) showing stable transgene expression were isolated (Figure 1b). DMACA staining of immature seeds revealed that the transgenic lines, particularly OE#10 and OE#23, exhibited intense blue staining within 20 min, indicating higher PA accumulation relative to wild-type controls (Figure 1c). Pathogenicity assays using detached leaves revealed significantly enhanced susceptibility to *B. cinerea* in transgenic lines relative to WT plants at the 4-week developmental stage, including increased leaf yellowing and larger lesion diameters at 72 hpi (Figure 2a,b). Furthermore, quantification of fungal colonization confirmed higher *B. cinerea* levels in the transgenic lines (Figure 2c). These results demonstrate that VvMYBPA1 overexpression enhances Arabidopsis susceptibility to *B. cinerea*.

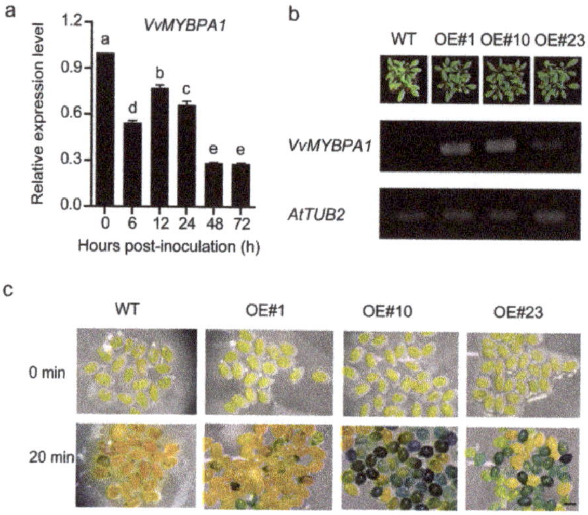

Figure 1. Heterologous overexpression of *VvMYBPA1* in *Arabidopsis thaliana*. (**a**) Temporal expression profile of *VvMYBPA1* in grape berry skins following *B. cinerea* infection. Different letters denote significant differences among groups (one-way ANOVA, $p < 0.05$). (**b**) Relative expression of *VvMYBPA1* in WT and three independent transgenic Arabidopsis lines (OE#1, OE#10, OE#23), normalized to *AtTUB2*. (**c**) Proanthocyanidin accumulation in seeds visualized by DMACA staining. Scale bar = 1 mm.

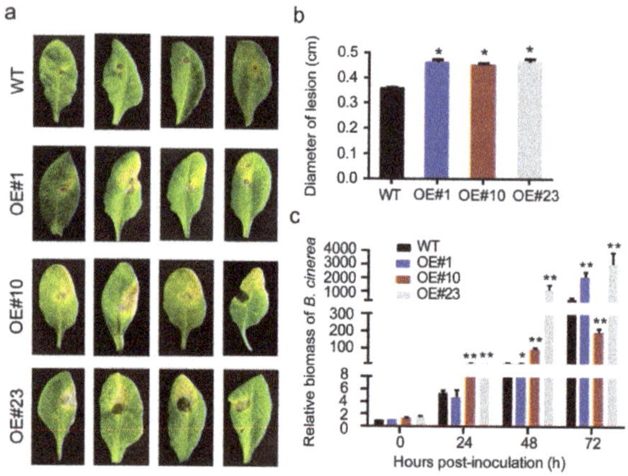

Figure 2. Ectopic expression of VvMYBPA1 enhances Arabidopsis sensitivity to *B. cinerea* infection. (**a**) Leaf symptoms observed in WT and three independent VvMYBPA1-overexpressing lines (#1, #10, #23) at 72 hpi. (**b**) Quantification of necrotic lesions on WT and VvMYBPA1-OE Arabidopsis leaves at 72 hpi. Data represent mean lesion diameters from three independent experiments ($n = 33$). (**c**) *B. cinerea* colonization was monitored by quantifying the ratio of *BcActin* to *AtActin* gene amplification in infected leaf samples collected at different infection stages. Error bars denote standard error of triplicate experiments. Statistical comparisons were made using two-tailed *t*-tests (* $p < 0.05$, ** $p < 0.01$).

2.2. VvMYBPA1 Stimulates ROS Production in A. thaliana Leaves upon B. cinerea Challenge

Given the critical role of ROS generation in plant interactions with *B. cinerea*, we assessed H_2O_2 and O_2^- accumulation in Arabidopsis leaves 72 hpi using DAB and NBT staining. In contrast to the wild-type (WT) controls, the overexpressing lines exhibited darker staining, indicating elevated ROS levels (Figure 3a,b). In *Arabidopsis thaliana*, the NADPH oxidase family members AtRBOHD and AtRBOHF serve as primary enzymatic sources of ROS generation [21]. We analyzed their transcription levels in overexpression lines and WT plants post inoculation. At 72 hpi, *AtRBOHD* expression was significantly higher in OE#1 and OE#23 compared to WT, while *AtRBOHF* transcript levels were markedly elevated in OE#10 (Figure 3c,d). Additionally, malondialdehyde (MDA) levels, indicative of ROS-mediated lipid peroxidation [22], were substantially higher in transgenic lines than in WT after inoculation (Figure 3f). Trypan blue staining further revealed increased cell death in transgenic leaves compared to WT at 72 hpi (Figure 3e). These findings suggest that in Arabidopsis VvMYBPA1 promotes *B. cinerea* proliferation by inducing oxidative burst-mediated cell death.

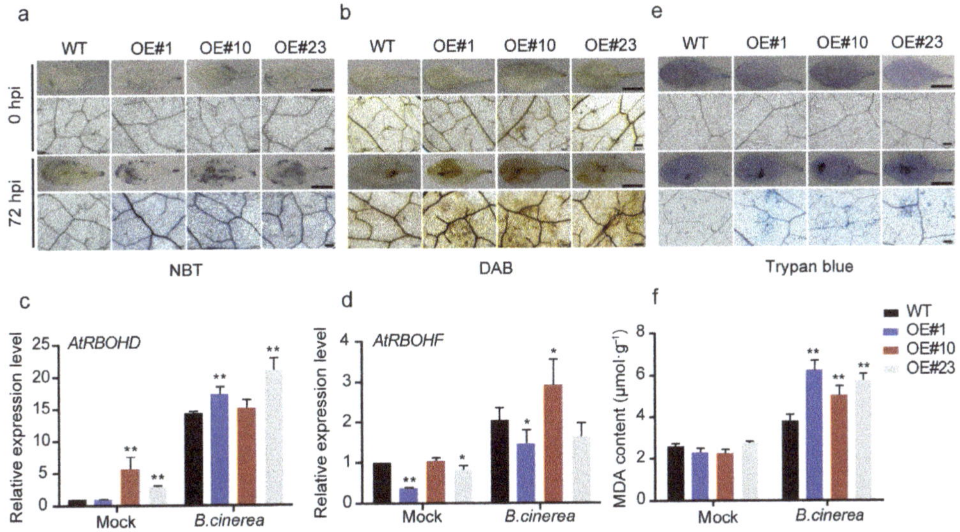

Figure 3. VvMYBPA1 overexpression exacerbates pathogen-triggered ROS production in Arabidopsis. (**a**) NBT staining reveals O_2^- distribution. (**b**) DAB staining indicates H_2O_2 accumulation. Analyses performed on WT and three independent VvMYBPA1-expressing lines at 72 hpi. Scale bars: 1 cm (whole leaf), 200 μm (magnified view). (**c**,**d**) Expression profiles of *AtRBOHD* and *AtRBOHF* in WT and transgenic lines during infection (0 and 72 hpi), normalized to *AtActin2*. (**e**) Trypan blue staining reveals cell death patterns at 72 hpi. Scale bars as in A-B. (**f**) MDA content as a lipid peroxidation marker at 0 and 72 hpi. Values shown are means ± SE from three biological replicates. Statistical significance was assessed by two-tailed Student's *t*-test (* $p < 0.05$, ** $p < 0.01$).

2.3. VvMYBPA1-Overexpressing Grape Berries Exhibit Decreased Resistance to B. cinerea Infection

Assessing the defensive function of VvMYBPA1 in grape berry–pathogen interactions, we generated transient VvMYBPA1-overexpressing grape berries via Agrobacterium-mediated transformation and inoculated them with *B. cinerea* (Figure 4a). VvMYBPA1 expression in transgenic berries was 8-fold higher than in controls, accompanied by significantly elevated proanthocyanidin levels both pre- and post-inoculation (Figure 4b,c). Three days post inoculation, lesion areas in the overexpression group were significantly larger, and *B. cinerea* colonization was markedly higher compared to controls (Figure 4d,e).

These results, consistent with heterologous overexpression in Arabidopsis, demonstrate that VvMYBPA1 overexpression compromises grape berry resistance against *B. cinerea*.

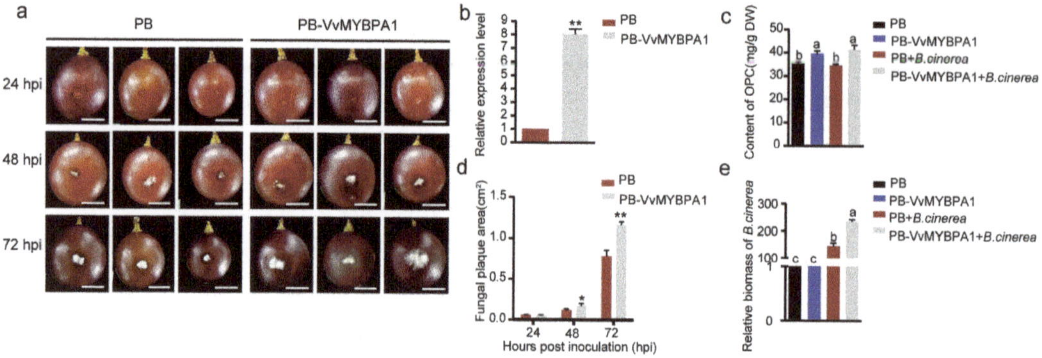

Figure 4. Transient overexpression of VvMYBPA1 increases susceptibility of 'Kyoho' grape berries to *B. cinerea* infection. (**a**) Disease symptoms on grape berries infiltrated with either empty vector (PB) or VvMYBPA1 overexpression construct (PB-VvMYBPA1) at 24, 48, and 72 h post inoculation with *B. cinerea* (spore suspension: 2×10^6/mL). Representative images shown were derived from three biological replicates (n = 30). Bars = 1 cm. (**b**) Relative expression of VvMYBPA1 in transiently transformed berries. (**c**) Oligomeric proanthocyanidin (OPC) content in PB and PB-VvMYBPA1 berries at 0 and 72 hpi. (**d**) Lesion area quantification at 24, 48, and 72 hpi (n = 30 from three independent experiments). (**e**) *B. cinerea* biomass was quantified via qPCR at 0 and 72 hpi. Bars represent mean ± SE. Significant variations between treatments marked by * (t-test: * $p < 0.05$, ** $p < 0.01$) or letters (ANOVA, $p < 0.05$).

We further monitored the enzymatic activity of defense-associated proteins spanning both phenylpropanoid pathway (PAL, PPO) and pathogen cell wall degradation (BGL, CHI) functions. Prior to *B. cinerea* inoculation, the enzymatic profiles of VvMYBPA1-overexpressing and control berries showed comparable activity levels (Figure 5a–d). Post inoculation, besides CHI activity, all other three enzyme activities significantly increased in the control berries (Figure 5a–d). In contrast, the VvMYBPA1-overexpression group exhibited distinct responses; while PAL and CHI activities showed comparable activities compared to the control (Figure 5b,c), BGL and PPO activities were significantly reduced (Figure 5a,d). These findings further demonstrate that VvMYBPA1 overexpression compromises grape berry resistance against *B. cinerea* infection.

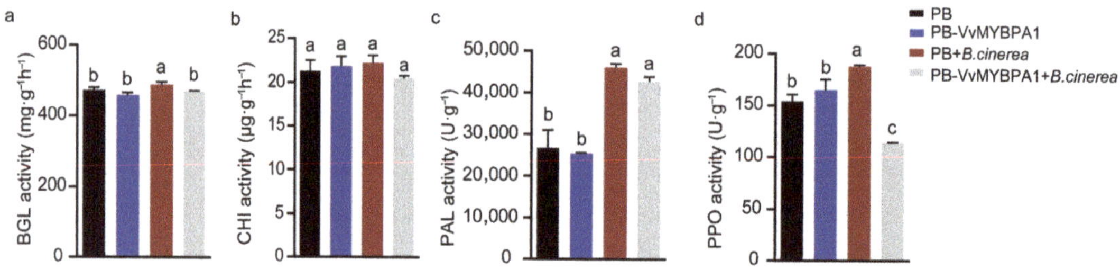

Figure 5. VvMYBPA1 overexpression alters defense enzyme responses to *B. cinerea* infection. (**a–d**) Activities of four defense-related enzymes (BGL, CHI, PAL, PPO) measured at 72 hpi. Data shown as mean ± SE from three biological replicates. Different letters denote statistical differences (ANOVA, $p < 0.05$).

2.4. *VvMYBPA1 Promotes ROS Accumulation in Mature Grape Berries During B. cinerea Inoculation*

To further investigate whether VvMYBPA1 in grape berries promotes *B. cinerea* proliferation by enhancing ROS accumulation, we analyzed the expression levels of two key *VvRBOH* genes, *VvRBOHA* and *VvRBOHB*, in VvMYBPA1-overexpressing and control berries two days post inoculation. The results revealed that both genes were significantly upregulated in the overexpression group relative to the control, with *VvRBOHB* expression approximately 18-fold higher (Figure 6a,b). Additionally, hydrogen peroxide (H_2O_2) levels were measured three days post inoculation. While no significant difference was observed prior to inoculation, H_2O_2 content in the overexpression group was markedly higher (approximately 1.5-fold) than in the control group post inoculation (Figure 6c). Furthermore, MDA levels were significantly elevated in the overexpression group (Figure 6d). VvVPE, a crucial enzyme in programmed cell death that facilitates vacuole rupture and subsequent release of hydrolytic enzymes, was also examined [23]. Its expression was low and comparable between groups before inoculation but increased approximately 2-fold in the overexpression group post inoculation (Figure 6e). These findings collectively demonstrate that VvMYBPA1 overexpression enhances H_2O_2 accumulation, exacerbates membrane lipid peroxidation, promotes host cell death, and ultimately facilitates *B. cinerea* proliferation.

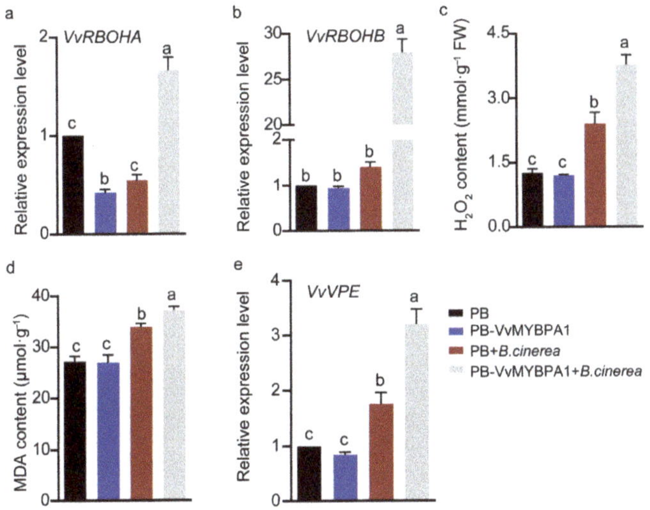

Figure 6. VvMYBPA1 overexpression alters *B. cinerea*-induced ROS accumulation in grape berries. (**a**,**b**) Expression profiles of NADPH oxidase genes (*VvRBOHA* and *VvRBOHB*) in control (PB) and VvMYBPA1-overexpressing (PB-VvMYBPA1) grape berries at 0 and 48 hpi. *Vvβactin* served as the internal reference gene. (**c**,**d**) H_2O_2 and MDA levels in PB and PB-VvMYBPA1 berries at 0 and 72 hpi. (**e**) Relative expression of *VvVPE* in PB and PB-VvMYBPA1 berries at 0 and 48 hpi, normalized to *Vvβactin*. Values are means ± SE of three biological replicates. Distinct letters indicate ANOVA-grouped differences ($p < 0.05$).

2.5. *Overexpression of VvMYBPA1 Altered the Activities of Antioxidant Enzymes in Grape Berries*

To elucidate the dual role of VvMYBPA1 in ROS metabolism during grape–*B. cinerea* interaction, we measured the transcript levels of key antioxidant enzymes (SOD, POD, and CAT) at 48 hpi and their corresponding enzyme activities at 72 hpi in both VvMYBPA1-overexpressing and control berries. Transcriptional analysis revealed a pathogen-induced upregulation of SOD in both genotypes, with no significant variation between genotypes (Figure 7a). Consistently, SOD activity remained comparable between transgenic and control fruits (Figure 7b). Notably, VvMYBPA1 overexpression led to a marked suppression of both POD transcription and enzymatic activity compared to the control (Figure 7c,d). While CAT transcript accumulation was enhanced in transgenic berries, its catalytic activity showed no significant alteration (Figure 7e,f). These findings demonstrate that VvMYBPA1 orchestrates antioxidant enzyme regulation during *B. cinerea* infection, extending beyond its role in ROS biosynthesis in grape berries.

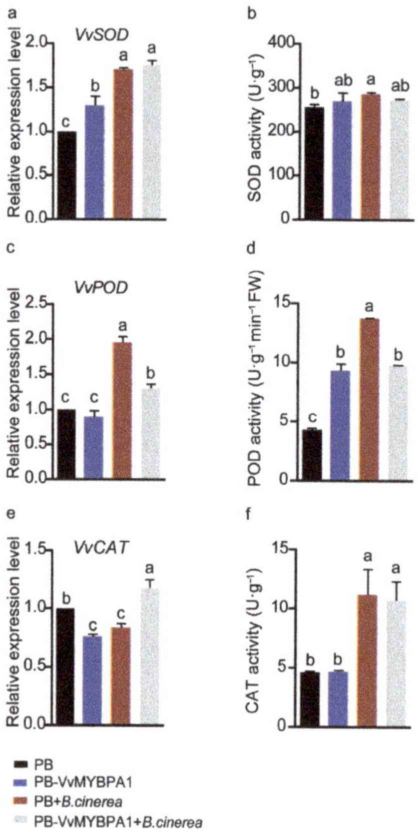

Figure 7. Transcriptional and enzymatic reprogramming of antioxidant systems in transgenic berries during *B. cinerea* infection. (**a**,**b**) SOD dynamics. (**c**,**d**) POD parameters. (**e**,**f**) CAT profiles. Transcriptional changes (assessed at 48 hpi) and enzymatic alterations (evaluated at 72 hpi) are shown. Bars show mean ± SE of three biological replicates. Letter-based significance groupings (ANOVA, $p < 0.05$).

3. Discussion

3.1. VvMYBPA1 Negatively Regulates B. cinerea Resistance in Mature Grape Berries

B. cinerea, a highly destructive fungal pathogen, poses a significant threat to the post-harvest storage of grape berries. Understanding the defense mechanisms against gray mold is therefore crucial for improving grape quality and extending shelf life during storage. Numerous secondary metabolites in plants contribute to fungal pathogen resistance, among which PAs play a pivotal role. For instance, heterologous expression of MnANR and MnLAR from *Morus alba* in *Nicotiana tabacum* has been demonstrated to confer enhanced defense against *B. cinerea* infection [24]. Additionally, catechin-treated cucumber seedlings exhibit increased pathogen resistance [13]. Previous investigations have well established the contribution of PAs to *Vitis vinifera* defense mechanisms against *Erysiphe necator* infection.

For example, VvPUB26 positively regulates resistance by ubiquitinating and degrading the PA synthesis repressor VvWRKY24 in grape leaves. Interestingly, VvPUB26-silenced plants show reduced powdery mildew resistance but elevated PA accumulation [15]. Genetic evidence from *Vitis quinquangularis* reveals that the VqWRKY56-VqbZIPC22 transcriptional module activates proanthocyanidin biosynthesis pathways, resulting in improved defense against *Erysiphe necator* infection [16].

The proanthocyanidin biosynthetic pathway requires coordinated action of multiple enzymes, including chalcone synthase catalyzing the initial condensation, chalcone isomerase mediating flavonoid ring formation, dihydroflavonol reductase reducing dihydroflavonols, and the terminal reductases LAR and ANR generating different PA subunits [24]. Several transcription factors, including VvMYBPAR, VvMYBPA1, VvMYBPA2, VvMYB5b, and VviMYB86, have been functionally identified as major determinants of PA biosynthesis in grapevines [12,25–28]. As previously reported, PA content in grape berry skins, along with the transcriptional level of *VvMYBPA1*, *VvLAR*, and *VvANR*, declines during berry ripening [12,29]. In this study, we investigated the function of VvMYBPA1 in gray mold resistance by ectopically expressing it in Arabidopsis and transiently overexpressing it in mature grape berries (Figures 2 and 4). Surprisingly, VvMYBPA1 overexpression in mature berries increased PA accumulation but significantly reduced resistance to *B. cinerea*. These findings expand the knowledge of the complex relationship between PA biosynthesis and gray mold resistance.

Plants employ two key immune strategies against pathogens: pattern-triggered immunity (PTI) and effector-triggered immunity (ETI) [30]. Upon *B. cinerea* infection, pathogen-associated molecular patterns are recognized, triggering host defenses such as transcriptional reprogramming, PR protein secretion, ROS burst, and activation of secondary metabolic pathways [4,10]. Key enzymes, including BGL, CHI, PAL, and PPO, play critical roles [31,32]. BGL and CHI degrade fungal cell walls, leading to pathogen death. PAL, central to the phenylpropanoid pathway, enhances lignin and flavonoid synthesis when activated. In grapes, 15 *PAL* genes contribute to *B. cinerea* resistance during the green berry stage [33,34]. PPO promotes lignification and generates antimicrobial quinones via phenolic oxidation [35]. In our study, pre-inoculation enzyme activities showed no significant differences between VvMYBPA1-overexpressing and control fruits. Post inoculation, however, enzyme activities declined in overexpressing fruits, with significant reductions in BGL and PPO (Figure 5). This suggests VvMYBPA1 negatively regulates *B. cinerea* resistance by suppressing key defense enzyme activities in secondary metabolism.

3.2. *ROS and PAs in Plant Defense: A Dual Role*

ROS are pivotal signaling molecules that mediate plant defense mechanisms against diverse environmental challenges, especially in immune responses. Studies have demonstrated that ROS production is closely associated with both PTI and ETI [36]. As secondary messengers, ROS can induce hypersensitive responses, ultimately leading to host cell death. This mechanism effectively restricts nutrient acquisition by biotrophic pathogens, thereby inhibiting their proliferation [37–39]. Nevertheless, an overabundance of ROS can trigger programmed cell death in host tissues, rendering plants more vulnerable to infection by hemi-biotrophic and necrotrophic pathogens [40,41]. Research has shown that in the interaction with the hemi-biotrophic pathogen *Phytophthora infestans*, the tomato 1R-type MYB transcription factor SlKUA1 enhances *SlRBOHD* expression, increasing ROS accumulation, while simultaneously reducing POD expression by binding to the *SlPrx1* promoter, thereby impairing ROS scavenging and ultimately affecting tomato resistance to late blight [41]. In contrast, during interaction with the necrotrophic pathogen *B. cinerea*, tomato SlWRKY3 promotes ROS accumulation and negatively regulates resistance by affecting *SlCAT1* expression [40]. Consistent with these findings, our study detected elevated ROS levels in both VvMYBPA1-overexpressing Arabidopsis leaves and grape fruits following *B. cinerea* infection (Figures 3 and 6), accompanied by significantly reduced POD expression and activity (Figure 7). The disruption of ROS homeostasis ultimately led to increased MDA content, elevated host cell death, and upregulation of the apoptosis-related gene *VvVPE*, which facilitated *B. cinerea* proliferation (Figure 6).

PAs, as major defensive phenolic compounds, exhibit antioxidant activity [42]. However, our study revealed that VvMYBPA1 overexpression not only increased proanthocyanidin content (Figures 1 and 4) but also enhanced ROS accumulation (Figures 3 and 6), affecting the expression and activity of antioxidant enzymes such as CAT, POD, and SOD (Figure 7), ultimately exacerbating membrane lipid peroxidation and host cell death (Figures 3 and 6). Parallel results have been documented in previous research. The WRKY transcription factor VqWRKY56, identified in wild grape *Vitis quinquangularis*, promotes anthocyanin and proanthocyanidin accumulation in grape leaves through interaction with VqbZIP22. Following powdery mildew infection, VqWRKY56-overexpressing lines exhibited elevated ROS levels and increased leaf cell death, which significantly inhibited fungal hyphal growth and spore production [16]. These findings unveil a sophisticated interplay between proanthocyanidin accumulation and defense activation during infection, where these processes exhibit dynamic and context-dependent coordination.

This work characterizes VvMYBPA1's function in mediating the grape–*B. cinerea* interaction, with particular emphasis on its role in maintaining ROS balance (Figure 8). We demonstrated that transient overexpression of VvMYBPA1 in berries enhances ROS accumulation and lipid peroxidation, increasing susceptibility to *B. cinerea*. The transient overexpression of MYBPA1, while capable of improving the content of PAs, concomitantly attenuates the established defense mechanisms in ripening grape berries against *B. cinerea* pathogenesis. This research significantly contributes to comprehending the intricate interplay between secondary metabolism and oxidative stress in grape–pathogen interactions. Future work will employ CRISPR-Cas9 knockout or RNAi-mediated knockdown of VvMYBPA1 in stable transgenic grape calli to definitively establish its negative regulatory role in Botrytis defense responses. These genetic studies will not only elucidate the molecular mechanism of susceptibility but also provide direct targets for breeding Botrytis-resistant grape cultivars through precision genome editing of MYB transcription factors.

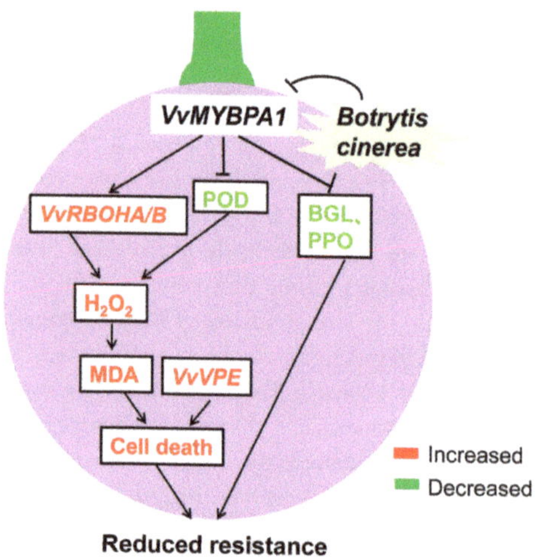

Figure 8. A proposed model illustrating how *VvMYBPA1* negatively regulates *Botrytis cinerea* resistance by modulating ROS homeostasis.

4. Materials and Methods

4.1. Plant Materials and Growth Conditions

Arabidopsis thaliana ecotype Columbia (Col-0) was maintained under laboratory conditions. Seeds were surface-sterilized in 15% (*v*/*v*) NaClO and then were treated at 4 °C for 3 days on half-strength Murashige and Skoog (MS) medium. Wild-type and transgenic seedlings were then transferred into an autoclaved soil mix at the two-leaf stage. Plants were cultivated in a controlled-environment growth chamber (22 °C, 16/8 h photoperiod, 60% RH). For this study, *Vitis vinifera* cv. 'Kyoho', a cultivar reported to show resistance to *B. cinerea* [43] and widely cultivated in Shanxi Province, was sampled from an orchard in Taiyuan City (112°33′ E, 37°51′ N). Post harvest, berries at E-L 38 stage were promptly transported to the laboratory for transient transformation and *B. cinerea* pathogenicity assays.

4.2. Vector Construction

To clone the *VvMYBPA1* gene into an overexpression vector, the VvMYBPA1 open reading frame (ORF) was PCR-amplified from cDNA derived from grape skins and VvMYBPA1 CDS F/VvMYBPA1 CDS R as primers and then cloned into pDONR221 (Invitrogen, Waltham, MA, USA) vector to generate a pENTRY-VvMYBPA1 construct using BP Clonase (Invitrogen), which was then sequenced. The correct pENTRY-VvMYBPA1 plasmid was then cloned into a GATEWAY-compatible vector pB7FWG2 utilizing LR Clonase (Invitrogen). The plasmid pB7FWG2-VvMYBPA1 was further validated by restriction endonuclease digestion via *Xba* I and *Xho* I sites.

4.3. Generation of *VvMYBPA1*-Overexpressing *Arabidopsis* Transgenics

Agrobacterium tumefaciens strain GV3101 carrying the VvMYBPA1 overexpression vector was used to transform *A. thaliana* by the floral dip method [44]. T0 seeds were harvested and sown on half-strength MS medium supplemented with 10 mg/L Basta. Three independent overexpression lines (OE#1, OE#10, OE#23) were selected from the T1 population based on transgene expression stability, with homozygous T3 progenies serving as the experimental materials.

4.4. *B. cinerea* Infection Assay

B. cinerea strain B05.10, isolated from decayed grapefruits in our lab, was preserved on PDA plates (22 °C). For inoculation of *A. thaliana*, fresh *B. cinerea* conidial suspensions were adjusted to an inoculum density of 5×10^6 conidia/mL as described by Xue and Yi [32]. For consistent infection, 10 µL inoculum droplets were applied to the abaxial surface of five mature leaves per plant at developmental stage 3.90 (4-week-old vegetative growth). For gene expression analysis, leaf samples were acquired at 0 and 72 h post inoculation (hpi). Histopathological evaluation was performed using trypan blue (cell viability), 3,3′-diaminobenzidine (DAB, H_2O_2 accumulation), and nitroblue tetrazolium (NBT, superoxide anion) staining at both pre-inoculation (0 hpi) and advanced infection (72 hpi) stages. For morphological assessment and lesion diameter measurement, detached leaves were infected with 10 µL of prepared conidial suspension using the droplet method. All measurements and photographic documentation were performed at 72 hpi.

For pathogen challenge assays, visually uniform 'Kyoho' fruits showing no signs of mechanical injury or pathological symptoms were subjected to biphasic surface decontamination (1% NaClO → 75% ethanol) with intermediate rinses in sterile ddH_2O, and then air-drying at room temperature. Precise epidermal trauma (2 mm transverse × 4 mm longitudinal) was generated in the equatorial plane of each berry using sterile needles, and 10 µL of *B. cinerea* conidial suspensions (2×10^6 conidia/mL) was inoculated into each wound. Post inoculation, berries were maintained in controlled conditions: primary incubation at 22 °C with saturated humidity (90–100% RH) under dark regime for 24 h, transitioning to diurnal cycles (12 h light/12 h dark) for disease progression monitoring. Three independent biological replicates were performed, each consisting of 10 berries. Pericarp tissues (approximately 1 cm in diameter) surrounding the inoculation site were used for physiological index determination and gene expression detection.

4.5. ROS Levels and Cell Death Assay

Hydrogen peroxide (H_2O_2) accumulation was detected using DAB staining [45]. Briefly, leaf samples were vacuum-infiltrated for 20 min in freshly prepared 1 mg/mL DAB solution and incubated in darkness overnight. Subsequently, leaf specimens underwent thermal destaining (100 °C, 10 min) in acetic acid: glycerol: ethanol (1:1:3, *v*/*v*/*v*), followed by immersion in 95% ethanol. Superoxide anion (O_2^-) accumulation was visualized by NBT staining [46]. Superoxide localization was achieved through vacuum-assisted infiltration (20 min, 25 inHg) of NBT reaction buffer (10 mM phosphate buffer (pH 7.8), 10 mM NaN_3, 0.1% (*w*/*v*) NBT), with subsequent dark incubation (60 min, 22 °C). Stained tissues were cleared by thermal treatment (100 °C, 10 min) in destaining solvent (acetic acid: glycerol: ethanol, 1:1:3 *v*/*v*/*v*) and archived in 95% ethanol for microscopic analysis. Histochemical detection of compromised cells was conducted via trypan blue uptake assay, according to Vogel and Somerville [47]. All specimens were imaged under controlled bright-field conditions under an Olympus light microscope (Olympus Corporation, Tokyo, Japan). The content of H_2O_2 and malondialdehyde (MDA) was monitored according to Han et al. [48]. For each treatment, a minimum of six leaves per biological replicate were analyzed, with three independent biological replicates performed.

4.6. Histochemical Staining of DMACA

Proanthocyanidin localization in *A. thaliana* seeds was visualized through dimethylaminocinnamaldehyde (DMACA) histochemistry [12]. Seed specimens were immersed in freshly prepared DMACA reagent (1% *w*/*v* DMACA in 6 M HCl-methanol, 1:99 *v*/*v*) for 20 min (25 °C), followed by sequential ethanol washes (75% *v*/*v*, 4 × 5 min) to eliminate nonspecific staining. PA deposition patterns were documented using stereomicroscopy with consistent illumination settings.

4.7. RNA Isolation and RT-qPCR

Total RNA extraction from Arabidopsis leaves was conducted utilizing Trizol reagent (TaKaRa, Beijing, China). At the same time, grape berry pericarp RNA was isolated through an adapted CTAB-based approach [49]. RNA quality was assessed via 2.0% agarose gel electrophoresis (Sigma-Aldrich, Hong Kong, China), with quantification performed using a Bio Spectrometer fluorescence spectrophotometer (Eppendorf, Hamburg, Germany). cDNA synthesis was performed using the Prime Script™ RT reagent Kit with a gDNA Eraser (TaKaRa, Osaka, Japan). Gene-specific primers (Supplementary Table S1) were obtained from Sangon Biotech (Shanghai, China). Quantitative PCR was carried out on a Bio-Rad CFX96™ system under standardized thermal cycling parameters: following a 30 s hot start at 95 °C, the reaction mixture underwent 40 amplification cycles comprising 5 s denaturation pulses at 95 °C and 30 s incubation periods at 60 °C. For Arabidopsis and grapevine, gene expression levels were standardized against *AtActin2* and *Vvβactin* as internal controls, with the $2^{-\Delta\Delta Ct}$ method employed for comparative quantification [50].

4.8. Transient Overexpression of VvMYBPA1 in Grape Berries

Transient overexpression of VvMYBPA1 in grape berries was conducted following the methodology established by Xie et al. [51]. *Agrobacterium tumefaciens* strains harboring either the pB7FWG2-VvMYBPA1 construct or the empty pB7FWG2 vector were first grown on LB agar plates containing 25 mg·L^{-1} rifampicin and 50 mg·L^{-1} spectinomycin. For pre-culture preparation, a selected colony was introduced into 2 mL of liquid LB medium and cultured at 28 °C (220 rpm) overnight (12 h). The bacterial culture was subsequently expanded to 20 mL in LB medium and allowed to grow until the OD_{600} reached 0.6–0.8. Cell pellets obtained by centrifugation (5 min, 4000× *g*, 25 °C) were subsequently resuspended in infiltration buffer of identical volume containing magnesium chloride (10 mM) and acetyleugenone (200 μM). The suspension was conditioned at 28 °C for a duration of 3 to 4 h before being used for infiltration. Transient expression was achieved by microinjecting 300 μL of Agrobacterium suspension into the stylar apex of grape berries using a 1 mL syringe, ensuring minimal leakage. Berries were sampled at 4 days post infiltration (DPI) for RNA extraction. The expression levels of VvMYBPA1 were quantified at 0 and 4 DPI using the RT-qPCR protocol described earlier. The agroinfiltrated grape fruits were kept for four days before being subjected to pathogenic infection tests.

4.9. Quantification of Oligomeric Proanthocyanidins Content

Oligomeric proanthocyanidins content (OPC) was quantified using a plant OPC assay kit (Beijing Solarbio Technology Co., Ltd., Beijing, China) following the manufacturer's protocol. A standard curve was established based on the linear relationship between OPC concentration and absorbance. The OPC concentration (X, mg·mL^{-1}) was calculated using the linear regression equation $X = (A_{500} - 0.0167)/0.1829$ ($R^2 = 0.9981$), where A_{500} represents the absorbance at 500 nm.

Grape fruit samples were freeze-dried to constant weight, homogenized using a mortar, and precisely weighed (0.1 g). The absorbance was recorded at 500 nm following the manufacturer's protocol. OPC content was determined by normalizing the standard curve-derived concentration (X) against sample dry mass (W), expressed as mg per gram dry weight (mg·g^{-1} DW).

4.10. Measurement of Enzyme Activity

The enzymatic activities of four defense-related proteins—phenylalanine ammonia-lyase (PAL), β-1,3-glucanase (BGL), chitinase (CHI), and polyphenol oxidase (PPO)—were assayed following established protocols [32]. Concurrently, antioxidant enzyme activities (catalase (CAT), peroxidase (POD), and superoxide dismutase (SOD)) were evaluated according to the methodology of Jiang et al. [35].

4.11. Statistical Analysis

Three independent biological replicates were included for each experimental treatment. Results are presented as mean values ± standard error (SE). For transgenic Arabidopsis lines, statistical comparisons of lesion diameter measurements, transcriptional profiles, enzymatic activities, and secondary metabolite accumulation were conducted using two-tailed paired Student's *t*-tests, with asterisks indicating significance levels (* $p < 0.05$; ** $p < 0.01$). For agro-infiltrated grape samples, one-way analysis of variance (ANOVA) was applied, with Duncan's multiple range test used for comparisons ($p < 0.05$, denoted by different letters). Data processing was conducted using Microsoft Excel (v2019, Microsoft Corp.). Statistical analyses were conducted using SPSS Statistics 16.0, while data visualization was implemented in GraphPad Prism 6.

Supplementary Materials: The following supporting information can be downloaded at: https://www.mdpi.com/article/10.3390/plants14162469/s1, Table S1: List of primers used in this study. Figure S1. Expression patterns of *VvMYBPA1* in grapevine cultivars responding to *Botrytis cinerea* infection.

Author Contributions: Conceptualization, L.H.; Data curation, Y.Z., Z.G., X.M. and Y.S.; Funding acquisition, L.H.; Methodology, Y.Z. and Z.G.; Supervision, H.Y.; Writing—original draft, L.H.; Writing—review and editing, H.Y. All authors have read and agreed to the published version of the manuscript.

Funding: This work was funded by grants from National Natural Science Foundation of China (grant No. 31800266), the Applied Basic Research Programs of Shanxi Province (No. 202203021221019), and the Open Project Program of Xinhuacun College of Shanxi (Shanxi Institute of Brewing Technology and Industry) (preparation) (No. XCSXU-KF-202315).

Data Availability Statement: The original contributions presented in this study are included in the article/Supplementary Materials. Further inquiries can be directed to the corresponding authors.

Conflicts of Interest: The authors declare no conflicts of interest.

References

1. Williamson, B.; Tudzynski, B.; Tudzynski, P.; van Kan, J.A. *Botrytis cinerea*: The cause of grey mould disease. *Mol. Plant Pathol.* **2007**, *8*, 561–580. [CrossRef] [PubMed]
2. Saito, S.; Michailides, T.J.; Xiao, C.L. Fungicide-resistant phenotypes in *Botrytis cinerea* populations and their impact on control of gray mold on stored table grapes in California. *Eur. J. Plant Pathol.* **2019**, *154*, 203–213. [CrossRef]
3. Shen, F.; Wu, W.; Han, X.; Wang, J.; Li, Y.; Liu, D. Study on the occurrence law and green control of grape gray mold from the perspective of ecological balance. *Bioengineered* **2021**, *12*, 779–790. [CrossRef]
4. Rahman, M.U.; Liu, X.; Wang, X.; Fan, B. Grapevine gray mold disease: Infection, defense and management. *Hortic. Res.* **2024**, *11*, uhae182. [CrossRef]
5. Camejo, D.; Guzman-Cedeno, A.; Moreno, A. Reactive oxygen species, essential molecules, during plant-pathogen interactions. *Plant Physiol. Biochem.* **2016**, *103*, 10–23. [CrossRef]
6. Deighton, N.; Muckenschnabel, I.I.; Goodman, B.A.; Williamson, B. Lipid peroxidation and the oxidative burst associated with infection of *Capsicum annuum* by *Botrytis cinerea*. *Plant J.* **1999**, *20*, 485–492. [CrossRef]
7. Castillo, L.; Plaza, V.; Larrondo, L.F.; Canessa, P. Recent advances in the study of the plant pathogenic fungus *Botrytis cinerea* and its interaction with the environment. *Curr. Protein Pept. Sci.* **2017**, *18*, 976–989. [CrossRef]
8. Liu, Y.; Yu, Y.; Fei, S.; Chen, Y.; Xu, Y.; Zhu, Z.; He, Y. Overexpression of Sly-miR398b compromises disease resistance against *Botrytis cinerea* through regulating ROS homeostasis and JA-related defense genes in Tomato. *Plants* **2023**, *12*, 2572. [CrossRef]
9. Shu, P.; Zhang, S.; Li, Y.; Wang, X.; Yao, L.; Sheng, J.; Shen, L. Over-expression of *SlWRKY46* in tomato plants increases susceptibility to *Botrytis cinerea* by modulating ROS homeostasis and SA and JA signaling pathways. *Plant Physiol. Biochem.* **2021**, *166*, 1–9. [CrossRef] [PubMed]
10. Wan, R.; Guo, C.; Hou, X.; Zhu, Y.; Gao, M.; Hu, X.; Zhang, S.; Jiao, C.; Guo, R.; Li, Z.; et al. Comparative transcriptomic analysis highlights contrasting levels of resistance of *Vitis vinifera* and *Vitis amurensis* to *Botrytis cinerea*. *Hortic. Res.* **2021**, *8*, 103. [CrossRef] [PubMed]
11. Yu, Y.; Guo, D.; Li, G.; Yang, Y.; Zhang, G.; Li, S.; Liang, Z. The grapevine R2R3-type MYB transcription factor VdMYB1 positively regulates defense responses by activating the *stilbene synthase gene 2* (*VdSTS2*). *BMC Plant Biol.* **2019**, *19*, 478. [CrossRef]
12. Bogs, J.; Jaffe, F.W.; Takos, A.M.; Walker, A.R.; Robinson, S.P. The grapevine transcription factor VvMYBPA1 regulates proanthocyanidin synthesis during fruit development. *Plant Physiol.* **2007**, *143*, 1347–1361. [CrossRef] [PubMed]
13. Li, J.; Luan, Q.; Han, J.; Chen, C.; Ren, Z. CsMYB60 confers enhanced resistance to *Fusarium solani* by increasing proanthocyanidin biosynthesis in cucumber. *Phytopathology* **2022**, *112*, 588–594. [CrossRef]
14. Wang, L.; Ran, L.; Hou, Y.; Tian, Q.; Li, C.; Liu, R.; Fan, D.; Luo, K. The transcription factor MYB115 contributes to the regulation of proanthocyanidin biosynthesis and enhances fungal resistance in poplar. *New Phytol.* **2017**, *215*, 351–367. [CrossRef]
15. Zhao, T.; Huang, C.; Li, N.; Ge, Y.; Wang, L.; Tang, Y.; Wang, Y.; Li, Y.; Zhang, C. Ubiquitin ligase VvPUB26 in grapevine promotes proanthocyanidin synthesis and resistance to powdery mildew. *Plant Physiol.* **2024**, *195*, 2891–2910. [CrossRef]
16. Wang, Y.; Wang, X.; Fang, J.; Yin, W.; Yan, X.; Tu, M.; Liu, H.; Zhang, Z.; Li, Z.; Gao, M.; et al. VqWRKY56 interacts with VqbZIPC22 in grapevine to promote proanthocyanidin biosynthesis and increase resistance to powdery mildew. *New Phytol.* **2023**, *237*, 1856–1875. [CrossRef]
17. Yu, D.; Wei, W.; Fan, Z.; Chen, J.; You, Y.; Huang, W.; Zhan, J. VabHLH137 promotes proanthocyanidin and anthocyanin biosynthesis and enhances resistance to *Colletotrichum gloeosporioides* in grapevine. *Hortic. Res.* **2023**, *10*, uhac261. [CrossRef]
18. Liu, M.; Zhang, Z.; Xu, Z.; Wang, L.; Chen, C.; Ren, Z. Overexpression of *SlMYB75* enhances resistance to *Botrytis cinerea* and prolongs fruit storage life in tomato. *Plant Cell Rep.* **2021**, *40*, 43–58. [CrossRef] [PubMed]

19. Lagreze, J.; Pajuelo, A.S.; Coculo, D.; Rojas, B.; Pizzio, G.A.; Zhang, C.; Tian, M.B.; Malnoy, M.; Vannozzi, A.; Costa, L.D.; et al. *PME10* is a pectin methylesterase driving PME activity and immunity against *Botrytis cinerea* in grapevine (*Vitis vinifera* L.). *Plant Biotechnol. J.* **2025**. [CrossRef] [PubMed]
20. Santiago, A.; Orduña, L.; Fernández, J.D.; Vidal, Á.; De Martín-Agirre, I.; Lisón, P.; Vidal, E.A.; Navarro-Payá, D.; Matus, J.T. The Plantae Visualization Platform: A comprehensive web-based tool for the integration, visualization, and analysis of omic data across plant and related species. *bioRxiv* **2024**. [CrossRef]
21. Torres, M.A.; Jones, J.D.; Dangl, J.L. Pathogen-induced, NADPH oxidase-derived reactive oxygen intermediates suppress spread of cell death in *Arabidopsis thaliana*. *Nat. Genet.* **2005**, *37*, 1130–1134. [CrossRef]
22. Mostofa, M.G.; Rahman, A.; Ansary, M.M.; Watanabe, A.; Fujita, M.; Tran, L.S. Hydrogen sulfide modulates cadmium-induced physiological and biochemical responses to alleviate cadmium toxicity in rice. *Sci. Rep.* **2015**, *5*, 14078. [CrossRef]
23. Yamada, K.; Basak, A.K.; Goto-Yamada, S.; Tarnawska-Glatt, K.; Hara-Nishimura, I. Vacuolar processing enzymes in the plant life cycle. *New Phytol.* **2020**, *226*, 21–31. [CrossRef]
24. Xin, Y.; Meng, S.; Ma, B.; He, W.; He, N. Mulberry genes *MnANR* and *MnLAR* confer transgenic plants with resistance to *Botrytis cinerea*. *Plant Sci.* **2020**, *296*, 110473. [CrossRef] [PubMed]
25. Deluc, L.; Bogs, J.; Walker, A.R.; Ferrier, T.; Decendit, A.; Merillon, J.M.; Robinson, S.P.; Barrieu, F. The transcription factor VvMYB5b contributes to the regulation of anthocyanin and proanthocyanidin biosynthesis in developing grape berries. *Plant Physiol.* **2008**, *147*, 2041–2053. [CrossRef] [PubMed]
26. Terrier, N.; Torregrosa, L.; Ageorges, A.; Vialet, S.; Verries, C.; Cheynier, V.; Romieu, C. Ectopic expression of VvMybPA2 promotes proanthocyanidin biosynthesis in grapevine and suggests additional targets in the pathway. *Plant Physiol.* **2009**, *149*, 1028–1041. [CrossRef]
27. Koyama, K.; Numata, M.; Nakajima, I.; Goto-Yamamoto, N.; Matsumura, H.; Tanaka, N. Functional characterization of a new grapevine MYB transcription factor and regulation of proanthocyanidin biosynthesis in grapes. *J. Exp. Bot.* **2014**, *65*, 4433–4449. [CrossRef]
28. Cheng, J.; Yu, K.; Shi, Y.; Wang, J.; Duan, C. Transcription factor VviMYB86 oppositely regulates proanthocyanidin and anthocyanin biosynthesis in grape berries. *Front. Plant Sci.* **2020**, *11*, 613677. [CrossRef] [PubMed]
29. Bogs, J.; Downey, M.O.; Harvey, J.S.; Ashton, A.R.; Tanner, G.J.; Robinson, S.P. Proanthocyanidin synthesis and expression of genes encoding leucoanthocyanidin reductase and anthocyanidin reductase in developing grape berries and grapevine leaves. *Plant Physiol.* **2005**, *139*, 652–663. [CrossRef]
30. Xiao, Y.; Wu, X.; Wang, Z.; Ji, K.; Zhao, Y.; Zhang, Y.; Wan, L. Activation and inhibition mechanisms of a plant helper NLR. *Nature* **2025**, *639*, 438–446. [CrossRef]
31. Ma, L.; He, J.; Liu, H.; Zhou, H. The phenylpropanoid pathway affects apple fruit resistance to *Botrytis cinerea*. *J. Phytopathol.* **2018**, *166*, 206–215. [CrossRef]
32. Xue, M.; Yi, H. Enhanced Arabidopsis disease resistance against *Botrytis cinerea* induced by sulfur dioxide. *Ecotoxicol. Environ. Saf.* **2018**, *147*, 523–529. [CrossRef] [PubMed]
33. Jiu, S.; Wang, C.; Zheng, T.; Liu, Z.; Leng, X.; Pervaiz, T.; Lotfi, A.; Fang, J.; Wang, X. Characterization of VvPAL-like promoter from grapevine using transgenic tobacco plants. *Funct. Integr. Genomics* **2016**, *16*, 595–617. [CrossRef]
34. Zhao, T.; Li, R.; Yao, W.; Wang, Y.; Zhang, C.; Li, Y. Genome-wide identification and characterisation of phenylalanine ammonia-lyase gene family in grapevine. *J. Hortic. Sci. Biotech.* **2021**, *96*, 456–468. [CrossRef]
35. Jiang, L.; Jin, P.; Wang, L.; Yu, X.; Wang, H.; Zheng, Y. Methyl jasmonate primes defense responses against *Botrytis cinerea* and reduces disease development in harvested table grapes. *Sci. Hortic.* **2015**, *192*, 218–223. [CrossRef]
36. Wu, B.; Qi, F.; Liang, Y. Fuels for ROS signaling in plant immunity. *Trends Plant Sci.* **2023**, *28*, 1124–1131. [CrossRef]
37. Coll, N.S.; Epple, P.; Dangl, J.L. Programmed cell death in the plant immune system. *Cell Death Differ.* **2011**, *18*, 1247–1256. [CrossRef]
38. Wang, X.; Guo, R.; Tu, M.; Wang, D.; Guo, C.; Wan, R.; Li, Z.; Wang, X. Ectopic expression of the wild grape WRKY transcription factor VqWRKY52 in *Arabidopsis thaliana* enhances resistance to the biotrophic pathogen powdery mildew but not to the necrotrophic pathogen *Botrytis cinerea*. *Front. Plant Sci.* **2017**, *8*, 97. [CrossRef] [PubMed]
39. Zhang, G.; Yan, X.; Zhang, S.; Zhu, Y.; Zhang, X.; Qiao, H.; van Nocker, S.; Li, Z.; Wang, X. The jasmonate-ZIM domain gene *VqJAZ4* from the Chinese wild grape *Vitis quinquangularis* improves resistance to powdery mildew in *Arabidopsis thaliana*. *Plant Physiol. Biochem.* **2019**, *143*, 329–339. [CrossRef]
40. Luo, D.; Cai, J.; Sun, W.; Yang, Q.; Hu, G.; Wang, T. Tomato SlWRKY3 negatively regulates *Botrytis cinerea* resistance via *TPK1b*. *Plants* **2024**, *13*, 1597. [CrossRef]
41. Wang, Z.; Lv, R.; Hong, Y.; Su, C.; Wang, Z.; Zhu, J.; Yang, R.; Wang, R.; Li, Y.; Meng, J.; et al. Transcription factor KUA1 positively regulates tomato resistance against *Phytophthora infestans* by fine-tuning reactive oxygen species accumulation. *Plant J.* **2025**, *121*, e70007. [CrossRef]

42. Yu, K.; Song, Y.; Lin, J.; Dixon, R.A. The complexities of proanthocyanidin biosynthesis and its regulation in plants. *Plant Commun.* **2023**, *4*, 100498. [CrossRef]
43. Rahman, M.U.; Ma, Q.; Ahmad, B.; Hanif, M.; Zhang, Y. Histochemical and microscopic studies predict that grapevine genotype "Ju mei gui" is highly resistant against *Botrytis cinerea*. *Pathogens* **2020**, *9*, 253. [CrossRef]
44. Clough, S.J.; Bent, A.F. Floral dip: A simplified method for Agrobacterium-mediated transformation of *Arabidopsis thaliana*. *Plant J.* **1998**, *16*, 735–743. [CrossRef]
45. Fryer, M.J.; Oxborough, K.; Mullineaux, P.M.; Baker, N.R. Imaging of photo-oxidative stress responses in leaves. *J. Exp. Bot.* **2002**, *53*, 1249–1254. [CrossRef]
46. Kim, S.H.; Woo, D.H.; Kim, J.M.; Lee, S.Y.; Chung, W.S.; Moon, Y.H. *Arabidopsis* MKK4 mediates osmotic-stress response via its regulation of MPK3 activity. *Biochem. Biophys. Res. Commun.* **2011**, *412*, 150–154. [CrossRef] [PubMed]
47. Vogel, J.; Somerville, S. Isolation and characterization of powdery mildew-resistant Arabidopsis mutants. *Proc. Natl. Acad. Sci. USA* **2000**, *97*, 1897–1902. [CrossRef] [PubMed]
48. Han, Y.; Wu, M.; Hao, L.; Yi, H. Sulfur dioxide derivatives alleviate cadmium toxicity by enhancing antioxidant defence and reducing Cd^{2+} uptake and translocation in foxtail millet seedlings. *Ecotoxicol. Environ. Saf.* **2018**, *157*, 207–215. [CrossRef]
49. Yan, D.; Yi, H. Transcriptome analysis provides insights into preservation mechanism of postharvest Muscat Hamburg grapes treated with SO_2. *Sci. Hortic.* **2024**, *331*, 113108. [CrossRef]
50. Livak, K.J.; Schmittgen, T.D. Analysis of relative gene expression data using real-time quantitative PCR and the $2^{-\Delta\Delta Ct}$ method. *Methods* **2001**, *25*, 402–408. [CrossRef]
51. Xie, J.; He, C.; Li, Z.; Li, M.; He, S.; Qian, J.; Tan, B.; Zheng, X.; Cheng, J.; Wang, W.; et al. A rapid and efficient Agrobacterium-mediated transient transformation system in grape berries. *Protoplasma* **2024**, *261*, 819–830. [CrossRef] [PubMed]

plants

MDPI

Article

Leaf Blight in *Ilex verticillata* Caused by *Alternaria alternata*: Mechanisms of Antioxidant Defense, Phytohormone Crosstalk, and Oxidative Stress Responses

Huijie Lu [1,2,†], Caixia Zhou [1,2,†], Peiwen Cheng [1,2,†], Liangye Huang [1,2], Qinyuan Shen [1,2], Ye Zheng [1,2], Yihui Li [1,2], Wenjun Dai [1,2], Jianhong Zhang [3], Dengfeng Shen [3], Anket Sharma [1,2], Muhammad Junaid Rao [1,2,*], Bingsong Zheng [1,2,*] and Huwei Yuan [1,2,*]

[1] National Key Laboratory for Development and Utilization of Forest Food Resources, Zhejiang A&F University, Hangzhou 311300, China; 19857999948@163.com (H.L.); 2022602122157@stu.zafu.edu.cn (C.Z.); chengpeiwen0826@163.com (P.C.); 2023602121036@stu.zafu.edu.cn (L.H.); 2024202011016@stu.zafu.edu.cn (Q.S.); 2024102012044@stu.zafu.edu.cn (Y.Z.); 2024602122054@stu.zafu.edu.cn (Y.L.); 2023602122015@stu.zafu.edu.cn (W.D.); anketsharma@gmail.com (A.S.)

[2] Provincial Key Laboratory for Non-Wood Forest and Quality Control and Utilization of Its Products, Zhejiang A&F University, Hangzhou 311300, China

[3] Ningbo Key Laboratory of Characteristic Horticultural Crops in Quality Adjustment and Resistance Breeding, Ningbo Academy of Agricultural Sciences, Ningbo 315041, China; nbjianhong@163.com (J.Z.); dengfengdz@163.com (D.S.)

[*] Correspondence: mjr@zafu.edu.cn (M.J.R.); bszheng@zafu.edu.cn (B.Z.); hwyuan@zafu.edu.cn (H.Y.); Tel.: +86-136-8576-1954 (H.Y.)

[†] These authors contributed equally to this work.

Abstract: *Ilex verticillata* (winterberry) is a valuable ornamental shrub increasingly threatened by leaf blight, a disease that compromises its aesthetic and economic value. While fungal pathogens like *Alternaria alternata* are known to cause leaf blight in horticultural crops, their role in *I. verticillata* and the host's defense mechanisms have not been fully characterized. Our study investigated the pathogen-host interaction by identifying the causal agent and examining the physiological and molecular defense mechanisms of *I. verticillata*. Through morphological and multi-locus molecular analyses (ITS, TEF1-α, G3PDH, RPB2), *A. alternata* was confirmed as the primary pathogen, fulfilling Koch's postulates. Pathogenicity assays revealed distinct disease progression stages, from necrotic lesions to tissue degradation. Transcriptomic profiling uncovered dynamic host responses, with early upregulation of pattern recognition receptors (PRRs) and transcripts encoding antioxidant enzymes (SOD, CAT), followed by downregulation of metabolic pathway genes. Phytohormone analysis highlighted intricate crosstalk, with salicylic acid (SA) peaking during mid-infection and jasmonic acid (JA) rebounding later, reflecting a coordinated defense strategy. Additionally, the oxidative stress marker malondialdehyde (MDA), an indicator of membrane lipid peroxidation, surged early, indicating membrane damage, while sustained induction of antioxidant enzymes suggested adaptive responses. The key finding was distinct phytohormone crosstalk, characterized by a mid-infection SA peak followed by a late JA rebound, alongside an early oxidative burst marked by MDA accumulation and sustained antioxidant enzyme activity. These findings provide a framework for understanding *I. verticillata's* defense mechanisms and offer insights for developing targeted disease management strategies, such as resistant cultivar breeding or hormone-mediated interventions.

Keywords: *Alternaria alternata*; *Ilex verticillata*; leaf blight; host–pathogen interaction; transcriptomics; antioxidant defense; phytohormones; oxidative stress

https://doi.org/10.3390/plants14193057

1. Introduction

Ilex verticillata (winterberry) is a deciduous shrub native to the northeastern United States [1], valued in horticulture for its persistent, vibrant red berries [2]. Introduced to China in 2006, it is now cultivated for landscaping and slope stabilization [1,2]. However, successful establishment outside its native range is challenged by adaptation issues, including significant susceptibility to pests and diseases such as spider mites, grey mold, and notably, leaf blight, which compromises its aesthetic and economic value [3,4]. However, field observations indicate that leaf blight causes severe foliar damage and fruit loss, potentially leading to plant death and constraining the cultivation of this valuable species [5].

Leaf blight, often favored by warm, humid conditions [6], is frequently caused by fungal pathogens in the genus *Alternaria* (species *A. alternata*, *A. tenuissima*, etc.) [7–9]. Fungal pathogens in the genus *Alternaria* are notorious for their broad host range, which includes numerous ornamental species (lilies and poplar), highlighting a significant threat to horticultural crops [10–13]. These fungi produce toxins that disrupt cellular functions, leading to lesion formation and tissue decay. Similarly, *Colletotrichum* species, such as *C. siamense*, have been identified as pathogens causing leaf blight in *I. verticillata* [14]. *Alternaria alternata*, a ubiquitous species, is particularly destructive, inducing leaf spots, blight, and rot through the production of phytotoxins and enzymatic degradation of host tissues [10,11]. While *A. alternata* has been implicated in leaf blight of other *Ilex* species and related plants, its role in *I. verticillata* leaf blight has not been definitively established. Accurate pathogen identification, coupled with an understanding of its interaction with the host, is critical for developing targeted control measures.

Plants defend against necrotrophic pathogens like *Alternaria* spp. through an integrated response involving structural reinforcement, antioxidant systems, and phytohormone signaling [15–18]. Early infection often triggers oxidative burst, leading to membrane damage marked by lipid peroxidation and malondialdehyde (MDA) accumulation [19], as observed in *Osmanthus fragrans* and *Phoebe zhennan* following *Alternaria* challenge [20,21]. To mitigate oxidative stress, plants activate their antioxidant enzymes such as superoxide dismutase (SOD), peroxidase (POD), and catalase (CAT) [13,22–24], though prolonged infection may suppress their activity [25]. Simultaneously, pathogen recognition via pattern recognition receptors (PRRs) can initiate a hypersensitive response (HR) and reinforce structural defenses through secondary metabolites like phenylpropanoids and pathogenesis-related (PR) proteins, including phenylalanine ammonia-lyase (PAL) and chitinase (CHI) [18,26,27]. At the molecular level, Transcriptomic studies in chrysanthemum and tree peony (*Paeonia suffruticosa* Andrews) have further highlighted the importance of jasmonic acid (JA) and salicylic acid (SA) pathways in regulating defense against *A. alternata* [28,29], with key regulators involved in JA signaling such as JAZ and MYC2 [28–31]. Similarly, WRKY and NAC transcription factors are differentially expressed in populus and tobacco (*Nicotiana attenuata*), in response to *A. alternata* infection [32,33]. However, despite these advances in related species, the defense response of *I. verticillata* to *A. alternata* remains uncharacterized. Examining these mechanisms in winterberry may identify key resistance traits applicable to breeding programs.

However, in contrast to well-studied pathosystems like *Alternaria*-tomato or *Alternaria*-apple tissues [10,11], the causal agent of leaf blight in *I. verticillata* and the host's defense responses are not well-defined [1,2]. This knowledge gap hinders the development of targeted and sustainable management strategies. Therefore, this study had three objectives: (1) to isolate and identify the causal agent of leaf blight in *I. verticillata* 'Meister' through morphological, molecular, and pathogenicity analyses; (2) to characterize the host's physiological response by measuring oxidative stress markers and antioxidant enzyme activities; and (3) to profile the transcriptional changes during infection to identify key defense-related pathways.

2. Results

2.1. Isolation and Purification of the Pathogen Causing Holly Leaf Blight

Healthy *I. verticillata* leaves showed no symptoms (Figure 1A) and yielded no pathogens (Figure 1B). In contrast, diseased leaves exhibited black spots, curling, and desiccation (Figure 1C). From these symptomatic tissues, we isolated three distinct fungal strains (Y9-2-1, Y10, Y16; Figure 1D). The clear contrast between healthy and diseased samples identified these isolates as putative causative agents. Subsequent purification enabled morphological and molecular characterization to identify the leaf blight pathogens.

Figure 1. Isolation of the pathogen of leaf blight disease in *I. verticillate*. (**A**): Healthy leaves, (**B**): Tissue isolation results of healthy leaves, (**C**): Diseased leaves, (**D**): Tissue isolation results of diseased leaves.

2.2. Morphological Characteristics and Phylogenetic Analysis of Whorled Holly Leaf Blight Pathogen

Morphological analysis revealed that strains Y9-2-1, Y10, and Y16 shared similar cultural and microscopic features (Figure 2). Colonies initially displayed yellow-brown pigmentation on the reverse side (Figure 2B,F,J), darkening to brown or black with maturity. They exhibited dense aerial hyphae (Figure 2A,E,I) and orderly marginal growth. Conidiophores were solitary or clustered, and light to dark brown in color (Figure 2D,H,L). Conidia were observed singly or in chains, with smooth surfaces, distinct transverse and longitudinal septa, and an inverted club-shaped to ovoid morphology (Figure 2C,G,K). The high degree of morphological similarity among isolates Y9-2-1, Y10, and Y16 supported their preliminary identification as *Alternaria* sp. and indicated they likely represent the same strain.

Phylogenetic analysis based on four genetic markers (G3PDH ~800 bp, ITS ~500 bp, TEF1-α ~250 bp, RPB2 ~1100 bp) were used to identify the fungal isolates and all isolates produced PCR amplicons of the expected sizes (800 bp, 500 bp, 250 bp, 1100 bp, respectively) for *Alternaria* species (Figure 3A–D). The concatenated sequence analysis placed isolates Y9-2-1, Y10, and Y16 within a strongly supported clade (100% bootstrap) alongside reference strains of *A. alternata* (Figure 3E). The high sequence similarity across all four loci and their consistent phylogenetic placement provide conclusive evidence for identifying the pathogen as *A. alternata*. Furthermore, the identical sequences for each locus among the three isolates indicate they represent a single strain responsible for the leaf blight symptoms on *I. verticillata*.

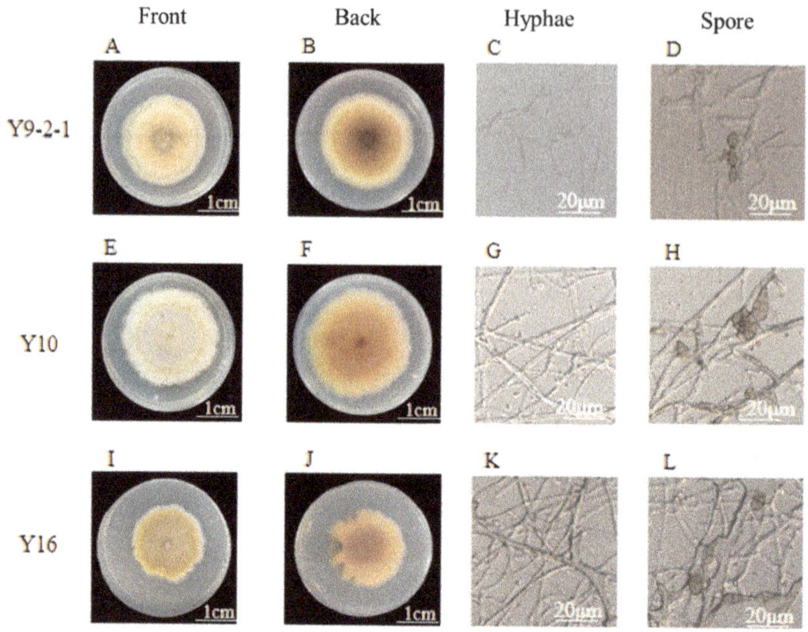

Figure 2. Cultural and morphological characteristics of *Alternaria* isolates causing leaf blight disease of *I. verticillata*. (**A**,**E**,**I**) Front view of colonies for strains Y9-2-1, Y10, and Y16, respectively, showing dense aerial hyphae with neat margins. (**B**,**F**,**J**) Reverse side of colonies displaying yellow-brown pigmentation in early growth, darkening to brown or black with maturity. (**C**,**G**,**K**) Microscopic features of hyphae. (**D**,**H**,**L**) Conidia observed singly or in chains, exhibiting smooth surfaces, distinct transverse and longitudinal septa, and an inverted club-shaped to ovoid morphology. The high degree of morphological similarity among the three isolates supports their preliminary identification as *Alternaria* sp. and indicates the same strain. Scale bars: 1 cm and 20 μm.

2.3. Pathogenicity Assessment of *Alternaria alternata* on *Ilex verticillata*

Our detached-leaf assays showed *A. alternata* caused distinct necrotic lesions within 3 days (Figure 4A). By day 5, the disease had progressed to 100% incidence, with lesions expanding to the entire leaf surface. The identity of the pathogen was confirmed by re-isolating fungi from the diseased tissue, which was morphologically identical to the original inoculant and confirmed as *A. alternata* by molecular analysis (Supplementary Figure S1A,B), thereby fulfilling Koch's postulates. Figure 4B shows healthy control leaves. Potted plant trials mirrored natural infections. Initial 2–5 mm black lesions expanded with characteristic leaf curling (Figure 4C), progressing to tissue collapse and perforation, identical to field symptoms. The consistent disease development across all replicates in both detached-leaf and potted plant assays demonstrates the pathogenicity of *A. alternata* in *I. verticillata*. The congruence of results from these complementary experimental systems provides strong evidence supporting *A. alternata* as the causative agent of leaf blight.

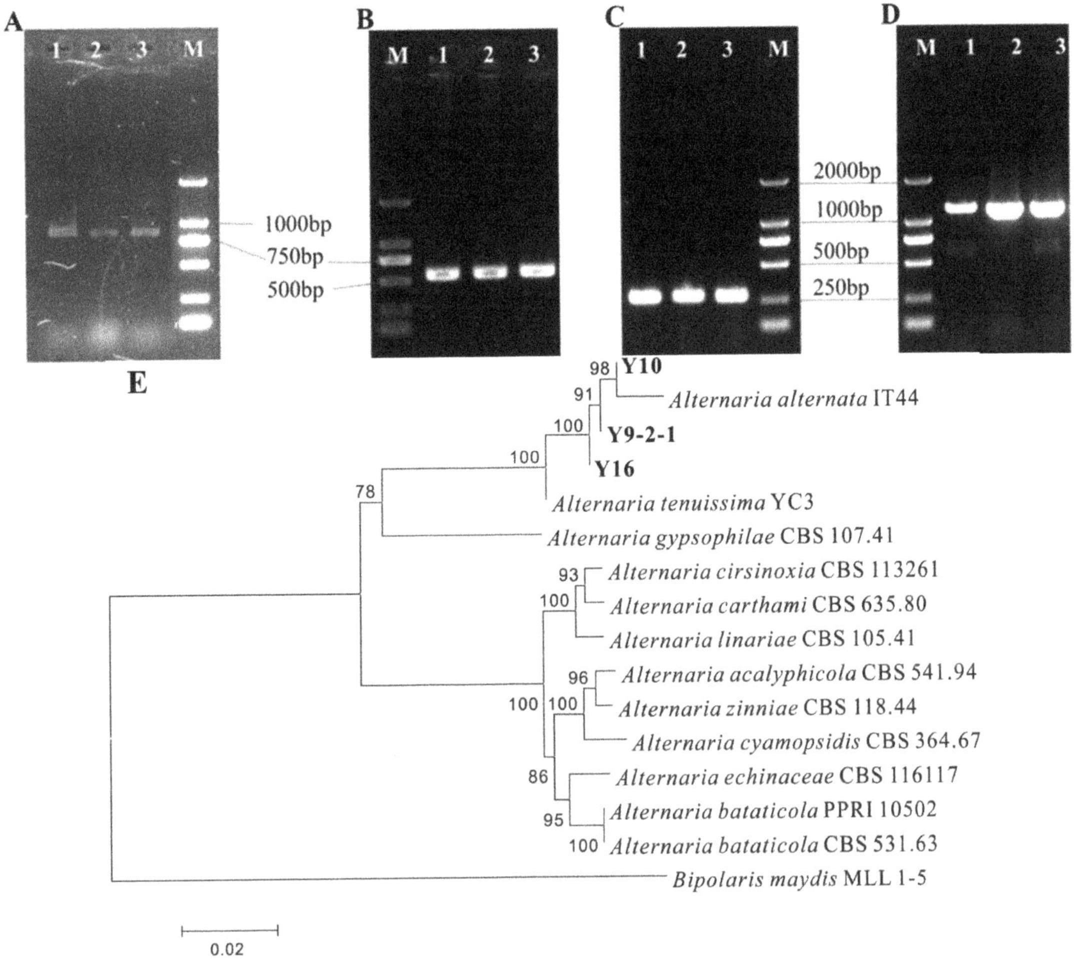

Figure 3. Molecular identification and phylogenetic analysis of *Alternaria* isolates causing leaf blight disease of *I. verticillate*. (**A–D**) Agarose gel electrophoresis of PCR amplicons for the (**A**) G3PDH, (**B**) ITS, (**C**) TEF1-α, and (**D**) RPB2 gene regions. Lanes: 1, isolate Y9-2-1; 2, isolate Y10; 3, isolate Y16; M, DL2000 DNA Marker. (**E**): Phylogenetic tree constructed from the concatenated sequences of the ITS, TEF1-α, RPB2, and G3PDH loci using the Neighbor-Joining method. The numbers at the nodes represent bootstrap percentages based on 1000 replicates; only values ≥ 50% are shown. The scale bar represents 0.02 nucleotide substitutions per site. Bold characters represent the strain isolated in *I. verticillate*.

In inoculated 'Meister' seedlings, *A. alternata* infection progressed through distinct stages (Figure 4D). Small black necrotic spots (1–3 mm) appeared by 3 dpi, expanding into rusty brown lesions with leaf curling by 12 dpi. By 19 dpi, severe tissue degradation caused dry, perforated lesions. This clear progression from localized spots to widespread damage confirms the pathogen's aggressive colonization pattern in whorled holly.

Figure 4. Disease progression, pathogenicity, and temporal progression on *I. verticillata* leaves following inoculation with *Alternaria alternata*. (**A**): *A. alternata* isolates Y9-2-1, Y10, and Y16 inoculation compared to non-inoculated control (CK). Photographs were taken at 3 days (3 d) and 5 days (5 d) post-inoculation, showing characteristic leaf blight symptoms including necrotic lesions and chlorosis. (**B**): Healthy, non-inoculated control leaves showing normal morphology. (**C**): Diseased leaves at 5 days post-inoculation with *A. alternata* isolates, exhibiting severe blight symptoms including necrotic lesions, chlorotic halos, and tissue collapse. (**D**): Temporal progression of *Alternaria alternata* infection on *I. verticillata* leaves. Leaves were inoculated with either sterile potato dextrose agar medium (PDA, control) or *A. alternata* conidial suspension (Aa). Initial symptoms (3 dpi) appeared as small necrotic spots that expanded into coalescing lesions with chlorotic halos by 12 dpi, culminating in extensive tissue necrosis by 19 dpi. PDA controls showed no pathological changes throughout the observation period.

2.4. *Transcriptomic Profiling of Host Responses During Alternaria alternata Infection*

2.4.1. Global Differential Gene Expression Patterns

RNA-seq revealed dramatic gene expression changes in *I. verticillata* leaves during *A. alternata* infection (Figure 5). Differentially expressed genes (DEGs) were identified using DESeq2 with a threshold of false discovery rate (FDR) < 0.05 and |log~2~Fold Change| ≥ 1. Using these criteria, we observed 1566 DEGs at 3 dpi (895 up- and 671 down-regulated), increasing to 4047 DEGs by 12 dpi (1503 up- and 2544 down-regulated) and 4322 DEGs at 19 dpi (1723 up- and 2599 down-regulated) (Figure 5A). This shifts from early upregulation (stress response) to late-stage downregulation (plant hormone signal transduction; Supplementary Figure S2). Comparing timepoints revealed major transcriptional shifts: 12 dpi introduced 1531 new upregulated and 2583 downregulated genes versus 3 dpi, while 19 dpi added 1737 up- /2745 down-regulated DEGs. Late-stage changes (12 vs. 19 dpi) were more moderate (631↑/360↓).

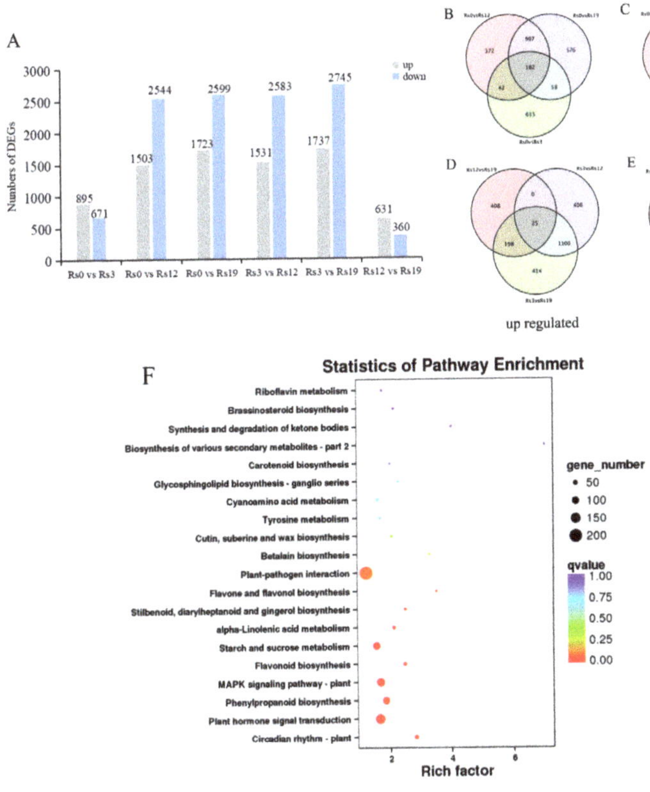

Figure 5. Temporal dynamics of DEGs (FDR < 0.05 and |log~2~Fold Change| ≥ 1) in *Ilex verticillata* leaves during *Alternaria alternata* infection. (**A**): Histogram showing the total number of DEGs identified at 3-, 12-, and 19 days post-inoculation (dpi) compared to uninoculated controls (0 dpi). (**B**,**C**) Venn diagrams illustrating (**B**) upregulated and (**C**) downregulated DEGs at each time point relative to 0 dpi controls, revealing stage-specific transcriptional responses. (**D**,**E**) Comparative Venn analyses of (**D**) upregulated and (**E**) downregulated DEGs across all infection timepoints (3 vs. 12 vs. 19 dpi), highlighting conserved and unique gene expression patterns during disease progression. (**F**): Top 20 significantly enriched KEGG pathways (FDR < 0.05) during *A. alternata* infection. The Rich Factor (x-axis) is the ratio of the number of DEGs to the total number of genes in a given pathway. The size of the bubble corresponds to the number of DEGs assigned to the pathway, and the color represents the -log10 transformed *p*-value from the enrichment analysis.

2.4.2. Stage-Specific Transcriptional Signatures

Venn analyses identified stage-specific responses (Figure 5B–E). Early infection featured 42 consistently upregulated and 63 downregulated genes (3–12 dpi), while late infection shared 907 up- /1546 down-regulated DEGs (12–19 dpi). Notably, 198 upregulated and 102 downregulated genes maintained their expression from early through mid-infection, suggesting their potential involvement in defense responses. The complete lack of overlapping DEGs between 3 and 19 dpi (using 12 dpi as reference) indicates a wholesale transcriptional reprogramming as infection progresses.

2.4.3. KEGG Enrichment Analysis of Differential Gene Expression

KEGG analysis revealed *I. verticillata's* coordinated defense against *A. alternata* through three key strategies (Figure 5F): First, pathogen recognition via Plant-pathogen interaction pathways (28 DEGs). Second, intracellular signaling through MAPK (19 DEGs) and hormone transduction (34 DEGs). Third, the production of antimicrobial compounds was activated, as indicated by the enrichment of the Phenylpropanoid and Flavonoid biosynthesis pathways, which are involved in synthesizing defensive secondary metabolites. Concurrently, the plant reshaped its primary metabolism, with changes in Starch and sucrose metabolism potentially indicating a resource reallocation to defense. The enrichment of the α-Linolenic acid pathway, a precursor for jasmonic acid, confirmed the activation of JA-mediated signaling. Furthermore, the induction of Cutin, suberine, and wax biosynthesis suggested a reinforcement of physical barriers at the leaf surface. Lastly, the modulation of Brassinosteroid pathways pointed to a complex hormonal coordination of the overall defense strategy. This multi-tiered response mirrors known plant defense mechanisms: rapid pathogen detection triggers signaling cascades that mobilize both chemical defenses (antimicrobials) and physical barriers (waxes). The conserved MAPK and jasmonate pathways highlight evolutionary convergence in fungal defense strategies. Together, these pathways reveal how whorled holly reconfigures its physiology to detect, signal, and combat fungal invasion at multiple levels.

2.5. *Transcriptional Dynamics of Plant-Pathogen Interaction Genes During Infection*

I. verticillata initiate defense against *A. alternata* through precisely timed gene regulation (Figure 6A–C). Key pattern recognition receptors showed phased activation—CERK1 and XA21 peaked early (3 dpi), while specific EIX1/2 genes gradually increased until 12–19 dpi (Figure 6A). This sequential PRR deployment suggests an evolving detection strategy matching pathogen progression. Hypersensitive response genes exhibited complex temporal patterns (Figure 6B). Calcium signaling components (CNGC/CPK) and R-protein regulators (RPM1) showed opposing trends, while the HR inhibitor KCS was consistently suppressed, a pattern that may facilitate the progression of cell death. Early responders like PBS1, PTI0, and FLS2 were rapidly induced and then downregulated, marking the transition from detection to defense execution. Genes associated with structural defenses also showed nuanced regulation, particularly those involved in calcium-mediated signaling (Figure 6C). Calcium signaling components, such as calmodulin-like proteins (CMLs) and calmodulins (CALMs), displayed divergent responses: seven CMLs and one CALM were transiently induced during early infection (0–3 dpi), potentially contributing to the initial activation of defense responses. In contrast, five CMLs and three CALMs were suppressed, while a single CALM maintained elevated expression, suggesting a complex and finely tuned modulation of calcium signals to orchestrate downstream defense processes. Cell death regulators indicated a carefully managed hypersensitive response. WRKY2, a transcription factor often associated with promoting programmed cell death, remained active until 12 dpi before declining. Conversely, EDS1, a central regulator of salicylic acid (SA)-mediated resistance and cell death, along with most WRKY1 homologs, was suppressed, which may serve to limit excessive tissue damage. One WRKY1 variant showed early activation followed by suppression, hinting at isoform-specific roles in the defense response. These patterns reveal a multi-layered defense strategy: (1) Immediate pathogen detection through phased PRR activation, (2) Precise control of hypersensitive response via opposing regulators, and (3) Dynamic modulation of structural defenses and cell death pathways. The coordinated yet flexible gene regulation allows *I. verticillata* to balance effective defense with resource conservation during prolonged infection.

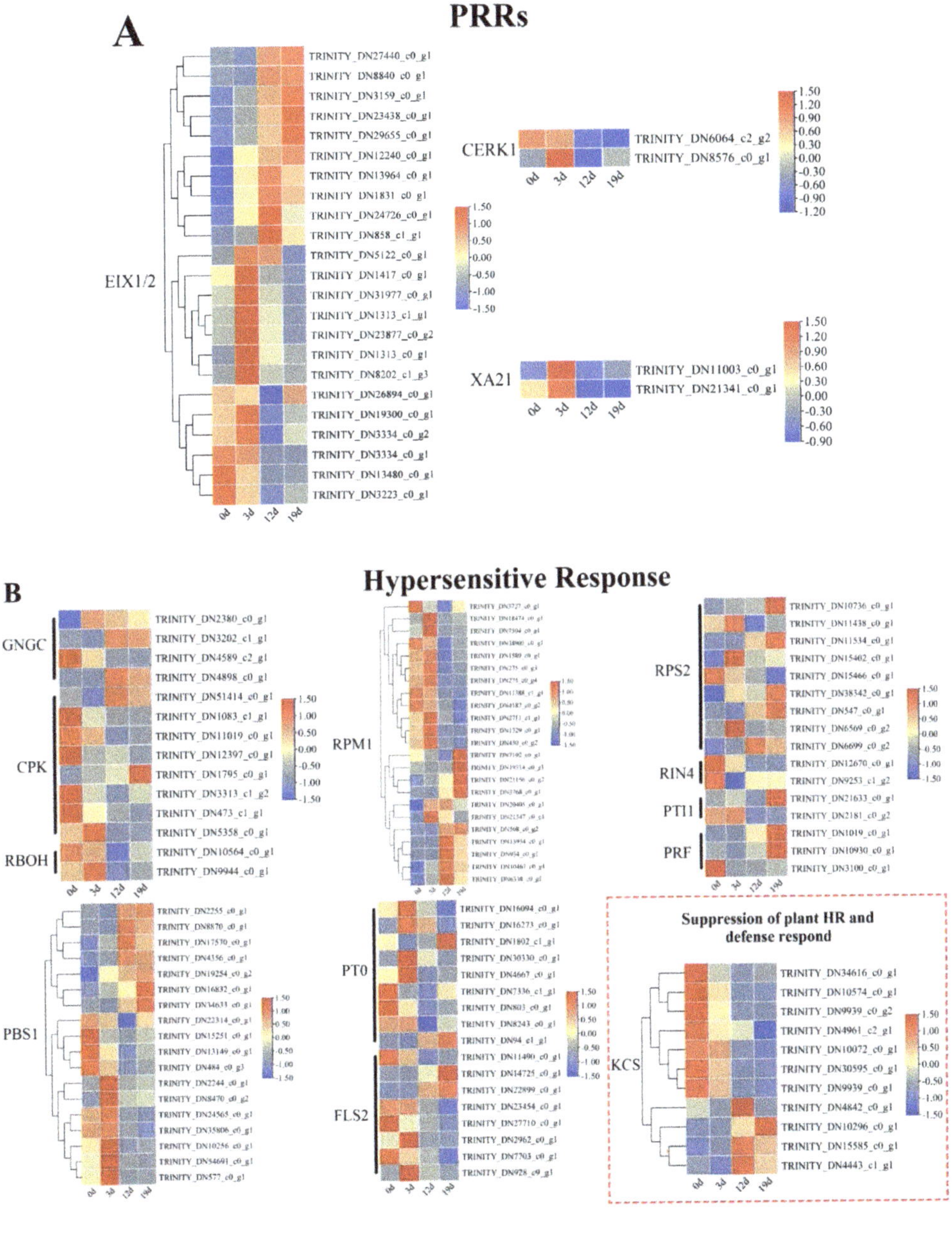

Figure 5. *Cont.*

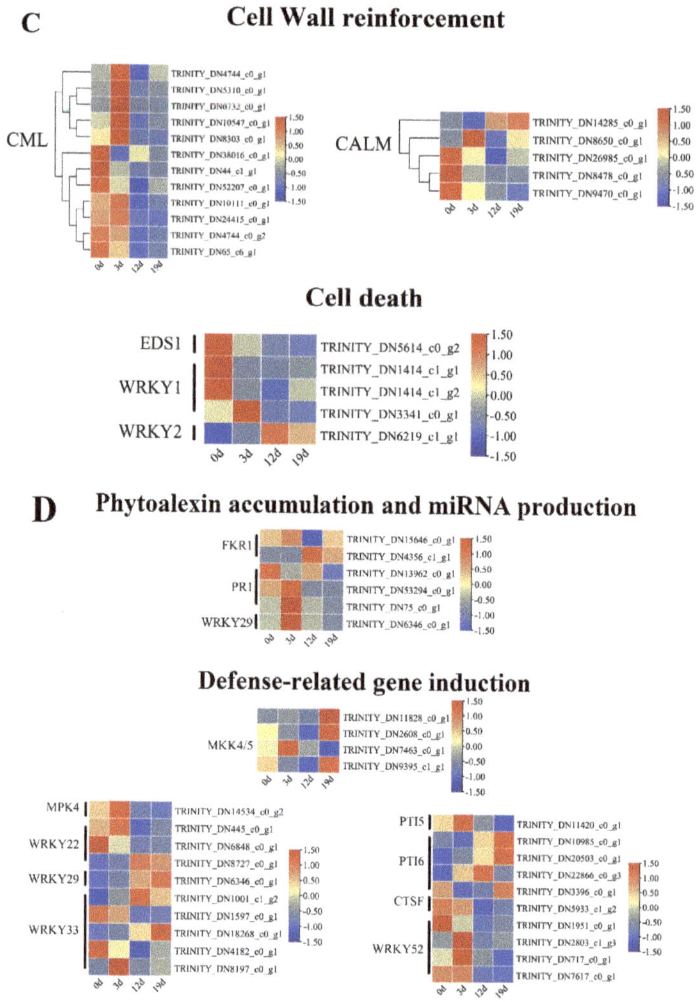

Figure 6. Expression analysis of DEGs (FDR < 0.05, |log2Fold Change| ≥ 1) related to the plant-pathogen interaction pathway. Heatmaps show the Z-score normalized expression of genes that were significantly differentially expressed (False Discovery Rate, FDR < 0.05 and |log2Fold Change| ≥ 1) at one or more time points during *A. alternata* infection (3, 12, 19 dpi) compared to the uninoculated control (0 dpi). (**A**): DEGs encoding pattern recognition receptors (PRRs). (**B**): DEGs associated with the hypersensitive response (HR). (**C**): DEGs involved in cell wall reinforcement and cell death regulation. (**D**): DEGs related to defense-related gene induction, phytoalexin accumulation, and miRNA production.

Early infection (0–3 dpi) triggered transient phytoalexin genes (FKR1, PR1, WRKY29), while defense signaling components (MKK4/5, MPK4, PT6) remain active longer (Figure 6D). This dual strategy balances immediate chemical defenses with sustained signaling, helping *I. verticillata* combat initial infection while maintaining resistance as the pathogen advances.

2.6. Dynamic Changes in Endogenous Hormone Levels During Pathogen Infection

2.6.1. Variation in Stress-Related Phytohormones During Pathogenesis

I. verticillata showed dynamic hormone changes during infection (Figure 7A–C). ABA decreased 12.6% by 3 dpi, partially recovering later. SA peaked at 12 dpi (+8.8%; for absolute values see Supplementary Table S2), then dropped sharply, suggesting transient resistance activation. JA showed opposite trends to SA, declining initially but rebounding by 19 dpi. These coordinated, opposing patterns reveal sophisticated hormonal crosstalk regulating defense strategies against *A. alternata*.

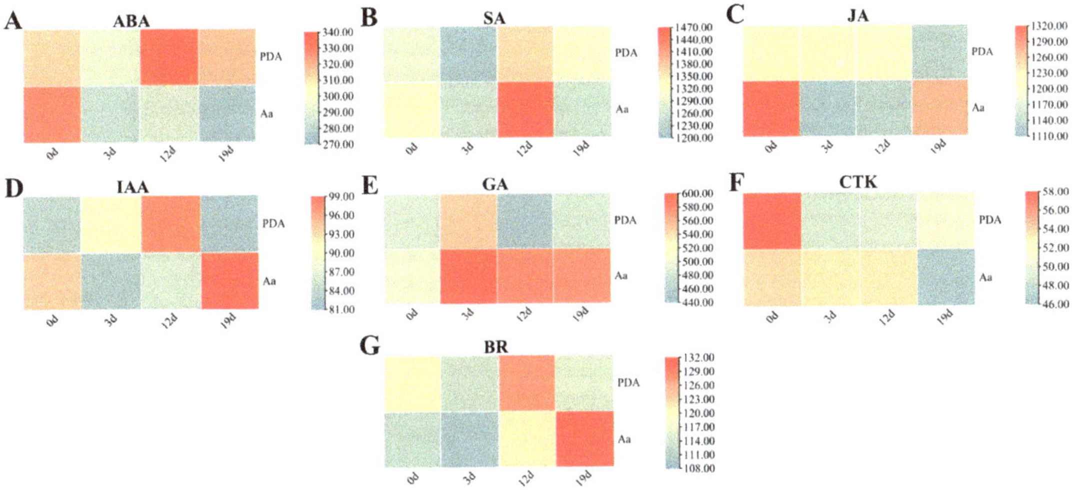

Figure 7. Changes in endogenous hormone levels in *Ilex verticillata* leaf during infection with the leaf blight pathogen. (**A**): ABA content (μg·g^{-1} FW). (**B**): SA content (pmol·g^{-1} FW). (**C**): JA content (pmol·g^{-1} FW). (**D**): IAA content (μg·g^{-1} FW). (**E**): GA content (pg·mL^{-1} FW). (**F**): CTK content (μg·L^{-1} FW). (**G**): BR content (ng·L^{-1} FW). The absolute hormone concentrations and mean values (n = 3) for each time point are provided in Supplementary Table S2. The gradient color from blue to red indicates the hormone content from small to large. PDA: *I. verticillata* leaves inoculated with sterile potato dextrose agar medium; Aa: *I. verticillata* leaves inoculated with *A. alternata*; 0 d, 3 d, 12 d, 19 d means 0, 3, 12, 19 days after infection. Note: Hormone levels were quantified using enzyme-linked immunosorbent assay (ELISA). Concentrations are reported in the specific units defined by the standard curves of the respective commercial ELISA kits used for each hormone; therefore, absolute values are not directly comparable across different hormones. The relative changes within each hormone over time are the critical metric.

2.6.2. Modulation of Growth-Related Hormones During Pathogenesis

A. alternata infection disrupted growth-related hormones in *I. verticillata* (Figure 7D–G). Auxin showed a V-shaped pattern, decreasing 7.9% by 12 dpi before rebounding 14.3% above mid-infection levels. GA peaked early (16.7% increase at 3 dpi) and remained elevated (+13.6% at 19 dpi). CTK levels remained stable initially but dropped sharply (11.6%) during late infection (12–19 dpi), potentially regulating senescence. BR displayed unique dynamics—increasing 7.4% in infected plants while decreasing in controls, suggesting pathogen manipulation of host defenses. These coordinated changes reveal: (1) Early GA-mediated growth responses; (2) Late-stage CTK-linked senescence; (3) Pathogen-induced BR manipulation; and (4) Recovery of auxin signaling. The distinct temporal patterns demonstrate how *A. alternata* alters host growth physiology while *I. verticillata* attempts to rebalance its hormonal defenses throughout infection progression.

with the activation of canonical defense signaling. Furthermore, the modulation of ABA biosynthesis and catabolism genes aligns with its suggested role in stress responses [44–46]. While the hormone quantification by ELISA indicates relative changes, these trends are corroborated by the congruent transcriptomic data. Collectively, these results suggest a model of a coordinated hormonal network contributing to the defense strategy of *I. verticillata*.

Our findings offer several specific avenues for managing *I. verticillata* leaf blight. The conclusive identification of *A. alternata* as the causal agent immediately informs the selection of effective fungicides known to target this pathogen. Furthermore, the key defense components (such as the early upregulation of the pattern recognition receptor *CERK1*, the sustained induction of SOD and CAT, and the distinct SA/JA hormonal crosstalk) provide concrete targets for molecular breeding. These genes could serve as biomarkers for screening resistant germplasm or be prioritized for genetic engineering. The temporal dynamics of the SA and JA pathways suggest that timed exogenous application of these hormones could potentially potentiate the plant's innate defense response, a hypothesis that warrants direct experimental testing [44]. Future work should focus on the functional validation of candidate genes like *CERK1* and *WRKY2* through transgenic approaches, and on investigating how the root microbiome influences the observed susceptibility to *A. alternata*. By building directly upon the molecular framework established here, future strategies can be developed to enhance the resilience of this valuable horticultural species.

4. Materials and Methods

4.1. Plant Material Collection and Preparation

Whorled holly (*Ilex verticillata* 'Meister') leaves were collected at 0-, 3-, 12-, and 19 days post-inoculation (dpi) with leaf blight pathogens, along with control leaves treated with sterile potato dextrose agar (PDA) medium. These time points were selected to correspond with distinct stages of disease progression observed in preliminary pathogenicity assays: the early symptom appearance (3 dpi), the active mid-stage (12 dpi), and the advanced tissue degradation phase (19 dpi). All collected samples were immediately flash-frozen in liquid nitrogen and stored at −80 °C until further analysis. The study utilized two-year-old cuttings of *I. verticillata* 'Meister', with healthy control plants provided by Hangzhou Runtu Horticulture Technology Co., Ltd (Hangzhou, China). Diseased plant materials were collected from field sites during November 2021 and October 2022 to enhance the representativeness of the pathogen isolates by sampling across different seasons (autumn) and years.

4.2. Experimental Design and Growth Conditions

The experiment was conducted using two-year-old *Ilex* 'Meister' cuttings maintained under controlled greenhouse conditions at Zhejiang Agricultural and Forestry University. Environmental parameters were maintained at 25–28 °C with 70–80% relative humidity throughout the study period. Plants were divided into two treatment groups: (1) control plants inoculated with sterile PDA medium and (2) experimental plants inoculated with purified pathogen strains. The inoculation procedure involved placing mycelial plugs (5 mm diameter) from actively growing cultures onto surface-sterilized leaves, with appropriate controls to validate the pathogenicity tests. Following inoculation, disease progression was monitored regularly, and leaf samples were collected at predetermined intervals for subsequent biochemical and molecular analyses.

4.3. Pathogen Isolation and Culture

Fungal pathogens were isolated from symptomatic leaf tissues collected at the junction of diseased and healthy areas. Tissue sections (0.5 × 0.5 cm) were surface sterilized in

75% ethanol for 1 min, followed by three rinses with sterile distilled water. The sterilized tissues were aseptically transferred to PDA plates and incubated at 28 °C for 2–3 days. Pure cultures were obtained through hyphal-tip subculturing onto fresh PDA medium. Three morphologically distinct isolates (designated Y9-2-1, Y10, and Y16) were selected for further characterization and maintained as stock cultures on PDA slants at 4 °C for short-term storage and in 20% glycerol at −80 °C for long-term preservation.

4.4. Morphological Characterization and Molecular Identification

Isolates were cultured on PDA at 28 °C for 4–6 days. Colony morphology (color, texture, growth pattern) was documented photographically. Spore structures were examined microscopically using the slide culture technique: sterile slides were inserted at 45° angles into PDA plates, allowing hyphal growth along the slide surface for direct microscopic observation. Microscopic examination was carried out using an upright microscope (Olympus BX53, Shanghai Lai's Electronic Technology Co., Ltd. [Shanghai, China]) at magnifications of 100×, 400×, and 1000× (using a 100× oil immersion objective lens for 1000× magnification). The identification of fungal structures (conidiophores, conidia) was based on standard taxonomic keys for *Alternaria* species, with key diagnostic criteria including: the morphology and coloration of conidiophores; and for conidia, their size, shape, the presence of both transverse and longitudinal septa, and their formation in chains.

For DNA extraction, fungal mycelium (100 mg) was ground in liquid nitrogen and processed using a modified CTAB protocol. Samples were lysed in CTAB buffer (with β-mercaptoethanol) at 65 °C for 20 min, followed by chloroform:isoamyl alcohol (24:1) extraction and isopropanol precipitation. DNA pellets were washed with 75% ethanol, air-dried, and resuspended in nuclease-free water. Quality was verified via spectrophotometry and 1.2% agarose gel electrophoresis. For phylogenetic identification of Alternaria isolates, a standard multi-locus sequence analysis (MLSA) approach was employed, targeting four genetic loci: the internal transcribed spacer (*ITS*) region, and portions of the translation elongation factor 1-alpha (*TEF1-α*), glyceraldehyde-3-phosphate dehydrogenase (*G3PDH*), and RNA polymerase II second largest subunit (*RPB2*) genes. This combination provides a balance between universal applicability (ITS) and higher phylogenetic resolution from protein-coding genes. These four loci were amplified (PCR conditions: 95 °C/5 min; 35 cycles of 95 °C/30 s, 55 °C/30 s, 72 °C/30 s; 72 °C/5 min). Primer sequences are listed in Table 1.

Table 1. Primers used for identification of *Alternaria*.

Gene	Primer Sequence (5′–3′)
ITS	ITS1: TCCGTAGGTGAACCTGCGG ITS4: TCCTCCGCTTATTGATATGC
TEF1-α	EF1-728F: CATCGAGAAGTTCGAGAAGG EF1-986R: TACTTGAAGGAACCCTACC
G3PDH	F: ATTGACATCGTCGCTGTCAACGA R: ACCCCACTCGTTGTCGTACCA
RPB2	F: GATGATCGTGATCATTTCGG R: CCCATAGCTTGCTTACCCAT

For phylogenetic analysis sequences were aligned against NCBI references using MEGA7.0. Concatenated datasets were analyzed in PhyloSuite v1.2.2, with Bayesian inference (MrBayes) and neighbor-joining (NJ) methods (1000 bootstrap replicates). Clades with ≥50% support were considered significant.

4.5. Pathogenicity Assays

For in vitro detached leaf assay, mycelial plugs (5 mm) from 7-day-old cultures were placed on surface-sterilized leaves wounded with sterile needles. Leaves were incubated on moist filter paper in Petri dishes (28 °C, darkness). Controls received sterile PDA plugs. Disease progression was monitored for 7 days, with re-isolation performed to fulfill Koch's postulates.

In vivo inoculation of potted plants was inoculated as above, with inoculated sites wrapped in sterile water-moistened cotton and parafilm. Plants were maintained at 25–28 °C, 80% RH under natural light cycles. Symptom development was recorded photographically.

4.6. Transcriptome Sequencing and Bioinformatics Analysis

Total RNA was extracted from infected and control leaves (three biological replicates per time point) by Biomike Biotechnology Co., Ltd. (Tianjin, China) and performed Illumina-based sequencing. cDNA libraries were prepared following quality checks, where RNA integrity was confirmed using an Agilent 2100 Bioanalyzer (Agilent Technologies Co., Ltd. [Beijing, China]; RNA Integrity Number, RIN > 8.0) and sequenced on the Illumina platform. Raw reads were processed to remove adapters and low-quality sequences, yielding clean data. Clean reads were aligned to the *I. verticillata* using StringTie and gene expression levels were quantified as FPKM (Fragments Per Kilobase of transcript per Million mapped reads). FPKM expression values of all genes are represented in Supplementary Table S1. Functional annotation was performed via alignment against GO (Gene Ontology) and KEGG (Kyoto Encyclopedia of Genes and Genomes) databases. Differentially expressed genes (DEGs) were identified using DESeq2 with thresholds of false discovery rate (FDR) < 0.05 and $|\log2\text{Fold Change}| \geq 1$. Functional annotation used GO and KEGG databases via Biomike Cloud Platform (www.biocloud.net, accessed on 23 September 2023).

4.7. Determination of Endogenous Plant Hormones

The levels of endogenous hormones were quantified using commercial enzyme-linked immunosorbent assay (ELISA) kits by Shanghai Fanke Industrial Co., Ltd. (Shanghai, China) according to the manufacturer's instructions. The specific catalog numbers for the kits used were as follows: abscisic acid (ABA): F4924-A; jasmonic acid (JA): F4976-A; salicylic acid (SA): F7911-A. Briefly, frozen leaf tissues were homogenized in the specific extraction buffer provided with each respective kit. The supernatant was collected after centrifugation. For the standard curves, the series of standard solutions provided in each kit were used, following the kit's instructions. Absorbance was measured using a microplate reader, and hormone concentrations in the samples were interpolated from their corresponding standard curves. Raw hormone values were represented in Supplementary Table S2.

4.8. Extraction of Crude Enzyme Solution

For antioxidant defense enzyme assays, approximately 0.2 g of powdered leaf tissue was homogenized in 2 mL of ice-cold 50 mM phosphate-buffered saline (PBS, pH 7.8) containing 0.2 mM EDTA and polyvinylpolypyrrolidone (PVPP). The homogenate was centrifuged at 12,000 rpm for 20 min at 4 °C, and the supernatant was collected as a crude enzyme extract.

4.8.1. Superoxide Dismutase (SOD) Activity Assay

SOD activity was determined by monitoring the inhibition of nitroblue tetrazolium (NBT) reduction. The reaction mixture (1.5 mL total volume) contained 870 μL PBS (pH 7.8),

150 μL each of 130 mM methionine, 750 μM NBT, 100 μM riboflavin, and Na_2-EDTA, along with 30 μL of crude enzyme extract. Control tubes replaced the enzyme extract with PBS. Samples were exposed to 150 $\mu mol/m^2 \cdot s$ light for 25–30 min, and absorbance was measured at 560 nm. SOD activity ($U \cdot g^{-1} \cdot min^{-1}$) was calculated using the formula:

$$\text{SOD activity } (U \cdot g^{-1} \cdot min^{-1}) = (Ack - AE) \times V/Ack \times 0.5 \times Fw \times a$$

where *Ack* and *AE* are the absorbances of the control and sample, respectively, *V* is the total volume of crude enzyme extract, *Fw* is the fresh weight of the sample, and *a* is the total reaction volume (1.5 mL).

4.8.2. Catalase (CAT) Activity Assay

CAT activity was measured by tracking the decomposition of H_2O_2 at 240 nm. The reaction mixture (1 mL) contained 875 μL PBS (pH 7.0), 100 μL 100 mM H_2O_2, and 25 μL crude enzyme extract. The decrease in absorbance was recorded over 30 s. CAT activity ($U \cdot g^{-1} \cdot min^{-1}$) was calculated as:

$$\text{CAT activity } (U \cdot g^{-1} \cdot min^{-1}) = \Delta A \times Vt/0.1 \times V1 \times t \times Fw$$

where ΔA is the absorbance change, *Vt* is the total enzyme extract volume, *V*1 is the volume used in the assay, *t* is the reaction time, and *Fw* is the sample fresh weight.

4.8.3. Peroxidase (POD) Activity Assay

POD activity was determined by measuring the oxidation of guaiacol at 470 nm according to the method of Kochba et al. [47] with modifications from Zhang et al. [48]. The reaction mixture (1 mL) consisted of 800 μL PBS (pH 7.0), 50 μL 20 mM H_2O_2, 50 μL 1% guaiacol, and 100 μL crude enzyme extract. The increase in absorbance was recorded for 30 s. POD activity was calculated using:

$$\text{POD activity} = (\Delta A \times Vt)/(0.001 \times Fw \times \text{vs.} \times t)$$

where ΔA is the absorbance change, *Vt* is the total enzyme volume, vs. is the volume used in the assay, *t* is the reaction time, and *Fw* is the sample fresh weight.

4.9. Malondialdehyde (MDA) Content Determination

MDA, a lipid peroxidation marker, was quantified using the thiobarbituric acid (TBA) method. A reaction mixture containing 300 μL 10% trichloroacetic acid (TCA), 100 μL 0.6% TBA, and 100 μL crude enzyme extract was incubated at 95 °C for 30 min, cooled, and centrifuged. Absorbance was measured at 532 nm, 600 nm, and 450 nm. MDA content (μmol/g) was calculated as:

$$\text{MDA content } (\mu mol/g) = [6.452 \times (A532 - A600) - 0.559 \times A450] \times V1/(Fw \times V2)$$

where *V*1 is the total extract volume, *V*2 is the volume used in the assay, and *Fw* is the sample fresh weight.

4.10. Statistical Analysis

All experimental data were analyzed using IBM SPSS Statistics (version 20). For physiological and biochemical measurements, the assumptions of normality (Shapiro–Wilk test) and homogeneity of variances (Levene's test) were verified prior to analysis. All data met these assumptions ($p > 0.05$). A one-way analysis of variance (ANOVA) was then performed to determine significant differences among treatment groups, followed by Duncan's multiple range test for post hoc comparisons at a significance level of $p < 0.05$. Three biological replicates were included for each experimental condition to ensure statistical reliability.

Differentially expressed genes (DEGs) were visualized as heatmaps using TBtools software (Version 1.123) [49], with hierarchical clustering based on normalized expression values (FPKM). The color gradient in heatmaps represented relative expression levels, with standardized z-scores used to facilitate comparison across genes and samples [49]. All statistical tests were two-tailed, and data were checked for normality and homogeneity of variance prior to analysis. Results are presented as mean values ± standard deviation (SD) unless otherwise specified.

5. Conclusions

This study establishes *Alternaria alternata* as the primary causative agent of leaf blight in *Ilex verticillata*, providing critical insights into the host's defense mechanisms. The integration of morphological, molecular, and pathogenicity analyses confirmed the pathogen's identity, while transcriptomic and physiological investigations revealed a multi-layered defense strategy. The host's response involved dynamic reprogramming of antioxidant systems, phytohormone signaling, and metabolic pathways, highlighting the complexity of its interaction with *A. alternata*. Notably, the antagonistic interplay between salicylic acid and jasmonic acid, along with sustained oxidative stress management, underscores the plant's adaptive resilience. These findings advance our understanding of *I. verticillata's* susceptibility to fungal pathogens and offer practical implications for disease management. The specific defense-related components identified in this study such as the early activation of pattern-recognition receptors (e.g., *CERK1*), the sustained antioxidant activity (e.g., SOD, CAT), and the dynamic crosstalk between salicylic acid and jasmonic acid signaling, provide potential targets for breeding resistant cultivars. By bridging fundamental and applied perspectives, this work contributes to the sustainable cultivation of *I. verticillata*, ensuring its ornamental and economic value in horticulture.

Supplementary Materials: The following supporting information can be downloaded at: https://www.mdpi.com/article/10.3390/plants14193057/s1, Figure S1: Pathogenicity assessment and microbial community analysis following Koch's postulates. (A) Corresponding purification and re-isolated fungal culture, [Symptoms on detached *Ilex verticillate* leaves inoculated with *Alternaria alternata* isolate Y9-2-1 (Figure 4C,D). (B) Heatmap showing the relative abundance of the dominant fungal genera in the leaf microbial community of control (CK) and *A. alternata*-inoculated (Y9-2-1) leaves. The color intensity indicates the relative abundance from low (blue) to high (red).; Figure S2: DEGs expression model of JA metabolic pathway in *I. verticillata* leaves after infection with *A. alternata*; Table S1: All gene expression; Table S2: Raw Hormone data.

Author Contributions: Conceptualization, H.Y. and B.Z.; methodology, H.L., H.Y., C.Z. and P.C.; software, H.L., H.Y. and L.H.; validation, H.L., H.Y., C.Z. and Q.S.; formal analysis, H.L. and M.J.R.; investigation, H.L., Y.Z., Y.L. and W.D.; resources, J.Z.; data curation, D.S.; writing—original draft preparation, H.Y. and M.J.R.; writing—review and editing, H.Y., B.Z. and M.J.R.; visualization, H.Y., B.Z. and A.S.; supervision, H.Y. and B.Z.; project administration, H.Y.; funding acquisition, H.Y. and B.Z. All authors have read and agreed to the published version of the manuscript.

Funding: This research was funded by Ningbo Key Research and Development Project (2024Z268, 2019B10014); Central Funded Forestry Science and Technology Promotion and Demonstration Project ((2023) TS 03-1, (2020) TS 07); Open Foundation of State Key Laboratory of Subtropical Silviculture, Zhejiang A&F University (SKLSS-KF2023-07); Open Subject of Ningbo Key Laboratory of Characteristic Horticultural Crops in Quality Adjustment and Resistance Breeding (NBYYL2023004, NBYYL2023003); Ningbo Forestry Development Center Project (2023); Forestry Science and Technology Development Project of the National Forestry and Grassland Administration (KJZXSA2019034); Overseas Expertise Introduction Project for Discipline Innovation (111 Project D18008).

Data Availability Statement: Data is contained within the article or Supplementary Materials.

Conflicts of Interest: The authors declare no conflicts of interest.

References

1. Yao, X.; Zhang, F.; Corlett, R.T. Utilization of the Hollies (*Ilex* L. spp.): A Review. *Forests* **2022**, *13*, 94. [CrossRef]
2. Yin, Y.; Hou, Z.; Sun, Q.; Zhu, B.; Liu, J.; Zou, Y.; Hao, M. Predicting the Habitat Suitability of *Ilex verticillata* (Aquifoliaceae) in China with Field-Test Validations. *PLoS ONE* **2025**, *20*, e0315908. [CrossRef] [PubMed]
3. Piassetta, R.d.R.L.; Mikos, A.P.; Auer, C.G. Fungal Diseases of Yerba-Mate Culture (*Ilex paraguariensis*) in Brazil. *BIOFIX Sci. J.* **2021**, *6*, 153–159. [CrossRef]
4. Bregant, C.; Carloni, F.; Linaldeddu, B.T.; Maddau, L.; Marcolongo, M.; Montecchio, L.; Murolo, S.; Piškur, B.; Ogris, N. First Report of *Phytophthora ilicis* Causing Leaf Spot, Shoot Blight and Bleeding Canker on *Ilex aquifolium* in Slovenia. *New Dis. Rep.* **2024**, *50*, e70005. [CrossRef]
5. Deng, W.; Zhang, W. Application of Ilex Plants in Urban Greening. *J. Landsc. Res.* **2021**, *13*, 121–123.
6. Kumar, N.; Dutta, R.; Ajay, B.C.; Radhakrishnan, T. Alternaria Leaf Blight (*Alternaria* spp.)–an Emerging Foliar Fungal Disease of Winter-Summer Groundnut (*Arachis hypogaea*): A Review. *Indian. J. Agric. Sci.* **2022**, *92*, 1043–1050. [CrossRef]
7. Xu, X.; Zhang, L.; Yang, X.; Cao, H.; Li, J.; Cao, P.; Guo, L.; Wang, X.; Zhao, J.; Xiang, W. *Alternaria* spp. Associated with Leaf Blight of Maize in Heilongjiang Province, China. *Plant Dis.* **2022**, *106*, 572–584. [CrossRef]
8. Ma, G.; Bao, S.; Zhao, J.; Sui, Y.; Wu, X. Morphological and Molecular Characterization of *Alternaria* Species Causing Leaf Blight on Watermelon in China. *Plant Dis.* **2021**, *105*, 60–70. [CrossRef]
9. Ahmad, T.; Nie, C.; Cao, C.; Xiao, Y.; Yu, X.; Liu, Y. First Record of *Alternaria tenuissima* Causing Aloe Barbadensis Leaf Blight and Leaf Spot Disease in Beijing, China. *Crop Prot.* **2024**, *175*, 106447. [CrossRef]
10. Schmey, T.; Tominello-Ramirez, C.S.; Brune, C.; Stam, R. *Alternaria* Diseases on Potato and Tomato. *Mol. Plant Pathol.* **2024**, *25*, e13435. [CrossRef]
11. Cao, C.; Gong, S.; Li, Y.; Tang, J.; Li, T.; Zhang, Q. Pathogenic Factors and Mechanisms of the *Alternaria* Leaf Spot Pathogen in Apple. *Horticulturae* **2024**, *10*, 212. [CrossRef]
12. Huang, Y.; Ma, H.; Yue, Y.; Zhou, T.; Zhu, Z.; Wang, C. Integrated Transcriptomic and Transgenic Analyses Reveal Potential Mechanisms of Poplar Resistance to *Alternaria alternata* Infection. *BMC Plant Biol.* **2022**, *22*, 413. [CrossRef] [PubMed]
13. Ping, H.; Rao, W.; Dong, Z.; He, X.; Zhao, J.; Chen, Z.; Liu, L. Physiological and Biochemical Responses of Tobacco at Different Growth Stages to *Alternaria alternata*. *J. Phytopathol.* **2024**, *172*, e13431. [CrossRef]
14. Feng, L.; Zhang, Y.; Chen, W.; Mao, B. Colletotrichum Siamense Strain Lvy 9 Causing Spot Anthracnose on Winterberry Holly in China. *Microorganisms* **2023**, *11*, 976. [CrossRef]
15. Rao, M.J.; Zheng, B. The Role of Polyphenols in Abiotic Stress Tolerance and Their Antioxidant Properties to Scavenge Reactive Oxygen Species and Free Radicals. *Antioxidants* **2025**, *14*, 74. [CrossRef]
16. Rao, M.J.; Duan, M.; Zhou, C.; Jiao, J.; Cheng, P.; Yang, L.; Wei, W.; Shen, Q.; Ji, P.; Yang, Y.; et al. Antioxidant Defense System in Plants: Reactive Oxygen Species Production, Signaling, and Scavenging During Abiotic Stress-Induced Oxidative Damage. *Horticulturae* **2025**, *11*, 477. [CrossRef]
17. Rao, M.J.; Duan, M.; Ikram, M.; Zheng, B. ROS Regulation and Antioxidant Responses in Plants Under Air Pollution: Molecular Signaling, Metabolic Adaptation, and Biotechnological Solutions. *Antioxidants* **2025**, *14*, 907. [CrossRef]
18. Schmid-Siegert, E.; Stepushenko, O.; Glauser, G.; Farmer, E.E. Membranes as Structural Antioxidants: Recycling of Malondialdehyde to Its Source in Oxidation-Sensitive Chloroplast Fatty Acids. *J. Biol. Chem.* **2016**, *291*, 13005–13013. [CrossRef]
19. Shanthi, K.C. Biochemical defense mechanisms in medicinal and aromatic plants: Early-stage responses to fungal infections. *Biochem. Cell Arch.* **2024**, *24*, 1759. [CrossRef]
20. Deng, B.; Jia, B.; Liu, G.; Zhang, X. Transcriptome Analysis to Elucidate the Enhanced Cold Resistance of *Phoebe zhennan* Pretreated with Exogenous Calcium. *Int. J. Agric. Biol.* **2021**, *25*, 151–159. [CrossRef]
21. Duan, M.; Bao, L.; Eman, M.; Han, D.; Zhang, Y.; Zheng, B.; Yang, S.; Rao, M.J. The Ectopic Expression of the MpDIR1(t) Gene Enhances the Response of Plants from *Arabidopsis thaliana* to Biotic Stress by Regulating the Defense Genes and Antioxidant Flavonoids. *Plants* **2024**, *13*, 2692. [CrossRef] [PubMed]
22. Rao, M.J.; Duan, M.; Eman, M.; Yuan, H.; Sharma, A.; Zheng, B. Comparative Analysis of Citrus Species' Flavonoid Metabolism, Gene Expression Profiling, and Their Antioxidant Capacity under Drought Stress. *Antioxidants* **2024**, *13*, 1149. [CrossRef] [PubMed]
23. Zhou, L.; Zhang, C.; Zhou, Q.; Liao, M.; Feng, Z.; Zhu, P.; Wang, M.; Luo, Y.; Luo, G. Transcriptome Analysis of Gardenia Jasminoides Ellis in Response to *Botryosphaeria dothidea*. *J. Plant Biochem. Biotechnol.* **2023**, *34*, 179–190. [CrossRef]
24. Appu, M.; Ramalingam, P.; Sathiyanarayanan, A.; Huang, J. An Overview of Plant Defense-Related Enzymes Responses to Biotic Stresses. *Plant Gene* **2021**, *27*, 100302. [CrossRef]

25. Bora, S.; Borah, R.; Giri, K. Role of Proteins and Enzymes in Plant Disease Control. In *Plant Protection: From Chemicals to Biologicals*; Walter de Gruyter GmbH & Co KG: Berlin, Germany, 2022; p. 395.
26. McCombe, C.L.; Greenwood, J.R.; Solomon, P.S.; Williams, S.J. Molecular Plant Immunity against Biotrophic, Hemibiotrophic, and Necrotrophic Fungi. *Essays Biochem.* **2022**, *66*, 581–593. [CrossRef]
27. Fang, X.; Xie, Y.; Yuan, Y.; Long, Q.; Zhang, L.; Abid, G.; Zhang, W. The Role of Salicylic Acid in Plant Defense Responses against Biotic Stresses. *Plant Horm.* **2025**, *1*, e004. [CrossRef]
28. Roychowdhury, R.; Hada, A.; Biswas, S.; Mishra, S.; Prusty, M.R.; Das, S.P.; Ray, S.; Kumar, A.; Sarker, U. Jasmonic Acid (JA) in Plant Immune Response: Unravelling Complex Molecular Mechanisms and Networking of Defence Signalling against Pathogens. *J. Plant Growth Regul.* **2025**, *44*, 89–114. [CrossRef]
29. Li, H.; Lu, Y.; Liu, Z.; Ren, Q.; Liu, Z.; Liu, S.; Ren, R.; Wang, F.; Liu, Y.; Zhang, Y. Transcriptomic Analysis Unveils Alterations in the Genetic Expression Profile of Tree Peony (*Paeonia suffruticosa* Andrews) Infected by *Alternaria alternata*. *BMC Genom.* **2024**, *25*, 861. [CrossRef]
30. Zhang, S.; Miao, W.; Liu, Y.; Jiang, J.; Chen, S.; Chen, F.; Guan, Z. Jasmonate Signaling Drives Defense Responses against *Alternaria alternata* in Chrysanthemum. *BMC Genom.* **2023**, *24*, 553. [CrossRef]
31. Ma, L.; Li, R.; Ma, L.; Song, N.; Xu, Z.; Wu, J. Involvement of NAC Transcription Factor NaNAC29 in *Alternaria alternata* Resistance and Leaf Senescence in *Nicotiana attenuata*. *Plant Divers.* **2021**, *43*, 502–509. [CrossRef]
32. Wang, W.; Bai, X.-D.; Chen, K.; Gu, C.-R.; Yu, Q.-B.; Jiang, J.; Liu, G.-F. Role of PsnWRKY70 in Regulatory Network Response to Infection with *Alternaria alternata* (Fr.) Keissl in Populus. *Int. J. Mol. Sci.* **2022**, *23*, 7537. [CrossRef]
33. Liu, W.; Xu, S.; Ou, C.; Liu, X.; Zhuang, F.; Deng, X.W. T2T Genomes of Carrot and *Alternaria dauci* and Their Utility for Understanding Host–Pathogen Interactions during Carrot Leaf Blight Disease. *Plant J.* **2024**, *120*, 1643–1661. [CrossRef] [PubMed]
34. Rajarammohan, S. Redefining Plant-Necrotroph Interactions: The Thin Line between Hemibiotrophs and Necrotrophs. *Front. Microbiol.* **2021**, *12*, 673518. [CrossRef] [PubMed]
35. Afzal, K.; Mujtaba, M.; Wang, Y.; Zhou, B. Effectors of Plants Pathogenic Fungi and Fungal like Microbes: A Comprehensive Review on Mechanisms, Roles, and Host Interactions. *Front. Plant Sci.* **2025**, *16*, 1626960. [CrossRef] [PubMed]
36. Devanna, B.N.; Jain, P.; Solanke, A.U.; Das, A.; Thakur, S.; Singh, P.K.; Kumari, M.; Dubey, H.; Jaswal, R.; Pawar, D. Understanding the Dynamics of Blast Resistance in Rice-Magnaporthe Oryzae Interactions. *J. Fungi* **2022**, *8*, 584. [CrossRef]
37. Liu, X.; Zhang, Z. A Double--edged Sword: Reactive Oxygen Species (ROS) during the Rice Blast Fungus and Host Interaction. *FEBS J.* **2022**, *289*, 5505–5515. [CrossRef]
38. Malvestiti, M.C.; Steentjes, M.B.F.; Beenen, H.G.; Boeren, S.; van Kan, J.A.L.; Shi-Kunne, X. Analysis of Plant Cell Death-Inducing Proteins of the Necrotrophic Fungal Pathogens *Botrytis squamosa* and *Botrytis elliptica*. *Front. Plant Sci.* **2022**, *13*, 993325. [CrossRef]
39. Nadarajah, K.K. Defensive Strategies of ROS in Plant–Pathogen Interactions. In *Plant Pathogen Interaction*; Springer: Berlin/Heidelberg, Germany, 2024; pp. 163–183.
40. Huang, Y.; Li, J.; Nong, C.; Lin, T.; Fang, L.; Feng, X.; Chen, Y.; Lin, Y.; Lai, Z.; Miao, L. *Piriformospora indica* Enhances Resistance to *Fusarium wilt* in Strawberry by Increasing the Activity of Superoxide Dismutase, Peroxidase, and Catalase, While Reducing the Content of Malondialdehyde in the Roots. *Horticulturae* **2024**, *10*, 240. [CrossRef]
41. Narsai, R.; Wang, C.; Chen, J.; Wu, J.; Shou, H.; Whelan, J. Antagonistic, Overlapping and Distinct Responses to Biotic Stress in Rice (*Oryza sativa*) and Interactions with Abiotic Stress. *BMC Genom.* **2013**, *14*, 1–21. [CrossRef]
42. Gupta, R.; Anand, G.; Bar, M. Developmental Phytohormones: Key Players in Host-Microbe Interactions. *J. Plant Growth Regul.* **2023**, *42*, 7330–7351. [CrossRef]
43. Wang, X.; Cui, J.; Li, X.; Wang, G.; Zheng, G.; Li, Y.; Li, X.; Ma, L.; Wang, L. Comparative Study on Growth, Physiological and Biotic Stress Resistance of Cucumber (*Cucumis sativus* L.) by Exogenous Application of Jasmonic Acid and Methyl Jasmonate under Greenhouse Conditions. *Russ. J. Plant Physiol.* **2022**, *72*, 176. [CrossRef]
44. Bharath, P.; Gahir, S.; Raghavendra, A.S. Abscisic Acid-Induced Stomatal Closure: An Important Component of Plant Defense against Abiotic and Biotic Stress. *Front. Plant Sci.* **2021**, *12*, 615114. [CrossRef]
45. Zhang, X.; Sun, X.; Zhang, S.; Yang, J.; Liu, F.; Fan, J. Comprehensive Transcriptome Analysis of Grafting onto *Artemisia scoparia* W. to Affect the Aphid Resistance of Chrysanthemum (*Chrysanthemum morifolium* T.). *BMC Genom.* **2019**, *20*, 776. [CrossRef]
46. Kochba, J.; Lavee, S.; Spiegel-Roy, P. Differences in Peroxidase Activity and Isoenzymes in Embryogenic Ane Non-Embryogenic 'Shamouti' Orange Ovular Callus Lines. *Plant Cell Physiol.* **1977**, *18*, 463–467. [CrossRef]
47. Zhang, R.; Peng, F.; Li, Y. Pecan Production in China. *Sci. Hortic.* **2015**, *197*, 719–727. [CrossRef]

48. Chen, C.; Chen, H.; Zhang, Y.; Thomas, H.R.; Frank, M.H.; He, Y.; Xia, R. TBtools: An Integrative Toolkit Developed for Interactive Analyses of Big Biological Data. *Mol. Plant* **2020**, *13*, 1194–1202. [CrossRef]
49. Chanana, S.; Thomas, C.S.; Zhang, F.; Rajski, S.R.; Bugni, T.S. HCAPCA: Automated Hierarchical Clustering and Principal Component Analysis of Large Metabolomic Datasets in R. *Metabolites* **2020**, *10*, 297. [CrossRef] [PubMed]

plants

Article

Correlation Analysis of Twig and Leaf Characteristics and Leaf Thermal Dissipation of *Hippophae rhamnoides* in the Riparian Zone of the Taohe River in Gansu Province, China

Qun Li [1,2,*], Min Ma [3], Yurui Tang [3], Tingting Zhao [3], Chengzhang Zhao [3,4,*] and Bo Li [2]

1 College of Resources and Environment, Xichang University, Xichang 615013, China
2 School of Life Sciences, Fudan University, Shanghai 200433, China
3 College of Geography and Environmental Science, Northwest Normal University, Lanzhou 730070, China; mamin@gsau.edu.cn (M.M.); 18419060862@163.com (Y.T.); zhtt123_56@163.com (T.Z.)
4 College of Geography and Environmental Science, Northwest Normal University, Research Center of Wetland Resources Protection and Industrial Development Engineering of Gansu Province, Lanzhou 730070, China
* Correspondence: liqunwv0203@163.com (Q.L.); zhaocz1710@163.com (C.Z.); Tel.: +86-18809464065 (Q.L.)

Abstract: Aims: The functional traits of twigs and leaves are closely related to the ability of plants to cope with heterogeneous environments. The analysis of the characteristics of twigs and leaves and leaf thermal dissipation in riparian plants is of great significance for exploring the light energy allocation and ecological adaptation strategies of plant leaves in heterogeneous habitats. However, there are few studies on the correlation between the twig–leaf characteristics of riparian plants and their heat dissipation in light heterogeneous environments. **Methods**: In this study, the riparian plant *Hippophae rhamnoides* in Taohe National Wetland Park was the research object. According to the differences in the canopy light environment of the *H. rhamnoides* population, three habitat gradients were set: I, the full sight zone; II, the moderate shade zone; and III, the canopy cover zone. We studied the relationship between the twig–leaf characteristics of *H. rhamnoides* and leaf thermal dissipation in a heterogeneous light environment. **Important Findings**: The results are as follows: from the full sight zone to the canopy cover zone, the population characteristics and the twig, leaf, and photosynthetic fluorescence physiological characteristics of *H. rhamnoides* demonstrated significant changes ($p < 0.05$). In the full sight zone, *H. rhamnoides* tended to have thick leaves with a smaller SLA on short and thick twigs, and the light energy absorbed by the leaves accounted for a higher proportion of thermal dissipation. In the moderate shade zone, *H. rhamnoides* tended to grow many thin leaves with high SLA on long and thick twigs, and the proportion of light energy absorbed by the leaves for heat dissipation was lower than that in the full sight zone. In the canopy cover zone, *H. rhamnoides* tended to grow a few large and thick leaves with a low SLA on slender and long twigs, and the proportion of light energy absorbed by the leaves for heat dissipation was the lowest. There was a significant correlation between the twig–leaf and leaf heat dissipation of *H. rhamnoides* in the three habitats ($p < 0.05$). The co-variation of plant branches and leaves and the timely adjustment of thermal dissipation in photoheterogeneous habitats reflect the phenotypic plasticity mechanism and self-protection strategy of riparian plants in adapting to heterogeneous environments.

Keywords: twig diameter; leafing intensity; thermal dissipation; electron transport rate; *Hippophae rhamnoides*; riparian zone

1. Introduction

Plant functional traits are the survival, growth, and reproduction strategies of plants adapted to the external environment, which determine how plants respond to environmental changes and optimize resource utilization and reflect the ability of plants to obtain, utilize, and preserve resources [1–3]. Plants cope with environmental stress through the coordination or trade-off of traits in different environments so that the harm of adverse environmental factors to individual plants is minimized [4]. As one of the structural units of plants (twig, hereafter), the annual twigs of trees and shrubs are the most active part of the plant branching system [5,6], which can directly reflect the life history countermeasures of plants adapting to the environment. The leaves on the twigs are the main components of plants and the main organs for carrying out physiological activities and exerting ecological functions. Their structural traits, to a certain extent, determine the photosynthetic activity of plants and the optimization of resource utilization in response to environmental changes [7]. These traits can reflect the utilization level of the light energy absorbed by the leaves, which are also an organ with strong plasticity and sensitivity to environmental changes [8]. The combination of branches and leaves can better reflect plant configuration shaping and space resource utilization ability, which is one of the main dimensions of plant ecological strategy [9]. Branch structure affects leaf size and photosynthetic efficiency, which determines the trade-off between leaf size and leaf number, affects the carbon acquisition of plants, and further affects the growth and survival of branches [10]. The characteristics of leaves and branches are highly coordinated in biomechanics and the morphological structure, which determines the utilization of light, water, and other resources and the adaptation strategies of plants [11]. Therefore, the relationship between twigs and leaves can effectively explain the adaptation strategies of plants to the environment [12].

Light is one of the most important environmental factors affecting plant growth and distribution [13,14]. Shading is a common phenomenon among naturally growing plants that can reduce intense solar radiation, promote the accumulation of soil organic matter, reduce the amplitude of ambient temperature, etc., and plays an important regulatory role in building the optimal physiological and functional state of plants [15,16]. Long-term exposure to heterogeneous light distribution inhibited the availability of resources among plant populations [14], thus affecting plant growth and development and dry matter accumulation efficiency [17]. To ensure sufficient light energy capture, maintain photosynthesis, and achieve the maximum investment and return of photosynthesis products [18], plants preferentially allocate biomass to the trait with the easiest access to light resources [19,20] by adjusting the number leaves on twigs, single leaf area, leaf thickness, and other functional traits [21], thus promoting plants to weigh resource input among various components [22]. The change in the characteristics of twigs and leaves and the balanced distribution of biomass ensures the coordination and unity of the effectiveness and safety of water transport in the process of leaf photosynthesis [23,24].

Plant photosynthesis is a complex metabolic pathway that can reflect the adaptive countermeasures formed by plants to adapt to environmental changes and reflect the law of interaction between plants and the environment [25,26]. Chlorophyll fluorescence is closely related to the characteristic process of photosynthesis [26,27]. Part of the light energy absorbed by plant leaves is used in photochemical processes to synthesize sugars for life activities; the other part is dissipated in the form of fluorescence or heat to protect itself [28]. Chlorophyll fluorescence parameters reflect the changes in the process of plant photosynthesis, so it becomes an indicator of photosynthetic activity and behavior [29]. Non-photochemical quenching (NPQ) is considered the fastest process by which plants relieve the excitation energy pressure in the photosynthetic membrane (the site of the photosystem II reaction), thus protecting the plant from stress damage [30]. Light is the driver of photosynthesis in plants, but it also impairs the photosynthetic process in many ways [31]. In strong light environments, the non-photochemical quenching (NPQ) mechanism maintains the normal photosynthesis of plant leaves by dissipating excess light energy absorbed by plant leaves in the form of heat, which is an important protection mechanism of plant leaves [32]. The environmental resources (such as light, water, nutrients, etc.) received by individual plants in different habitats can significantly change the structural morphology and photosynthetic fluorescence characteristics of plant twigs and leaves [33]. In the shaded environment, the far red light is enriched [34–36], the photochemical efficiency of PSII is higher, the biomass distribution of plant branches and stems is increased, and the light distribution between the photosystems is conducive to PSII, which can promote photosynthesis and plant growth [37]. However, less attention has been paid to the correlation and response mechanism of chlorophyll fluorescence characteristics in plants in natural shade environments. Therefore, studies on the correlation between the twig–leaf traits and chlorophyll fluorescence characteristics of plants in natural shade environments can reflect the ecological strategies adopted by plants to cope with heterogeneous light environments and have important significance for understanding the mechanisms of phenotypic plasticity and self-protection of plants.

The riparian zone is the riverbed between high and low water levels and the area where the influence of the river transitions from the high water level to the river [38]. It is also an important ecological transition zone for the material, energy, and information exchange between the river and the terrestrial ecosystem [39]. It has rich biodiversity and provides an extremely wide range of ecosystem services and functions [40,41]. The seasonal and interannual changes in river hydrological conditions lead to the alternating cycle of flood and drought in riparian areas [39] and the changes in topography and soil structure. The particularity, complexity, and spatial heterogeneity of their habitats affect the resource acquisition and spatial distribution of riparian plant communities, thus changing the light, soil water, and nutrient resources available to populations within the communities. To improve the fitness of the population, the individual plants in the population can adjust their morphology and physiology [42,43]. *Hippophae rhamnoides* is a deciduous plant belonging to the Elaeagnaceae family, Hippophae. It grows rapidly, has strong root spreader ability, and has a wide range of adaptability. It is an excellent tree species for windbreak and sand fixing, soil and water conservation, and ecological environment improvement [44]. The strong habitat adaptability and environmental modification ability of *H. rhamnoides* can alleviate the impact of floods on the ground, create favorable conditions for the restoration of riparian biodiversity, and play a very important role in the survival and development of shoreline forests. At present, studies on *H. rhamnoides* mainly focus on leaf photosynthetic characteristics [45,46], morphological and physiological responses to salt stress [47], root studies [48], and ecological stoichiometry studies [49]. However, the correlation between the twig–leaf traits and photosynthetic fluorescence physiological characteristics of *H. rhamnoides* in heterogeneous habitats remains unclear.

Given this, this paper took *H. rhamnoides*, a riparian plant in the Taohe National Wetland Park, Gansu Province, as the research object and proposed the following hypothesis: *H. rhamnoides* would demonstrate adaptive changes in twig–leaf traits and photosynthetic fluorescence physiological characteristics in photoheterogeneous habitats. We aim to answer the following questions: (1) Under photoheterogeneous conditions, what is the change rule of the twig–leaf traits of *H. rhamnoides*? (2) What is the distribution of light energy absorbed by *H. rhamnoides* leaves in terms of photosynthesis and heat dissipation? What is the relationship between the twig–leaf characteristics of *H. rhamnoides* and leaf heat dissipation? This study helps shed light on the photosynthetic and photosynthetic fluorescence physiological processes of riparian plants in photoheterogeneous habitats and provides a scientific theoretical basis for the stable maintenance of riparian ecosystems and the management of riparian plants.

2. Materials and Methods

2.1. Study Sites

The study area is located in the south of Lintao County, Dingxi City, Gansu Province (103°45′43″–103°50′55″ E, 35°05′27″–35°15′58″ N; Figure 1), which is located in the intersection area of the Qinghai-Tibet Plateau and Loess Plateau. On the northwest margin of the Qinling fold system, there are undulating mountains in the east, west, and south and a deep gully with an altitude of 2200–2500 m. This region is a semi-arid and semi-humid climate zone in the middle and temperate zone. The average annual temperature is 7.2 °C, the average annual precipitation is 519.2 mm, and the annual solar radiation is 139.76 kCal/cm^2. The Taohe River is a first-level tributary of the Yellow River, originating from the Henan Mongolian Autonomous County of Qinghai Province, flowing from the east through Luqu, Lintan, and Zhuoni County, south to Minxian County Chabu, and then turning to the northwest through Jiudian Gorge, Haidian Gorge, Lintao Basin, and Yongjing County, into the Liujiaxia Reservoir. The soil is mainly red soil, tidal soil, and sandy soil. The main plants are *H. rhamnoides*, *Periploca sepium*, *Populus simonii*, *Berberis thunbergii*, *Ligustrum lucidum*, and *Hedysarum multijugum* [50].

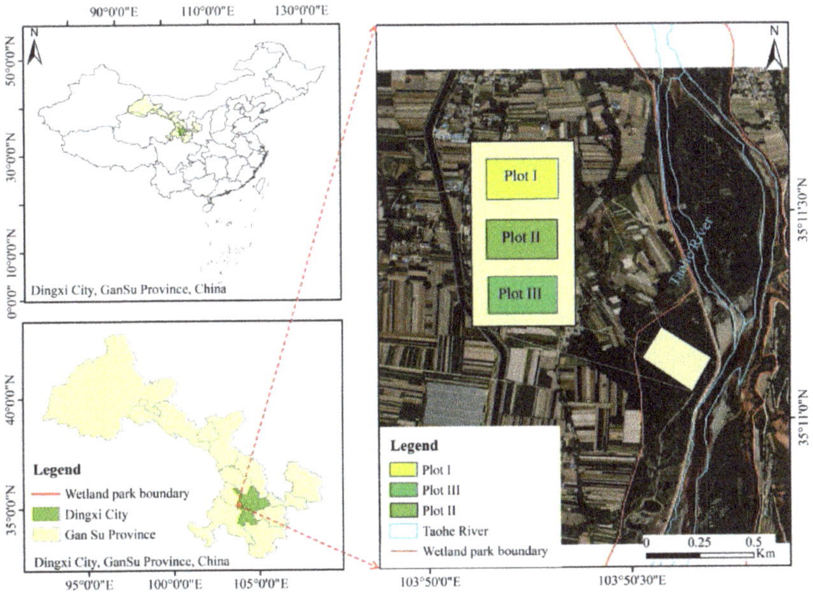

Figure 1. Study area and locations of the plots.

2.2. Experimental Methods and Design

Based on many field investigations and the accessibility of plant communities on both sides of the Taohe National Wetland Park, a 300 m × 200 m transect was set up along the river terrace in the middle part of the Taohe National Wetland Park from 20 to 31 August 2019, and a representative Qiaoguan community was selected from the transect. Due to the influence of canopy light resources, light acquisition by understory shrubs was limited, and the biological characteristics of *H. rhamnoides* populations were significantly different in various microhabitats. According to the canopy shelter and shrub distance of the *H. rhamnoides* population, the riparian tree–shrub community is divided into three light levels: plot I, the full sight zone, which is an open area more than 1 times the tree height from the forest margin; plot II, the moderate shade zone, is a forest clearance with a size of 1 times the height of the trees or a canopy gap with a diameter of not less than 5 m and direct light; plot III, the canopy cover zone, which is an environment in which the canopy is fully closed or intersecting with another canopy [44,51]. First, according to the habitat gradient of *H. rhamnoides*, six 5 m × 5 m quadrats were set for each light level, for a total of 18 quadrats. Then, plant community characteristics and the properties of the *H. rhamnoides* population (canopy density, plant height, crown width, and density) were recorded, and the altitude, longitude, and latitude of each quadrat were measured using a portable handheld GPS. Finally, six *H. rhamnoides* strains with good growth were selected as test plants in each quadrat, and the following indexes were measured.

2.2.1. Measurement of the Leaf Photosynthetic Parameters of *H. rhamnoides*

From 09:00 to 12:00 on a sunny day, during the same period, from each *H. rhamnoides*, we selected four annual twigs from the middle and upper part of the canopy and marked them. The branch angle and leaf inclination angle of the twigs were measured with a protractor, and the average value was obtained. Two healthy, intact leaves were selected from the labeled twigs and wiped clean with dry gauze. The photosynthetic parameters were determined using a portable gas exchange system (GFS-3000 Heinz Walz GmbH, Effeltrich, Germany). During measurement, an artificial red–blue light source was used. The leaf portion inside the clamp-on leaf chamber was exposed to a photosynthetic photon flux density (PPFD) of 1200 μmol (photons) m^{-2} s^{-1}. The experiments were performed outdoors with a controlled CO_2 concentration (340 ± 10 μmol·mol^{-1}), flow rate (750 ± 10 μmol·s^{-1}), temperature (20 ± 5 °C), and humidity [relative humidity (RH) = 70 ± 5%]. The parameters of net photosynthetic rate (P_n,) and transpiration rate (T_r) were measured under an atmospheric CO_2 concentration of 340 μmol·mol^{-1}, repeated three times for each marked *H. rhamnoides* plant. The water use efficiency is determined by calculating the P_n/T_r ratio [52].

2.2.2. Measurement of Photosynthetically Active Radiation (PAR) in the *H. rhamnoides* Population

During the same period of photosynthetic parameter determination, the photoquantum meter (3415F, Walz, Plainfield, IL, USA) was installed above the *H. rhamnoides* (15 cm from the canopy), in the middle (1/2 of the plant height), and on the surface (15 cm from the surface). The photosynthetically active radiation of *H. rhamnoides* was measured at 3 light levels [26].

2.2.3. Measurement of Chlorophyll Fluorescence Parameters

Chlorophyll fluorescence parameters were measured using an IMAGING-PAM chlorophyll fluorescence system (Heinz Walz GmbH, Effeltrich, Germany) following a previously described method [53,54]. All chlorophyll fluorescence kinetics parameters were measured on the leaf after dark adaptation at ambient temperature for 30 min in each plot. The effective quantum yield of PSII photochemistry [Y(II)], the quantum yield of non-regulatory energy dissipation [Y(NO)], the quantum yield of regulatory energy dissipation [Y(NPQ)], photochemical quenching (QP), electron transport rate (ETR), and non-photochemical quenching (NPQ) were determined when steady-state chlorophyll fluorescence was attained after 5 min of actinic irradiance [26,52,55].

$$QP = (Fm' - Fs)/Fv' = 1 - (Fs - Fo')/(Fm' - Fo') \quad (1)$$

$$NPQ = (Fm - Fm')/Fm' = Fm/Fm' - 1 \quad (2)$$

$$Y(II) = \Phi PSII = (Fm' - Ft)/Fm' \quad (3)$$

$$Y(NO) = 1/(NPQ + 1 + qL\ (Fm/Fo - 1)) \quad (4)$$

$$Y(NPQ) = 1 - Y(II) - Y(NO) \quad (5)$$

$$ETR = 0.5 \times Y(II) \times PAR \times 0.84 \quad (6)$$

Here, Fm' is the maximum fluorescence under light, Fo' is the minimum fluorescence under light, Fv' is the fluorescence yield caused by saturated pulse light, Fs is steady-state fluorescence, Ft is real-time fluorescence, qL is the photochemical quenching coefficient based on the "lake model", and PAR is photosynthetically active radiation [26].

2.2.4. Measurement of Twig and Leaf (Twig Diameter, Twig Length, Number of Leaves, Fresh Leaf Weight, Leaf Area, Leaf Thickness, Leaf Dry Weight, Specific Leaf Area, and Chlorophyll Content) Traits of *H. rhamnoides*

The twigs with the epidermis of the mother branch after the measurement of leaf photosynthesis were cut and stored in a sealed plastic bag and brought back to the laboratory to complete the following measurements: First, the number of leaves on each twig was recorded, the leaves were removed, the fresh leaf weight was measured using an electronic balance (accuracy 0.0001 g) 3 times, and the average value was obtained. Then, a portable laser leaf area meter (CI-202, Walz, Camas, WA, USA) was used to determine the leaf area. The chlorophyll content was measured with a portable chlorophyll meter (SPAD-502, Minolta, Osaka, Japan) in different habitats [56]. The twig diameter and leaf thickness were measured using a vernier caliper with an accuracy of 0.02 mm. The leaf thickness was measured at different parts of the tested leaves, and the measurement was repeated 3 times to find the average value. The twig length was measured with steel tape (accuracy 0.1 mm). Finally, the leaves were put into an envelope, put into the oven at 105 °C for 30 min, and then dried at 85 °C until the constant weight was the dry weight of the leaves (accuracy 0.0001 g). The leaf water content was calculated using the following formula: (fresh leaf weight − dry leaf weight)/fresh leaf weight. SLA is expressed as the ratio of leaf area to leaf dry weight [26] (Li et al., 2022b). Leafing strength was calculated using the following formula: number of leaves/(length of twig × diameter of twig) [57].

2.2.5. Determination of Soil Water Content, Soil Bulk Density, Soil Salinity, and the Contents of Sand, Silk, and Clay

First, a soil profile of 1 m × 1 m × 0.5 m was randomly excavated in the same square where the community investigation was conducted. Soil samples were taken with a ring knife (200 cm^3) and divided into 5 layers 10 cm apart. The sample collection was repeated 3 times. Then, it was taken back to the laboratory to bake in the oven at 105 °C for 12 h. The mass was taken out and weighed, and the soil mass water content and soil bulk density of the 0–50 cm soil layer were calculated. The soil salt content was determined using the electrical conductivity method, and the electrical conductivity of the leaching solution was measured using a portable electrical conductivity meter (DDS-11C, Shanghai Lei Magnetic Instrument Factory, Shanghai, China). The average value was obtained after 3 repetitions. The mechanical composition of the soil was determined using the traditional screening method, and each soil sample was repeated three times. The weight of sand, silt, and clay were recorded separately (accuracy 0.0001 g). Finally, the proportion of sand, silt, and clay in each soil sample was calculated separately.

2.3. Statistical Analysis

All the original data of the experiment were sorted using Microsoft Excel 2021, and the water use efficiency, specific leaf area, and leafing intensity of *H. rhamnoides* were calculated. The biological traits and soil properties of the wetland communities in the three habitats were statistically analyzed. A one-way ANOVA was used to compare the twig–leaf characteristics, photosynthetic and fluorescent characteristics, and population traits of *H. rhamnoides* among the different habitats ($\alpha = 0.05$). Then, R4.0.5 was used for a correlation analysis and mantel test. Other charts of the experiment were drawn using origin 2022.

3. Results

3.1. Physical and Chemical Properties of Soil with Different Habitats

The physical and chemical properties of soil were significantly different in different habitats (Figure 2; Supplementary Table S1, $p < 0.05$). From the full sight zone (plot I) to the canopy cover zone (plot III), the soil bulk density, soil water content, soil silk content (Csilt), and soil clay content (Cclay) displayed increasing trends and increased by 79.09%, 59.89%, 1.04 times, and 2.77 times, respectively (A, B, D, F). The electric conductivity of the soil and the soil sand content (Csand) showed decreasing trends and decreased by 28.15% and 75.90%, respectively (B, E).

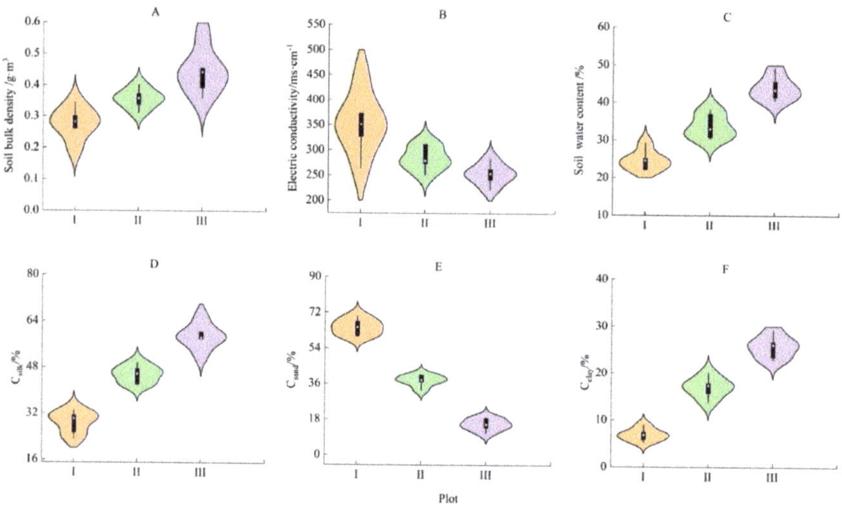

Figure 2. The physical and chemical properties of soil in each plot. I, Full sight zone; II, Moderate shade zone; III, Canopy cover zone. Different lowercase letters indicate significant differences among plots ($p < 0.05$). (**A**), the soil moisture content (SMC) in each plot; (**B**), soil electrical conductivity (EC) in each plot; (**C**), the soil bulk density (SBD) in each plot; (**D**), the content of silt (Csilt) in each plot; (**E**), the content of sand (Csand) in each plot; (**F**), the content of clay (Cclay) in each plot.

3.2. Biological Characteristics of Different Plant Community Habitats and Population Traits of *H. rhamnoides*

The photosynthetically active radiation (PAR) and crown density of the community were significantly different under different habitats ($p < 0.05$). From the full sight zone (plot I) to the canopy cover zone (plot III), the PAR of the community exhibited decreasing trends and decreased by 37.63%. The crown density displayed an increasing trend and increased by 1.42 times from the full sight zone to the canopy cover zone. Meanwhile, the population traits of *H. rhamnoides* were also significantly different from the full sight zone to the canopy cover zone. The crown width and average height and density of the *H. rhamnoides* population displayed increased trends, increasing by 1.67, 4.61, and 1.44 times, respectively (Table 1).

Table 1. Community and population traits of *H. rhamnoides* in different plots (mean ± SE).

Plot	Community Traits			Population Traits	
	PAR/μmol·m^{-2}·s^{-1}	Crown Density/%	Crown Wide/cm	Average Height/cm	Density/Strain·m^2
I	1457.33 ± 45.41 a	58.33 ± 4.65 c	55.50 ± 9.86 c	40.00 ± 5.99 c	20 ± 0.92 c
II	1167.33 ± 38.32 b	90.88 ± 6.84 b	99.80 ± 9.31 b	103.75 ± 4.62 b	29 ± 1.25 b
III	909.00 ± 22.35 c	141.02 ± 15.00 a	148.10 ± 10.12 a	216.58 ± 14.74 a	48 ± 2.42 a

Different lowercase letters in the same column indicate significant differences among plots ($p < 0.05$, n = 36). I, Full sight zone; II, Moderate shade zone; III, Canopy cover zone.

3.3. *Leaf and Twig Traits of H. rhamnoides in Each Habitat*

With the habitat change from the full sight zone to the canopy cover zone, the twig–leaf traits of *H. rhamnoides* were significantly different ($p < 0.05$; Table 2). The leaf water content and twig length displayed increasing trends and increased by 18.01% and 5.03%; meanwhile, the leaf area, leaf thickness, and chlorophyll content decreased first and then increased and ultimately increased by 1.83%, 6.78%, and 5.12%, respectively. Moreover, the specific leaf area showed the opposite trend and ultimately increased by 26.96%. The leaf inclination, branch angle, and leafing intensity displayed a decreasing trend and decreased by 62.17% and 17.12%, respectively; the twig diameter and the number of blades increased first and then decreased and ultimately decreased by 6.68% and 24.17%, respectively. The leaf fresh weight and leaf dry weight decreased first and then increased, and there was no significant change on the whole from the full sight zone to the canopy cover zone.

Table 2. Main leaf and twig characteristics of *H. rhamnoides* in each habitat (mean ± SE).

Plot	I	II	III	Plasticity Index
Leaf fresh weight/g	0.05 ± 0.002 a	0.03 ± 0.001 b	0.05 ± 0.001 a	0.52
leaf dry weight/g	0.02 ± 0.001 a	0.01 ± 0.001 b	0.02 ± 0.001 a	0.58
leaf water content/%	57.15 ± 1.13 c	63.86 ±1.01 b	67.45 ± 0.73 a	0.15
leaf area/cm^2	1.98 ± 0.08 a	1.49 ± 0.07 b	2.02 ± 0.07 a	0.26
leaf thickness/mm	0.16 ±0.002 b	0.14 ± 0.002 c	0.17 ± 0.003 a	0.18
Chlorophyll content/SPAD	63.16 ± 1.87 b	58.59 ± 0.99 c	66.39 ± 1.10 a	0.12
leaf inclination/°	82.78 ±1.79 a	32.50 ± 0.68 b	31.31 ± 0.65 c	0.62
specific leaf area/cm^2·g^{-1}	98.74 ± 5.64 c	177.11 ± 11.17 a	125.36 ± 6.62 b	0.44
Branch Angle/°	74.02 ± 2.25 a	76.82 ± 1.66 a	61.35 ± 2.44 b	0.20
Twig length/cm	9.26 ± 0.58 a	9.50 ± 0.57 a	9.74 ± 0.70 a	0.05
Twig diameter/cm	1.66 ± 0.04 b	2.12 ± 0.20 a	1.55 ± 0.05 c	0.27
Number of blades/NO.	26 ± 2 a	28 ± 2 a	20 ± 2 b	0.30
Leafing intensity	1.69 ± 0.08 a	1.56 ± 0.08 a	1.32 ± 0.06 b	0.22

Different lowercase letters in the same row indicate significant differences among plots ($p < 0.05$, n = 36). I, Full sight zone; II, Moderate shade zone; III, Canopy cover zone.

The plasticity index of the twig and leaf characteristics of *H. rhamnoides* was the leaf inclination, leaf dry weight, leaf fresh weight, specific leaf area, number of blades, twig diameter, leaf area, leaf intensity, branch angle, leaf thickness, leaf water content, chlorophyll content, and twig length in order from large to small.

3.4. *Photosynthetic and Fluorescence Characteristics of H. rhamnoides in Each Habitat*

The leaf fluorescence and photosynthetic characteristics of *H. rhamnoides* showed significant changes in different habitats ($p < 0.05$, Table 3). The actual photochemical efficiency of PSII (Y(II)) and photochemical quenching (QP) showed increasing trends from the full sight zone to the canopy cover zone and increased by 72.07% and 41.61%, respectively. The quantum yield of regulated energy dissipation (Y(NPQ)), the quantum yield of non-regulated energy dissipation (Y(NO)), and the net photosynthetic rate (P_n) exhibited the largest values in the full sight zone and the lowest values in the canopy cover zone. From the full sight zone to the canopy cover zone, the Y(NPQ), Y(NO), and P_n decreased by 20.66%, 19.93%, and 25.30%, respectively. Meanwhile, the quantum yield of regulated energy dissipation (NPQ) and water use efficiency (WUE) increased first and then decreased and ultimately decreased by 5.28% and 22.70%, respectively. However, the electronic transport rate (ETR) presented the largest values in the moderate shade zone and the lowest values in the full sight zone and ultimately increased by 1.68 times. Moreover, the transpiration rate (T_r) showed the lowest value in the moderate shade zone and the largest values in the canopy cover zone.

Table 3. Photosynthetic and fluorescence characteristics of *H. rhamnoides leaves* in each habitat (mean ± SE).

Plot	I	II	III
Y(II)	0.22 ± 0.01 c	0.31 ± 0.01 b	0.40 ± 0.01 a
Y(NPQ)	0.46 ± 0.01 a	0.44 ± 0.01 a	0.37 ± 0.01 b
Y(NO)	0.32 ± 0.02 a	0.25 ± 0.01 b	0.25 ± 0.01 b
NPQ	0.40 ± 0.02 b	0.47 ± 0.02 a	0.38 ± 0.01 c
Photochemical quenching	0.41± 0.02 c	0.53 ± 0.01 b	0.59 ± 0.02 a
Electronic transport rate/$\mu mol \cdot m^{-2} \cdot s^{-1}$	37.08 ± 2.50 c	105.75 ± 10.71 a	99.3 ± 8.51 b
Transpiration rate/mmol $H_2O \cdot m^{-2} \cdot s^{-1}$	1.18 ± 0.01 b	1.04 ± 0.04 c	1.21 ± 0.10 a
Net photosynthetic rate/$\mu mol\ CO_2 \cdot m^{-2} \cdot s^{-1}$	7.53 ± 0.03 a	7.16 ± 0.24 a	5.63 ± 0.48 b
Water use efficiency/$\mu mol\ CO_2 \cdot mmol\ H_2O^{-1}$	6.41 ± 0.09 b	6.90 ± 0.11 a	4.95 ± 0.20 c

Different lowercase letters in the same row indicate a significant difference among plots ($p < 0.05$. n = 36). I, Full sight zone; II, Moderate shade zone; III, Canopy cover zone. Y(II), the actual photochemical efficiency of PSII; Y(NPQ), the quantum yield of regulated energy dissipation; Y(NO), the quantum yield of non-regulated energy dissipation; NPQ, the quantum yield of regulated energy dissipation.

3.5. Conversion of Quantum Yields in PSII of H. rhamnoides Leaves in Each Habitat

The proportion of actual photochemical efficiency of PSII (Y(II)) exhibited the largest values in the canopy cover zone and the lowest values in the full sight zone. The proportion of non-regulated energy dissipation (Y(NO)) presented the largest values in the full sight zone and the lowest values in the canopy cover zone. The quantum yield of regulated energy dissipation (Y(NPQ)) showed the largest values in the full sight zone and no significant changes from the moderate zone to the canopy cover zone. With the habitat change from the full sight zone to the canopy cover zone, the proportion of Y(NO) and Y(NPQ) decreased, whereas the proportion of Y(II) increased. The results indicated that the distribution of photosynthetically active radiation absorbed by *H. rhamnoides* leaves through photosynthesis, heat dissipation, and chlorophyll fluorescence changed significantly with the change in the light environment of the *H. rhamnoides* population ($p < 0.05$, Figure 3).

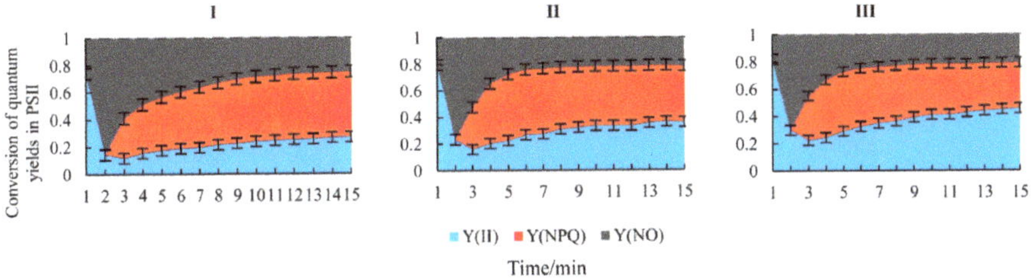

Figure 3. Conversion of quantum yields in photosystem II (PSII) in leaves of *H. rhamnoides* in each habitat (mean ± SD). Y(II), the photochemical quantum yield in PSII; Y(NO), the quantum yield of fluorescence and light-independent constitutive thermal dissipation; Y(NPQ), the quantum yield of thermal dissipation used in regulatory energy dissipation. Photosynthetically active radiation = 1200 μmol·m^{-2}·s^{-1}. **(I)**, Full sight zone; **(II)**, Moderate shade zone; **(III)**, Canopy cover zone.

3.6. Changes in Electron Transfer Rate in H. rhamnoides Leaves in Each Habitat

There were significant differences in the fast optical response curves of the PSII electron transport rate in *H. rhamnoides* leaves in different light environments ($p < 0.05$, Figure 4). As the habitat changed from the full sight zone to the canopy cover zone, the electronic transport rate (ETR) of *H. rhamnoides* leaves showed the following trend: full sight zone (plot I) < canopy cover zone (plot III) < moderate shade zone (plot II). When the saturated pulse light intensity was higher than 1000 μmol·m^{-2}·s^{-1}, the ETR of *H. rhamnoides* leaves in Plot I and Plot II gradually changed from an increasing trend to a stable trend. However, when the saturated pulse light was stronger than 1200 μmol·m^{-2}·s^{-1}, the ETR of *H. rhamnoides* leaves showed a slow growth trend in the moderate shade zone. The results showed that the photosynthetic active radiation in the canopy of the plant community directly affected the ETR of *H. rhamnoides* leaves and then affected the photosynthesis of *H. rhamnoides* leaves. The ETR presented in plot I < plot III < plot II indicates that the moderate shade can promote electron transfer in riparian plant *H. rhamnoides* leaves to a certain extent.

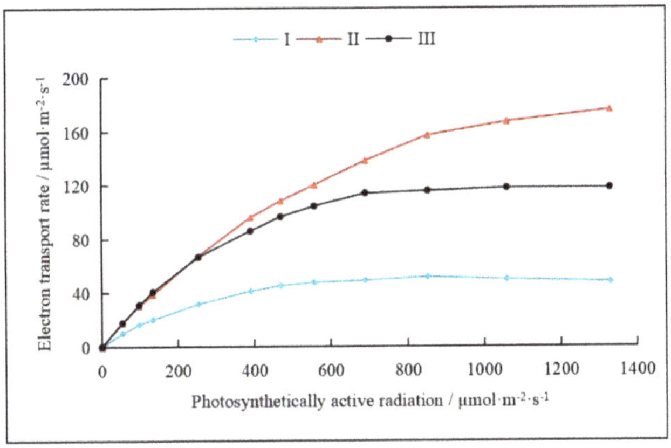

Figure 4. The electronic transport rate of *H. rhamnoides* leaves in each habitat (mean ± SD) I, I, Full sight zone; II, Moderate shade zone; III, Canopy cover zone.

3.7. Correlation Between the Twig–Leaf Characteristics and the Fluorescence Characteristics of H. rhamnoides

The Pearson correlation analysis results are shown in Figure 5 (Supplementary Table S2). There were highly significant negative correlations ($p < 0.01$, Figure 5, Supplementary Table S2) between the quantum yield of regulated energy dissipation (NPQ) and leaf fresh weight (LFW), leaf dry weight (LDW), leaf area (LA), and leaf thickness (LT). Meanwhile, the same correlation was observed between the proportion of the actual photochemical efficiency of PSII (Y(II)) and LDW, leaf inclination (LI), branch angle (BA), and leafing intensity (LGI), as well as between the quantum yield of regulated energy dissipation (Y(NPQ)) and leaf water content (LWC), LA, LT, and CHL, the proportion of non-regulated energy dissipation (Y(NO)) and LWC and LI, and the photochemical quenching (QP) and LDW and LI. In addition, there was a less significant negative correlation ($p > 0.05$, Figure 5, Supplementary Table S2) between NPQ and CHL and between Y(NO) and twig length (TL) in the three habitats. Moreover, there were highly significant positive correlations ($p < 0.01$, Figure 5, Supplementary Table S2) between Y(II) and LWC, as well as between Y(NO) and LDW. In addition, the same correlations were detected between Y(NPQ) and LI, BA, and the number of blades (NB); between NPQ and NB; and between QP and LWC. In addition, there was a less significant positive correlation ($p > 0.05$, Figure 5, Supplementary Table S2) between NPQ and twig length (TL) and LGI among the three habitats.

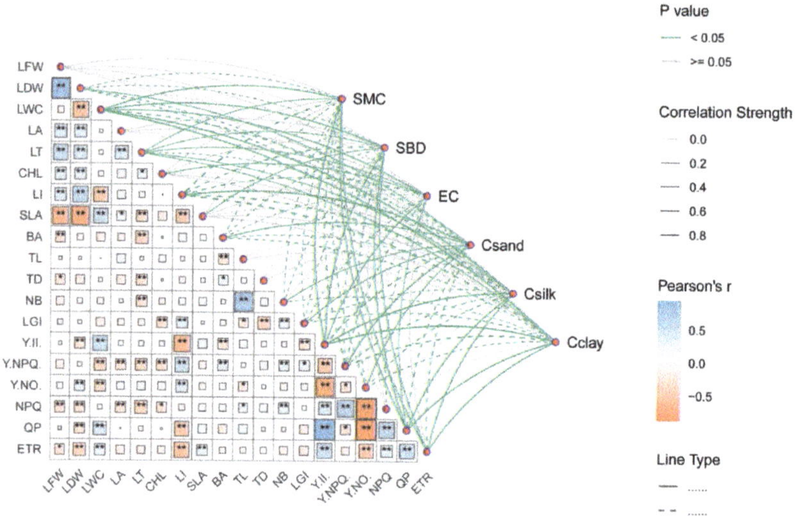

Figure 5. Correlation analysis between twig–leaf characteristics and fluorescence characteristics of *H. rhamnoides*. LFW, leaf fresh weight; LWC, leaf water content; CHL, chlorophyll content; LT, leaf thickness; LDW, leaf dry weight; LA, leaf area; LI, leaf inclination, SLA, specific leaf area; BA, branch angle; TL, twig length; TD, twig diameter; NB, the number of blade; Y(II), the actual photochemical efficiency of PSII; Y(NPQ), the quantum yield of regulated energy dissipation; Y(NO), the quantum yield of non-regulated energy dissipation; NPQ, the quantum yield of regulated energy dissipation; QP, photochemical quenching; T_r, transpiration rate; P_n, net photosynthetic rate; WUE, water use efficiency; ETR, electronic transport rate. * $p < 0.05$ (significant at the 0.05 level, bilateral; the null hypothesis is rejected at the 95% confidence level, and the sample shows a linear correlation); ** $p < 0.01$ (significant at the 0.01 level, bilateral; the null hypothesis is rejected at the 99% confidence level, and the sample shows a linear correlation). Blue represents a positive correlation between the twig–leaf characteristics and fluorescence characteristics, and tangerine represents a negative correlation between twig–leaf characteristics and fluorescence characteristics. The deeper the color, the more significant the correlation. The lighter the color, the weaker the correlation.

The mantel test showed that the soil moisture content (SMC), the content of silt (Csilt), the content of clay (Cclay), and the soil bulk density (SBD) were significantly correlated with Y(II), LWC, QP, ETR, LA, and leaf length, while the content of sand (Csand) was significantly correlated with Y(NPQ), WUE, and leaf inclination (LI). Moreover, soil electrical conductivity (EC) was significantly correlated with Y(II) and Tr. There is also a significant correlation between the fluorescence characteristics of *H.rhamnoides* and the six environmental factors in three density gradients (Figure 5; Supplementary Table S2).

4. Discussion

The functional traits of branches and leaves determine how plants respond to environmental changes and optimize resource utilization [7]. They are organs that are more sensitive to environmental changes and have strong plasticity [8], and they are also the main organs and energy converters for plants to obtain carbon through photosynthesis. In this study, it was found that the number of leaves, leaf inclination, branch angle, Y(NPQ), Y(NO), NPQ, Pn, and WUE of *H. rhamnoides* gradually decreased with the change in the light environment from the full sight zone (I) to the canopy cover zone (III). Leaf water content, leaf thickness, specific leaf area, chlorophyll content, Y(II), QP, Tr, and ETR were increased overall and tended to form thin and long twigs (Tables 2 and 3). There was a significant correlation between the twig–leaf and leaf fluorescence parameters of *H. rhamnoides* ($p < 0.05$, Figure 5, Supplementary Table S2).

4.1. Correlation Between the Twig–Leaf Characteristics and the Fluorescence Characteristics of H. rhamnoides in the Full Sight Zone

The resource allocation and configuration characteristics of twigs are important topics in the study of plant life history strategies [58]. In the full sight zone, the photosynthetically active radiation was the highest, the canopy density of the plant community was the lowest, and the crown width, average height, and density of the *H. rhamnoides* population were the lowest (Table 1). The soil water content, bulk density, and proportion of silt and clay particles were the smallest, and the soil electrical conductivity and sand content were the highest (Figure 2; Supplementary Table S1). In this habitat, *H. rhamnoides* tended to be built on short and thick twigs, with thicker leaves and less SLA. The light energy absorbed by the leaves also accounted for a higher proportion of heat dissipation, and there was a significant correlation between the twig–leaf traits and chlorophyll fluorescence parameters ($p < 0.05$, Figure 5, Supplementary Table S2). The main reasons are as follows: (1) The habitat had abundant light resources, low soil bulk density and water content, high soil salinity and sand content, and poor soil water holding capacity. Compared with large leaves, small leaves with a low SLA have stronger thermal conductivity and smaller transpiration areas and can accommodate more chloroplasts, which is conducive to the adaptation of *H. rhamnoides* to a strong light environment [59,60]. While ensuring the effective capture of light energy, it can prevent water loss of leaves caused by overheating in strong light environments so that *H. rhamnoides* can maintain a high water use efficiency (Table 3) and avoid the phenomenon of holes and embolism caused by *H. rhamnoides* in strong light environments. Therefore, in this habitat, *H. rhamnoides* did not need to build a large photosynthetic area and formed a low-SLA strategy by reducing the leaf area to adapt to the strong light environment. (2) Under the combined stress of water and salt, the effective supply of water resources is insufficient, and the short and thick twigs shorten the effective transport distance of water from the root system to the leaf of *H. rhamnoides* [61], enhancing the water transport ability and maintaining the water transportation in *H. rhamnoides*. This is basically consistent with the research results of Wang et al. [62], which state that "water stress will significantly limit the longitudinal growth of plants while having little effect on radial growth, which is beneficial for increasing water utilization efficiency". (3) In the full sight zone, *H. rhamnoides* had the lowest electron transfer rate (Figure 4). In order to prevent excessive light from damaging the photosynthetic structure of its leaves, *H. rhamnoides* used the acquired light energy for heat dissipation, and Y(NPQ) was the largest. Y(II) is the smallest (Table 3; Figure 3). This heat dissipation ability of leaves helps to prevent excessive water loss in plant leaves and improve plant adaptation to high-light habitats [59,60]. Therefore, the leaf heat dissipation of *H. rhamnoides* was significantly negatively correlated with leaf area and leaf thickness ($p < 0.05$, Figure 5, Supplementary Table S2).

4.2. Correlation Between the Twig–Leaf Characteristics and Fluorescence Characteristics of H. rhamnoides in the Moderate Shade Zone

Annual branches are the most dynamic part of the plant body, which is the support body of the leaves that determines the spatial distribution of the leaves. These branches allow the plant body to make full use of light energy for photosynthesis and reflect the response of the plant to the environment better than the whole plant can [63]. In the moderate shade zone, the SMC and SBD were higher than that in the full sight zone but lower than the canopy cover zone; the EC, Csilk, and Cclay were higher; and the Csand was lower than that in the full sight zone (Figure 2; Supplementary Table S1). The photosynthetic active radiation was lower than that in the full sight zone, the canopy density of the plant community was higher than that in the full sight zone, and the crown width, average height, and density of *H. rhamnoides* population were higher than that in the full sight zone (Table 1). In this habitat, *H. rhamnoides* tended to grow many thin leaves with high SLA on long and thick twigs, and the proportion of light energy absorbed by the leaves for heat dissipation was lower than that in the full sight zone. There was a significant correlation between the twig–leaf traits and chlorophyll fluorescence parameters ($p < 0.05$, Figure 5, Supplementary Table S2). The main reasons are as follows: (1) in the moderate shade zone, *H. rhamnoides* is in the excessive zone of the strong and weak light environments and has low survival pressure. To avoid the risk of water stress caused by excessive transpiration due to excessive leaf area, *H. rhamnoides* only needs to slowly increase the photosynthetic area to meet its growth demand. Therefore, it has a small Pn (Table 3). The investment of *H. rhamnoides* in leaf resources in this habitat was smaller than that in the strong light environment, resulting in thin leaves with high SLA, which helped to increase the effective utilization of interforest light resources by *H. rhamnoides* and maximize leaf WUE (Table 3). (2) In this habitat, the hydrothermal conditions were moderate, the soil water holding capacity was good, and there were no obvious resource constraints. *H. rhamnoides* was an associated species. To improve the interspecific competitiveness of *H. rhamnoides*, large and thin leaves were grown on long and thick twigs with the largest branch angle (Table 2). This could not only maintain a higher water supply capacity in the leaves but could also increase the vertical light resource competition ability of the *H. rhamnoides* population to maintain higher photosynthetic efficiency. (3) Chlorophyll fluorescence is closely related to the photosynthetic process of leaves [26,27,46]. In this habitat, the light intensity was moderate, and the proportion of light energy absorbed by *H. rhamnoides* leaves for heat dissipation decreased in the full sight zone. In addition, Y(NPQ) was lower, which increased the proportion of energy used for photosynthesis, and Y(II) was higher (Table 3; Figure 3). The decrease in light intensity was accompanied by a decrease in the light inhibition, heat dissipation, and transpiration water consumption of leaves, which resulted in an increase in the electron transfer rate of *H. rhamnoides* leaves (Table 3; Figure 4). *H. rhamnoides* tablets maintained a high water use efficiency (Table 3), thereby ensuring the smooth reproduction and renewal of the *H. rhamnoides* population.

4.3. Correlation Between the Twig–Leaf Characteristics and Fluorescence Characteristics of H. rhamnoides in the Canopy Cover Zone

The degree of crowding and resource competition within the population was high, and the available space occupied by leaves was relatively narrow [64]. In the canopy cover zone, the SMC, SBD, Csilk, and Cclay were highest, and the Csand and EC were lowest (Figure 2; Supplementary Table S1). The photosynthetic active radiation was the lowest, the vegetation community was the highest, and the crown width, average height, and density of the *H. rhamnoides* population were the highest (Table 1). In this habitat, *H. rhamnoides* tended to grow large and thick leaves with a small amount of SLA on long and thin twigs, and the proportion of light energy absorbed by leaves for heat dissipation was the lowest. There was also a significant correlation between the twig–leaf and chlorophyll fluorescence parameters ($p < 0.05$, Figure 5, Supplementary Table S2). The main reasons are as follows: (1) In the canopy cover zone, the soil water content was sufficient. The primary problem *H. rhamnoides* faced was the limitation of light resources caused by the shading of poplar and other trees in the canopy layer. To improve the competitiveness of light resources, *H. rhamnoides* formed thick leaves with a high SLA. The water content and chlorophyll content of leaves were increased (Table 2), and the normal photosynthetic physiological process of *H. rhamnoides* was maintained. This is different from the large and thin leaves formed by reeds under low light conditions reported by Li et al. [26], mainly because the leaf type of *H. rhamnoides* is different from that of reed leaves. (2) This habitat had the largest canopy density, and the construction of thin and long twigs was conducive to improving the interspecific competitiveness of *H. rhamnoides.* Placing the leaves on the twig at a higher vertical position was conducive to the acquisition and utilization of limited light resources by *H. rhamnoides.* The higher chlorophyll content in thick leaves maintained a higher electron transfer rate in *H. rhamnoides* leaves. However, compared with the moderate shade zone and full sight zone, a higher electron transfer rate in *H. rhamnoides* leaves was observed. The transpiration rate, net photosynthetic rate, and water use efficiency of *H. rhamnoides* leaves were significantly decreased (Table 3). (3) Chlorophyll fluorescence is the fastest and most efficient way to perceive external environmental stress [26]. In the canopy cover zone, light intensity decreased significantly, and *H. rhamnoides* leaves were exposed to weak light stress. Therefore, there was no need to use too much absorbed light energy for leaf heat dissipation, and Y(NPQ) and NPQ were the lowest, but the photosynthetic energy proportion was increased. Qp and Y(II) were the highest (Table 3; Figure 3). The results showed that *H. rhamnoides* PSII had higher photochemical efficiency in shaded environments [35,36], reflecting the phenotypic plasticity mechanism of *H. rhamnoides,* which adjusts the morphology structure and physiological characteristics of the twigs–leaves in time with the change in habitat conditions.

5. Conclusions

This study found that in the full sight zone, *H. rhamnoides* built thick leaves with a smaller SLA on short and thick twigs. A relatively high proportion of the light energy absorbed by the leaves was used for heat dissipation, which was an effective way for *H. rhamnoides* to adapt to the strong light environment. In the moderate shade zone, *H. rhamnoides* tended to grow many thin leaves with high SLA on long and thick branchlets, and the proportion of light energy absorbed by the leaves for heat dissipation was low. In the canopy cover zone, *H. rhamnoides* tended to grow a few large and thick leaves with a low SLA on slender and long twigs. The proportion of the light energy absorbed by the leaves used for heat dissipation was the lowest. This is a survival strategy of *H. rhamnoides* that enhances interspecific competition, alleviates intraspecific resource competition, and allows the plant to adapt to the weak light environment. In heterogeneous light habitats, the coordinated variation of the plant's twig–leaf traits and the timely adjustment of the leaf photosynthetic fluorescence enhance the fitness of riparian plants, reflecting the ecological adaptation strategies of plants to improve interspecific competition and maintain normal photosynthetic physiological processes within the population. This study only analyzed the response patterns of twig–leaf traits and photosynthetic fluorescence physiological characteristics to the light environment from the perspective of light resources. The functional traits of other plant components in the riparian zone also play a crucial role in responding to heterogeneous environments, and we will conduct further exploration and research on this topic.

Supplementary Materials: The following supporting information can be downloaded at: https://www.mdpi.com/article/10.3390/plants14020282/s1, Table S1: The physical and chemical properties of soil in different plot (mean ± SE); Table S2: Correlation analysis between twig-leaf traits and fluorescence characteristics of *H. rhamnoides*.

Author Contributions: Q.L., M.M., Y.T., T.Z., C.Z. and B.L. conceptualized and designed the experiments. Q.L., M.M., Y.T. and T.Z. performed the experiments and the investigation; Q.L. contributed to the methodology, analyzed the data, and drafted the manuscript. C.Z. and B.L. provided support in the areas of funding acquisition, resources, and the critical revision of the article. All authors have read and agreed to the published version of the manuscript.

Funding: This work was supported, in part, by the National Natural Science Foundation of China (Grant. No. 41461013 and No. 41861009) and the Xichang University Doctoral Start Foundation Program (YBZ202237).

Data Availability Statement: The data used in the present work have been listed in the Supplementary Materials.

Acknowledgments: We would like to acknowledge the professional manuscript services of American Journal Experts (https://secure.authorservices.springernature.com). In addition, we are very grateful to Xiaoya Li and Manping Kang for their assistance in the field and laboratory work.

Conflicts of Interest: None. The manuscript has been approved by all authors. I declare on behalf of all authors that the work described is original research that has not been published previously and is not under consideration for publication elsewhere, in whole or in part.

References

1. Sweeney, C.J.; de Vries, F.T.; van Dongen, B.E.; Bardgett, R.D. Root traits explain rhizosphere fungal community composition among temperate grassland plant species. *New Phytol.* **2021**, *229*, 1492–1507. [CrossRef] [PubMed]
2. Fu, D.; Wu, X.; Duan, C.; Smith, A.R.; Jones, D.L. Traits of dominant species and soil properties co-regulate soil microbial communities across land restoration types in a subtropical plateau region of Southwest China. *Ecol. Eng.* **2020**, *153*, 105897. [CrossRef]
3. Eller, C.B.; Meireles, L.D.; Sitch, S.; Burgess, S.S.O.; Oliveira, R.S. How Climate Shapes the Functioning of Tropical Montane Cloud Forests. *Curr. For. Rep.* **2020**, *6*, 97–114. [CrossRef]
4. Sun, M.; Tian, K.; Zhang, Y.; Wang, H.; Guan, D.X.; Yue, H.T. Research on leaf functional traits and their environmental adaptation. *Plant Sci. J.* **2017**, *35*, 940–949. [CrossRef]
5. Minden, V.; Kleyer, M. Internal and external regulation of plant organ stoichiometry. *Plant Biol.* **2014**, *16*, 897–907. [CrossRef]
6. Wu, Z.; Fu, Y.H.; Crowther, T.W.; Wang, S.; Gong, Y.; Zhang, J.; Zhao, Y.; Janssens, I.; Penuelas, J.; Zohner, C.M. Poleward shifts in the maximum of spring phenological responsiveness of *Ginkgo biloba* to temperature in China. *New Phytol.* **2023**, *240*, 1421–1432. [CrossRef]
7. Yang, J.; Gao, Y.; Zhao, C.; Chen, H. Leaf phenotypic plasticity and integration balance plant adaptation to water table decline: A mesocosm experiment. *Plant Soil* **2024**, *497*, 611–627. [CrossRef]
8. Felix, J.A.; Stevenson, P.C.; Koricheva, J. Plant neighbourhood diversity effects on leaf traits: A meta-analysis. *Funct. Ecol.* **2023**, *37*, 3150–3163. [CrossRef]
9. Westoby, M.; Wright, I.J. Land-plant ecology on the basis of functional traits. *Trends Ecol. Evol.* **2006**, *21*, 261–268. [CrossRef]
10. Xiang, S.; Liu, Y.; Fang, F.; Wu, N.; Sun, S. Stem architectural effect on leaf size, leaf number, and leaf mass fraction in plant twigs of woody species. *Int. J. Plant Sci.* **2009**, *170*, 999–1008. [CrossRef]
11. Kawai, K.; Okada, N. Coordination of leaf and stem traits in 25 species of Fagaceae from three biomes of East Asia. *Botany* **2019**, *97*, 391–403. [CrossRef]
12. Westoby, M.; Falster, D.S.; Moles, A.T.; Vesk, P.A.; Wright, I.J. Plant ecological strategies: Some leading dimensions of variation between species. *Annu. Rev. Ecol. Syst.* **2002**, *33*, 125–159. [CrossRef]

13. Tang, G.-L.; Li, X.-Y.; Lin, L.-S.; Li, L.; Lu, J.-R. Change of different shading on moisture conditions and the physiological response in *Alhagi sparsifolia*. *Chin. J. Plant Ecol.* **2013**, *37*, 354–364. [CrossRef]
14. Dang, J.J.; Zhao, C.Z.; Li, Y.; Hou, Z.J.; Dong, X.G. Variations with slope in stem and leaf traits of *Melica przewalskyi* in alpine grassland. *Chin. J. Plant Ecol.* **2014**, *38*, 1307–1314. [CrossRef]
15. Zhang, Q.M.; Zhang, C.; Liu, M.S.; Yu, W.; Xu, C.; Wang, H.J. The influences of arboraceous layer on spatial patterns and morphological characteristics of herbaceous layer in an arid plant community. *Acta Ecol. Sin.* **2007**, *27*, 1265–1271. [CrossRef]
16. Chen, Y.T.; Xu, Z.Z. Review on research of leaf economics spectrum. *Chin. J. Plant Ecol.* **2014**, *38*, 1135–1153. [CrossRef]
17. Zou, C.M.; Wang, Y.Q.; Liu, Y.; Zhang, X.H.; Tang, S. Responses of photosynthesis and growth to weak light regime in four legume species. *Chin. J. Appl. Ecol.* **2015**, *39*, 909–916. [CrossRef]
18. Brodribb, T.J.; Bowman, D.J.M.S.; Nichols, S.; Delzon, S.; Burlett, R. Xylem function and growth rate interact to determine recovery rates after exposure to extreme water deficit. *New Phytol.* **2010**, *188*, 533–542. [CrossRef]
19. Funk, J.L.; Cornwell, W.K. Leaf traits within communities: Context may affect the mapping of traits to function. *Ecology* **2013**, *94*, 1893–1897. [CrossRef]
20. Westoby, M.; Reich, P.B.; Wright, I.J. Understanding ecological variation across species: Area-based vs mass-based expression of leaf traits. *New Phytol.* **2013**, *199*, 322–323. [CrossRef]
21. Navas, M.-L.; Roumet, C.; Bellmann, A.; Laurent, G.; Garnier, E. Suites of plant traits in species from different stages of a Mediterranean secondary succession. *Plant Biol.* **2010**, *12*, 183–196. [CrossRef] [PubMed]
22. Brodribb, T.J.; Jordan, G.J. Water supply and demand remain balanced during leaf acclimation of *Nothofagus cunninghamii* trees. *New Phytol.* **2011**, *192*, 437–448. [CrossRef] [PubMed]
23. Cai, J.; Tyree, M.T. The Impact of Vessel Size on Vulnerability Curves: Data and Models for Within-Species Variability in Saplings of Aspen, *Populus tremuloides* Michx. *Plant Cell Environ.* **2010**, *33*, 1059–1069. [CrossRef] [PubMed]
24. Wu, T.; Geng, Y.F.; Chai, Y.; Hao, J.B.; Yuan, C.M. Response of leaf anatomical structure and photosynthesis characteristics of *Partheno-cissus himalayana* to three habitat types. *Ecol. Environ. Sci.* **2014**, *23*, 1586–1592. [CrossRef]
25. Sello, S.; Meneghesso, A.; Alboresi, A.; Baldan, B.; Morosinotto, T. Plant biodiversity and regulation of photosynthesis in the natural environment. *Planta* **2019**, *249*, 1217–1228. [CrossRef]
26. Li, Q.; Wen, J.; Zhao, C.-Z.; Zhao, L.-C.; Ke, D. The relationship between the main leaf traits and photosynthetic physiological characteristics of *Phragmites australis* under different habitats of a salt marsh in Qinwangchuan, China. *AoB Plants* **2022**, *14*, plac054. [CrossRef]
27. Zha, T.-S.; Wu, Y.J.; Jia, X.; Zhang, M.Y.; Bai, Y.J.; Liu, P.; Ma, J.Y.; Bourque, C.P.-A.; Peltola, H. Diurnal response of effective quantum yield of PSII photochemistry to irradiance as an indicator of photosynthetic acclimation to stressed environments revealed in a xerophytic species. *Ecol. Indic.* **2017**, *74*, 191–197. [CrossRef]
28. Antala, M.; Juszczak, R.; Rastogi, A. Nonphotochemical quenching does not alter the relationship between sun-induced fluorescence and gross primary production under heatwave. *New Phytol.* **2025**, *245*, 927–930. [CrossRef]
29. Wu, Y.J.; Zha, T.S.; Jia, X.; Qin, S.G.; Li, Y.; Wang, B. Temporal variation and controlling factors of photochemical efficiency and non-photochemical quenching in *Artemisia ordosica*. *Chin. J. Ecol.* **2015**, *34*, 319–325. [CrossRef]
30. Ware, M.A.; Belgio, E.; Ruban, A.V. Photoprotective capacity of non-photochemical quenching in plants acclimated to different light intensities. *Photosynth. Res.* **2015**, *126*, 261–274. [CrossRef]
31. Levin, G.; Yasmin, M.; Liran, O.; Hanna, R.; Kleifeld, O.; Horev, G.; Wollman, F.-A.; Schuster, G.; Nawrocki, W.J. Processes independent of nonphotochemical quenching protect a high-light-tolerant desert alga from oxidative stress. *Plant Physiol.* **2024**, *197*, kiae608. [CrossRef] [PubMed]
32. Demmig-Adams, B.; Winter, K.; Krüger, A.; Czygan, F.-C. Zeaxanthin Synthesis, Energy Dissipation, and Photoprotection of Photosystem II at Chilling Temperatures. *Plant Physiol.* **1989**, *90*, 894–898. [CrossRef]
33. Yan, C.; Wang, Z.S.; An, S.Q.; Chen, S.N.; Wei, N.; Lu, X.M. Differences in photosynthetic capacity among different diameter-classes of *Parrotia subaequalis pop-ulations* and their implications to regeneration limitation. *Acta Ecol. Sin.* **2008**, *28*, 4153–4161.
34. Roig-Villanova, I.; Martínez-García, J.F. Plant Responses to Vegetation Proximity: A Whole Life Avoiding Shade. *Front. Plant Sci.* **2016**, *7*, 236. [CrossRef]
35. Cutolo, E.A.; Guardini, Z.; Dall'Osto, L.; Bassi, R. A paler shade of green: Engineering cellular chlorophyll content to enhance photosynthesis in crowded environments. *New Phytol.* **2023**, *239*, 1567–1583. [CrossRef]
36. Martinez-Garcia, J.F.; Rodriguez-Concepcion, M. Molecular mechanisms of shade tolerance in plants. *New Phytol.* **2023**, *239*, 1190–1202. [CrossRef]
37. Leschevin, M.; Ksas, B.; Baltenweck, R.; Hugueney, P.; Caffarri, S.; Havaux, M. Photosystem rearrangements, photosynthetic efficiency, and plant growth in far red-enriched light. *Plant J.* **2024**, *120*, 2536–2552. [CrossRef]
38. Nilsson, C.; Berggren, K. Alterations of riparian ecosystems caused by river regulation. *BioScience* **2000**, *50*, 783–792. [CrossRef]
39. Han, L.; Wang, H.Z.; Yu, J. Research progress and prospects on riparian zone ecology. *Ecol. Environ. Sci.* **2013**, *22*, 879–886. [CrossRef]

40. Garssen, A.G.; Baattrup-Pedersen, A.; Voesenek, L.A.C.J.; Verhoeven, J.T.A.; Soons, M.B. Riparian plant community responses to increased flooding: A meta-analysis. *Glob. Chang. Biol.* **2015**, *21*, 2881–2890. [CrossRef]
41. Popoff, N.; Jaunatre, R.; Le Bouteiller, C.; Trinquier, M.; Paillet, Y.; Evette, A. Ecological succession and fine sediment accretion influence local patch dynamics of a pioneer riparian species (*Typha minimaHoppe*). *Freshw. Biol.* **2021**, *66*, 2351–2363. [CrossRef]
42. Li, Q.; Zhao, C.Z.; Wen, J.; Kang, M.P.; Li, X.Y. Fractal analysis of root architecture responses of *Saussurea salsa* to a gradient of flooding intensity and salinity. *Plant Soil* **2022**, *471*, 669–683. [CrossRef]
43. Wen, J.; Li, Q.; Zhao, C.; Kang, M. Morphological Traits and Biomass Allocation of *Leymus secalinus* along Habitat Gradient in a Floodplain Wetland of the Heihe River, China. *Agronomy* **2024**, *14*, 1899. [CrossRef]
44. Tang, Y.R.; Zhao, C.Z.; Zhao, H.; Hou, G.; Ma, M.; Zhao, T.T.; Wang, Y.F.; Zeng, H.X. The relationship between leaf dry mass and leaf area, leaf thickness of *Hippophae rhamnoides* under different light conditions in Taohe River riparian forest. *Chin. J. Ecol.* **2021**, *40*, 2745–2753. [CrossRef]
45. Cai, H.X.; Wu, F.Z.; Yang, W.Q. Effects of drought stress on the photosynthesis of *Salix paraqplesia* and *Hippophae rhamnoides* seed-lings. *Acta Ecol. Sin.* **2011**, *31*, 2430–2436.
46. Wu, X.; Tang, Y.K.; Chen, C.; Jia, C.; Chen, Y.M. Photosynthesis light response characteristics and environmental adaptability of *Hippophae rhamnoides*, *Pinus tabuliformis*, and *Robinia pseudoacacia* in the Loess hilly region of China. *Acta Ecol. Sin.* **2019**, *39*, 8111–8125. [CrossRef]
47. Urbinati, G.; Nota, P.; Frattarelli, A.; Lucioli, S.; Forni, C.; Caboni, E. Morpho-physiological responses of sea buckthorn (*Hippophae rhamnoides*) to NaCl stress. *Plant Biosyst. Int. J. Deal. All Asp. Plant Biol.* **2019**, *154*, 827–834. [CrossRef]
48. Zhang, C.; Zhang, Q.; Jiang, J.; Yang, Q. Pullout properties of *Hippophae rhamnoides* L. roots in the loess area. *Arch. Agron. Soil Sci.* **2023**, *69*, 2170–2186. [CrossRef]
49. Du, Z.; Bai, H.; Liu, M.; Liu, Y.; Zhu, G.; Chai, G.; He, Y.; Shi, J.; Duan, Y. Response of ecological stoichiometry and homeostasis characteristic to nitrogen addition in *Hippophae rhamnoides* L. *Sci. Total Environ.* **2024**, *951*, 175591. [CrossRef]
50. Li, Q.; Zhao, H.; Zhao, C.Z.; Ma, M.; Tang, Y.R.; Zhao, T.T. Response of main plant community diversity to soil environmental factors in Taohe national wetland park. *Acta Ecol. Sin.* **2022**, *42*, 2647–2684. [CrossRef]
51. Cifuentes, L.; Moreno, F. Trait coordination at leaf level explains the resistance to excess light stress in shade-tolerant tropical tree species. *Tree Physiol.* **2022**, *42*, 1325–1336. [CrossRef] [PubMed]
52. Li, Q.; Zhao, C.Z.; Zhao, L.C.; Wang, J.L.; Zhang, W.T.; Yao, W.X. Empirical relationship between specific leaf area and thermal dissipation of *Phragmites australis* in salt marshes of Qinwangchuan. *Chin. J. Plant Ecol.* **2017**, *41*, 985–994. [CrossRef]
53. Rascher, U.; Hütt, M.T.; Siebke, K.; Osmond, B.; Beck, F.; Luttge, U. Spatiotemporal variation of metabolism in a plant circadian rhyme: The biological clock as an assembly of coupled individual oscillators. *Proc. Nat. Acad. Sci. USA* **2001**, *98*, 11801–11805. [CrossRef] [PubMed]
54. Siebke, K.; Weis, E. Assimilation images of leaves of Glechoma hederacea; Analysis of non-synchronous stomata related oscillations. *Planta* **1995**, *196*, 155–165. [CrossRef]
55. Genty, B.; Briantais, J.-M.; Baker, N.R. The relationship between the quantum yield of photosynthetic electron transport and quenching of chlorophyll fluorescence. *Biochim. Biophys. Acta (BBA) Gen. Subj.* **1989**, *990*, 87–92. [CrossRef]
56. Netto, A.T.; Campostrini, E.; de Oliveira, J.G.; Bressan-Smith, R.E. Photosynthetic pigments, nitrogen, chlorophyll a fluorescence and SPAD-502 readings in coffee leaves. *Sci. Hortic.* **2005**, *104*, 199–209. [CrossRef]
57. Kleiman, D.; Aarssen, L.W. The leaf size/number trade-off in trees. *J. Ecol.* **2007**, *95*, 376–382. [CrossRef]
58. Bazzaz, F.A.; Grace, J. *Plant Resource Allocation*; Academic Press: San Diego, CA, USA, 1997; pp. 1–37.
59. Hu, W.H.; Zhang, S.S.; Xiao, Y.A. Physiological responses and photo-protective mechanisms of two *Rhododendron* plants to natural sunlight after long term shading. *Chin. J. Plant Ecol.* **2015**, *39*, 1093–1100. [CrossRef]
60. Hu, Y.S.; Mo, X.Y.; Liu, Y.H. Specific leaf area and its influencing factors of forests at different succession stages in Changbai Mountains. *Acta Ecol. Sin.* **2015**, *35*, 1480–1487.
61. Tyree, M.T.; Ewers, F.W. The hydraulic architecture of trees and other woody plants. *New Phytol.* **1991**, *119*, 345–360. [CrossRef]
62. Wang, G.G.; Bauerle, W.L.; Mudder, B.T. Effects of light acclimation on the photosynthesis, growth, and biomass allocation in American chestnut (*Castanea dentata*) seedlings. *For. Ecol. Manag.* **2006**, *226*, 173–180. [CrossRef]
63. Wang, X.Y.; Cao, J.J.; Zhang, X.F.; Zhang, X.F.; Kong, Y.Y.; Tian, H.; Li, M.T.; Xu, X.Y.; Gong, Y.F. Effects of topographic factors on leaf traits of apricot in the Loess Plateau, Northwest China. *Chin. J. Appl. Ecol.* **2019**, *30*, 2591–2599. [CrossRef]
64. Li, Q.; Zhao, C.Z.; Kang, M.P.; Li, X.Y. The relationship of the main root-shoot morphological characteristics and biomass allocation of *Saussurea salsa* under different habitat conditions in Sugan lake wetland on the northern margin of the Qinghai-Tibet Plateau. *Ecol. Indic.* **2021**, *128*, 107836. [CrossRef]

MDPI AG
Grosspeteranlage 5
4052 Basel
Switzerland
Tel.: +41 61 683 77 34

Plants Editorial Office
E-mail: plants@mdpi.com
www.mdpi.com/journal/plants

www.ingramcontent.com/pod-product-compliance
Lightning Source LLC
Chambersburg PA
CBHW061618210926
51850CB00009B/2362
* 9 7 8 3 7 2 5 8 7 4 6 0 6 *